Rural Analysis and Management

Francisco J. Tapiador

Rural Analysis and Management

An Earth Science Approach to Rural Science

With 166 Figures and 24 Tables

Springer

Author
Francisco J. Tapiador
Institute of Environmental Sciences, ICAM
Avda. Carlos III s/n, 45004
Toledo
Spain
Francisco.Tapiador@uclm.es

Library of Congress Control Number: 2007932916

ISBN 978-3-540-73342-3 Springer Berlin Heidelberg New York

Springer is a part of Springer Science+Business Media
springer.com

Typesetting: Integra Software Services Pvt. Ltd., India

Cover design: deblik, Berlin

Printed on acid-free paper SPIN: 11314608 30/2132/Integra 5 4 3 2 1 0

This book is dedicated to Raquel

Foreword

This handbook possesses two virtues that are becoming increasingly difficult to find together. The first one is providing a novel and original scheme to analyze the rural areas of the planet in the context of Earth sciences. The handbook is so original in its plan that it actually defines what modern rural science is. The scheme followed, albeit new, is rooted in a venerable geographical tradition and presents a sensible balance between the academic and the applied viewpoints. This novel way of studying rural areas represents by itself a valuable contribution to this complex field.

The second virtue of the book is the ability of synthesizing the most important aspects of the issues tackled, and there are many issues this book tackles, from physical to human geography and from deontology to matrix algebra. When writing a handbook in such a vast field as rural science, it is indeed tempting to accumulate meaningless information assisted by the wealth of available on-line information. On the other side, the danger of forgetting essential aspects is always present. None of these problems seems to have affected the making of this book. The pictures and tables always have a place, the text never looses its coherence, and no comment seems superfluous. Also, the ambitious scope of the book is seemly effortlessly solved by a rational, clear, and logical scheme that dissects all the relevant issues up to the required point. The result is a logical and uniform handbook in what was previously a daunting attempt, given the breadth of the topic.

Another aspect that favourably adds to the interest of this book is the careful choice of literature the author has done. The final contents embody what could be considered as rural science and nothing else. This is a coherent and thoughtfully thought out text on rural science, which is also a handbook on rural analysis and management.

This book will be of use to both higher education instructors and students. It is indeed difficult to find a handbook that can be read either as a continuous text or in a random way, but even rarer to have an academic book that so easily conveys a discipline: students of environmental sciences, geography, and rural or agrarian engineering studies will find how effortlessly this book helps to understand and remember essential topics. This is partly thanks to the way the text is presented. Confining most of the large number of works consulted to the end of each chapter is a clever and increasingly rare way of acknowledging the sources without distracting the reader. This absence of references in the main text contributes to a limpid reading, benefiting the educational side of the book.

This is not only a book for students, however. For the researcher and the practitioner, this handbook is an invaluable guide to carry out rural planning studies in an ordered and scientifically based manner. Just by considering all the topics presented in the first part of the book, valuable analysis will surely follow. And by taking into account the regional specificities (part II), and the management issues (part III), a proper rural planning and management can be carried out encompassing much of the current knowledge on this undoubtedly difficult but worthy task.

Jordi Martí-Henneberg

Professor in Human Geography
University of Cambridge, United Kingdom
January 2007

Prologue

Earth is a rural landscape. In spite of cities, both the rural and the natural areas of the world cover most of the planet. Today, the rural areas of the planet nourish us and provide us with most of the energy we use. Until the nineteenth century, the history of humanity was a rural history. Nonetheless, rural analysis is not a science of the past but a living discipline with a profound effect on our daily lives. How forests are managed, or how mass tourism is handled has implications for urbanites too. The agricultural policies are intertwined with national taxation systems, and even the geopolitics of large geographical areas is related to irrigation and rural interests. Global change has profound effects in rural areas, and even those apparently natural areas are so anthropized that they need to be analyzed in the framework of rural science.

Modern rural analysis is a multi-disciplinary geoscience which has its origins in geographical analysis studies. Today, it comprises many different areas and several different and often disparate expertises, both in social and in natural sciences. Whereas there are excellent journals and societies devoted to rural research, I felt that there was a gap in the literature between the academic point of view and how theoretical methods and studies are applied in a real practical work. This was one reason to start writing this handbook.

Another motivation to embark on this two year endeavour was my geographical passion. There are a few sciences that provide such a variety of scales, methods and rewards as geography does. The geographer can develop methods of rainfall estimation from space; analyze the making of the American railroads; or the morphology of the cities. He is concerned with both biodiversity and social deprivation and his analysis can directly help to make a better world for everyone. Instead of focusing on a particular aspect of the planet, the geographer starts acquiring a deep understanding on many particular subjects to eventually build a territorial synthesis, a complex task that can only be achieved using an objective methodology. The more we learn about the intricate interactions between humankind and nature, and of these elements by themselves, the more we will be able to control and manage our environment for our benefit. In this aspect, I believe that the geosciences contribution to rural analyses and how rural areas are analysed and managed is crucial. Geography is just one of the sciences involved in rural analysis, but its own perspective and methods are central in rural studies.

The first part of the book presents a comprehensive description of the elements of rural analysis, providing the basis for a synthetic view of rural landscapes that are explored in the second part. In the third part of the handbook, management issues arising in rural development projects are discussed. This part covers topics such as global change, the role of information technologies in rural areas, gender relations and how to bridge the gap between scientific analysis and decision-making processes. The final part of the book comprises of a few real examples to illustrate how all the analyses and techniques shown in the previous parts can be applied in commissioned rural project.

This book is a handbook and as such there is little room for originality, apart from the selection of topics and the links between them. Therefore, most of the material found here is an elaboration of the established academic literature, which is acknowledged at the end of each chapter. Nonetheless, I have widely used my personal experience in carrying out geographical analyses, both in academia and as a consultant, and I have organized existing material in a consistent, and I think novel way, by linking specific topics to the actual rural analysis work done in the field. Thus, many of the tables and figures are updates of my own work on several topics.

A couple of additional notes may help to contextualize the book. I have consciously shied away from any controversy on terminology (such as how 'counter-urbanization' or 'rural' can be defined), since they are not appropriated for a handbook and may cause more confusion than the information they convey. I have also reduced to a minimum the citations in the running text in order to avoid interrupting the reader. Only when I quote, or when either the opinion of the experts differs from mine or a special credit is due, have I added the corresponding specific citation.

Finally, an epistemological note: it is clear that no piece of scientific work can hide from its social and intellectual constraints. I would like to explicitly state the ones in which I think this book is inserted. The style, aim and contents of this handbook are consciously rooted in the democratic and open European society tradition. A rational view on society and the prevalence of logic and reason over other forms of understanding life and reality are also present. Also, a broad but clear positivistic approach to science intertwines the treatment of the topics: I do believe that if something cannot be thought of but has to be apprehended, or it is deemed as ineffable or as subjective, that is not suitable for scientific scrutiny. It is within those parameters that this book was developed.

Acknowledgments

Many people and institutions have to be acknowledged both in the motivation and in the making of this book. I will always be in debt to Professor José-Luis Casanova for making my academic career possible in the beginning, and continuing to help me since. My friend and colleague Professor Jordi Martí-Henneberg also needs to be thanked for his help in every aspect of my career, and in being an academic role model with his *seny*. The research generating this book was also greatly influenced by Professor Horacio Capel: it was a privilege to be able to discuss science with such a prominent figure in the international academy. The time spent in Barcelona University deeply contributed to consolidating many of my views on science and society.

I am also in debt with Dr. Chris Kidd. The three years I spent as research fellow at Birmingham University contributed to the formulation of the first ideas of this handbook. This book would have never been possible without the campus's main library. Also, the library and the librarians of Imperial College at London have to be acknowledged: the two years I spent doing the research for this book at the Kensington Campus were amongst the most productive of my career thanks to their competent and silent work.

My closest colleagues at the Earth Sciences group in the University of Castilla-La Mancha (UCLM), Professor Manuel Castro, Dr. Miguel Angel Gaertner, Dr. Clemente Gallardo, and Dr. Enrique Sánchez deserve my gratitude for making my life easier and happier in our daily routine. The same applies to Dr. Josu Mezo. I am also grateful for the efforts of my research team: Ana Mateos, Eva Salgado, Kerstin Burckhart and Alex Simón Colina. Thanks are also due to Kelsey Boast who proofread and carefully revised the final manuscript.

A final special thanks is due to Professor Richard Smith and to the Department of Geography at Cambridge University for being so kind and inviting me to join the department whilst preparing the final layout of this book. It was an enormous privilege to interact with Professor Sir Tony Wrigley and Dr. Roger Schofield, amongst many others.

Francisco J. Tapiador

Cambridge, England (winter of 2006)

Contents

Introduction. The Rural Areas of the World

The importance of the rural areas for humankind is apparent. Rural areas have hosted the main part of the human population until the twentieth century, and today they generate the bulk of the food, energy and materials required by the whole society. At the end, they represent the product of a continuing struggle between humankind and nature, illustrating how humans have prevailed over a harsh environment.

Rural areas are one of the most notorious early fingerprints of humankind in nature. They represent the best (and the worst) that humans do to strive and progress, and thus represent a way to better know our selves. Landscapes such as Somerset in England, and most of the Dutch landscape, are the result of human contribution to the beauty of the planet. Making habitable the inhospitable can be regarded as a conquest of human ingenuity, and rural landscapes worldwide show how successful societies have been in doing so. But rural landscapes are also the product of the environment, and much knowledge can be extracted about the fabric of reality by studying this part of nature. The subtle links between climate and rural land use have only recently been fully demonstrated by modelling, to name just one example.

What is a rural area? If we follow the definition of the European Commission a rural area is 'a spatial phenomenon that extends across regions, landscapes, natural areas, agricultural land, villages and other larger urban centres, pockets of industrialisation and regional centres. It encompasses a diverse and complex economic and social fabric. It is the home of a great wealth of natural and cultural resources and traditions. It is becoming more important as a place for relaxation and leisure activities'. Also, in order to help to classify areas into rural land, the European Commission provided a set of spatial features in terms of landscapes: rural areas close to highly urbanised areas are those characterised by a residential, recreational and industrial overspill; a growth in population, intensive agriculture and growth in traffic flows; rural areas used for tourism are characterised by being predominantly coastal and mountain areas, often equipped for mass tourism, with a reduction of agricultural activity, development of agglomerations and fragmentation of habitats; rural areas where access is difficult could be characterised as being areas of mountains, islands, forests, with a high out-migration and inward migration of retirees/second home owners; and the classical rural areas are those predominantly agricultural, characterised by agriculture that is highly productive or efficient, traditional or weak, with low productivity. Within this categorization, most rural areas host diverse activities, characterised by a high dependency on agriculture, and by the development of complimentary activities.

A more precise definition can be provided using quantitative measurements. Population density is an objective value that can be used to decide if an area is rural or not. The final result depends on the threshold used, which can vary depending on the country. For instance, the OECD define rural areas as those with less than 150 inhabitants/sq. km while EUROSTAT use a value of 100 inhabitants/sq. km. But the population density by itself is just the first step to define rurality. A finer calculation can be used to provide not two but three classes. An area is said to be predominantly rural if more than half of the population lives in rural communities (being those with less than 10,000 inhabitants); significantly rural if between 15%–50% of the population lives in rural communities; and predominantly urban if less than 15% of the population lives in rural communities. This working definition has its problems but can be used as a guide to decide what is rural and what is not.

The previous notes provide an idea of the processes involved in rural analysis and the resulting landscapes, and shows the difficulty in defining such an apparently obvious subject. In the rest of this book, we will deal with this diversity. The aim is to provide a methodology to dissect the different components in which rural areas can be analyzed, providing a guide to further study. Given the broad nature of the field, it is impossible to delve into every element at a professional level; a complete discussion on rural climatology or ecology would be a handbook in itself. Moreover, it is unnecessary to perform such a deep analysis since this book is aimed to those coordinating or planning rural analyses, or performing rural management tasks at a higher stage than the technical level. It is for these professionals (and those studying to become professionals) for whom this handbook has been written. Even if they are masters in one particular area, for instance graduated geologists or sociologists, they will find it useful to know what to expect from other professionals' reports, or what should be asked for when commissioning a study in a particularly relevant aspect of a rural area. Also, they will find how its particular expertise makes links with the rest of the disciplines involved in rural earth science, in both the analysis and the management components.

References and Bibliography Used

Canadian Rural Information Service. The OECD definition of 'rural' (2000)
European Commission. The Future of Rural Society. European Commission (1988)
European Commission. Europe 2000: Co-operation for European Territorial Development (1994)
European Commission. Rural Development Regulation (No. 1257/99) European Commission (1999)
European Commission. New Rural Development Regulation (No. 490/2004). European Commission (2004)
European Commission. Third Cohesion Report - A new Partnership for Cohesion, Convergence (2004)
European Commission, Directorate-General for Agriculture. CAP Reform: Rural Development (2000)

European Rural Exchange. Integrated Rural Development, Policy and Practice, A discussion paper by the European Rural Exchange. European Rural Exchange (2002)
NESC. New Approaches to Rural Development (1994)
NIAO. Report on the Rural Development Programme 1994-99 (2000)
NISRA Final Report of the Inter-Departmental Urban-Rural Definition Group: Classification and Delineation of Settlements (2004)

Chapter 1
Methodological Notes on Analysis, Synthesis and Management

Quantitative information is the basis of any scientific analysis of geographical space, and thus is the basis of rural analysis and management. While qualitative approaches can help to organize information in the initial stages of any study in a meaningful way, the ultimate scientific analysis would require a contrastable support to make decisions objectively-based, in spite of the political orientation of the final decisions.

In that context, the role of the rural analyst is to provide the decision-makers with the (best available) objective analysis. The decision agents would then decide their policies based upon this scientific assessment. Since in democracy politicians represent the people's will, the ultimate role of the analyst is to extract knowledge from information using scientific tools, and let the people decide by themselves in light of this knowledge.

A further step is the management stage of rural areas: which is the process of decision-making based upon the best scientific advice. This task must also be based on the best available information, and therefore the role of the analyst is instrumental too.

Rural analysis is a very complicated subject indeed. To deal with this truth there have been two approaches in the past: one is to avoid to actually doing a rural analysis and endlessly discussing on the difficulty of the task, usually unfairly criticizing pioneering techniques and thus incomplete approaches; or to attempt to divide the actually difficult task into more manageable divisions and painstakingly start solving the puzzle by using quantitative methods. To understand how an Airbus A350 flies is also a very complicated task for a layman, but no hypothetical flight analyst would only discuss the difficulty of the task without providing some technical insight. On the other hand, he would propose contrastable hypotheses based upon the available information, he will test these hypotheses with experiments or simulations, and he will provide explanations and proposals. A rural area is neither more or less complicated that an airplane once the appropriate methodology is designed and put into practice.

1.1 The Analysis Team

The size of a rural analysis team is highly variable. It may comprise of a single person doing a preliminary study of limited scope, or many professionals working together in a very complicated area.

The actual work of the coordinator can be compared with that of an orchestra conductor. Even if the conductor is himself a competent violinist, he would harmonize different expertise to bring together the best of each instrument. He will know the emphasis required for each phrase, and the combination to produce the best result. A geologist could be an expert on hydrology, but if he is trained as a rural analyst he will act more on the analysis-synthesis side of the whole work rather than as a proper hydro-geologist. Otherwise, the rural analysis work will be doomed from the start.

The coordinator does not need to have a specific background. It is true that some profiles such as 'geographer' or 'environmentalist' have a longer tradition in analyzing territory, but there are not specific needs to act as a coordinator as long as the professional has the required previous experience and a suitable formation in both natural and social sciences.

The coordinator will be the only person in the team directly dealing with the client/s, having the ultimate responsibility on the outcome of the process. His coordination task within the analysis team will be orientated to define the guidelines of the work, hiring the appropriate expertise when required, writing down the conclusions and policy recommendations, and having the last word in the whole technical process. It is not recommended for the coordinator to collaborate in a particular topic within the project, no matter how experienced he is in the subject. Rather, he is intended to maintain a balanced view on every aspect of the analysis.

If the scale of the work is large, the coordinator post can be split into a technical and a managing post. The first will directly deal with the analysis team, reporting to the managing director. The final responsibility however, should always be borne by the managing director, who is the person in contact with the client.

A small core rural analysis team may comprise of the coordinator and a natural scientist, a social scientist and an economist or policy analyst. Thematic experts (depending on the problem) may join the core team in a permanent or temporal task devoted manner. This includes soil scientists, naturalists, geologists, natural risks analysts, ecologists, agricultural experts, civil engineers, sociologists, GIS experts, economists, remote sensing experts, PR, managers, publicists, etc. They will provide the specific knowledge to develop the scheduled tasks, and the expertise required to let the core team know the critical aspects of their respective fields in the case under scrutiny.

The field work side of the project is also essential. It is almost always unwise to resort to students or part-time, underpaid staff. Professional free-lance people working either full or part-time can be expensive to hire, but the results pay off in terms of productivity and in the final quality of their products. At the end, economic logic prevails and as in other economic activities what you pay for is what you get.

1.2 Ethics and Deontology

The same caution about underpaid staff applies to the budgeting of the work. A good professional work is expensive to carry out and the client must acknowledge this fact. If the budget proposed by the commissioner is clearly insufficient to carry the analysis in a professional way the team may propose a downscaled version of the work: that is, a work with either a limited field work, or just a description of the area without a deep analysis. What is not advisable is to accept a large project without the required funding. The results will never be up to the expectations of the commissioners, and the credibility and prestige of the team will thus be affected.

1.3 Work Breakdown

Rural analyses are usually commissioned by public administration, either at local, regional or national level. In the EU, the task can be linked to specific programs (ERDF, EQUAL, etc.) as a part of the assessment of the programs, or as preparation for bids. In any of these situations, the scheme of a rural analysis study remains the same, as described in Table 1.1. The description of each phase is as follows:

Table 1.1 Work breakdown in a rural analysis project

PRELIMINARY FEASIBILITY STUDY
1- Client specifications, aims and objectives
2- Problem definition
3- Budgeting and agreement on a preliminary study
4- Field work, phase 1
5- Need and requirements of specific analyses
6- Preliminary feasibility study. Basic indicators analysis
7- Project definition meeting and budgeting. Client approval
8- Acceptance. Commissioning of external specific analyses
DATA GATHERING
9- Database, archives and information sources identification and exploration
10- GIS data model definition, version 1
11- Field work, phase 2
12- Meeting with the social agents, clients, expert panel and team
13- Additional information gathering
PRELIMINARY ANALYSES
14- Integration of data from specific analyses fieldwork
15- GIS redefinition and GIS data model, version 2
16- Data processing
17- Cross-relationships identification
18- Models development
19- Preliminary analyses
20- Users and agents feedback
FINAL ANALYSIS
21- Proposals and scenario development
22- Third field work visit
23- Draft report and tools for public participation
24- Final meeting with the client
25- Public presentation and participation
26- Integration of results from public participation
27- Delivery of the final report

1. **Client specifications, aims and objectives.** It is crucial to discuss the aims, objectives, constraints and motivations of the people commissioning the study. It is advisable to keep a record of the first meeting (as meeting minutes), circulate this document after the meeting, and explicitly ask for the approval of the participants on what was written down in the minutes. This will save a lot of work afterwards, preventing likely misunderstandings and false expectations, and focusing the work on what is actually expected. The aim of this task is to have a first contact between the commissioners and the analysts, but it is critical to keep a formal record of the meeting.
2. **Problem definition.** The first task is to define the geographical area of study. This may seem trivial, but sometimes it is required to consider a larger area than the one defined by the client to fully capture the processes in operation, and to provide workable development guidelines. It is up to the analysts to decide if that is the case and propose any adjustment to the client/s. It is also imperative to explicitly state the aim and objectives of the work in the light of the first meeting and consider the possibilities and likely needs of the work.
3. **Budgeting and agreement on a preliminary study.** It is advisable to propose a preliminary study to the clients, with a budget between 5–15% of the final full project. The aim of this preliminary study would be to fully identify the complexity of the problem under study, and to provide a first glimpse of the final product to the clients. Even if the client prefers a direct study, it is advisable to divide the work in this way, since it helps the analysis work.
4. **Field work: Phase 1.** The first technical visit to the area under study can be carried out once the preliminary project has been accepted (and not before, since it is a relevant cost that the team does not need to assume before a formal agreement). The purpose of this visit is so the core analysis team physically gets to know the place and use their expertise to assess the scale of the project, the potential problems, and the main directions of the main research.
5. **Need and requirements of specific analyses.** At this stage, the core team should decide if additional expertise is needed. If an area presents a specific problem (if there are, for example, clear natural risks to be considered in a mountain area), it would be advisable to integrate an expert in the analysis team or to commission a specific study. Since this may be costly and time consuming, it is vital that this need is addressed in the first stages of the project.
6. **Preliminary feasibility study.** The preliminary study should comprise of a schematic assessment of the rural area, without any recommendations, diagnoses or planning ideas. It can be based on indicators (cf. Sect. 2.1 below). The full scale of the work should then be stated here. It is dangerous to either propose an oversized work for simple cases and an overambitious project for complex areas. It is preferable to define and accept small work and have the work done in three months rather than oversell and delay a small project for longer than required. The need of specific studies should be justified in the preliminary report taking into account the basic indicators.
7. **Project definition meeting and budgeting: Client approval.** The preliminary study should be the basis for defining the whole project with the client: to define the economic needs and to have the project budget and structure fully accepted.

If this is not the case, the analysis stops here and the preliminary study is billed to the client.

8. **Acceptance: Commissioning of external specific analyses.** If the client accepts the proposal it is then up to the team to decide if they take the commission. Ethical –and not only economical grounds- are issues to be considered by the team. If the commission is finally accepted, it is the time to start the specific studies since they are likely to be the more time consuming tasks and they do not directly depend on the core team. Enough time should be allocated to be able to re-commission any study if the first choice team fails to provide a timely report for any reason.
9. **Database, archives and information sources identification and exploration.** At the same time the specific study starts, the core team has to carry out the data gathering. An analysis should be based upon quantitative data and not on opinions or subjective appraisals. It is necessary to identify the data sources, the relevant archives (if any) and start storing the information.
10. **GIS data model definition: Version 1.** At the same time, and in a feedback process with the previous task, the data model of the GIS should be built. This data model should be flexible enough to incorporate forthcoming changes and should always be seen as a preliminary product.
11. **Field work: Phase 2.** The second phase of the fieldwork should then be carried out. This is when the intensity of the field work is greater, and when most of the field work should be done, now that it is becoming increasingly clear what is significant and what is not in the area. A visit to the immediate neighbouring areas is also advisable.
12. **Meeting with the social agents, clients, expert panel and team.** The second fieldwork visit can be used to have a first meeting with the agents involved. This phase is crucial to assess the self-perception of the rural inhabitants on their own area in order to find potential problems.
13. **Additional information gathering.** It is likely that new data sources will appear both in the second field visit and in the contact with the agents. It is thus advisable to have a second round of information gathering. It is noticeable that this newly available information is usually inaccessible, and thus it should be properly valued. As in the rest of the phases, the new information must be properly described, documented and filtered of errors.
14. **Integration of fieldwork data from specific analyses.** By the time the core team arrives to this phase, the specific analyses should have been finished. (To make this possible, the schedule of such analyses would have to be defined accordingly). It is now time to integrate any field work or data recorded in those analyses.
15. **GIS redefinition and GIS data model: Version 2.** In light of the new information, it should be required to redefine the GIS data model, the scale of the GIS, the GIS software to use, and the analytical tools used in the GIS.
16. **Data processing** Once the final data model has been decided, it is time to link all the databases and process the information to help the analysis.
17. **Cross-relationships identification.** As will be discussed in Chaps. 4 and 6, one of the most difficult tasks in rural analysis is to establish the cross-relationships

between environmental and anthropic elements. It is therefore sensible to dedicate a complete task to this topic.

18. **Models development.** If there is enough information to set up a model it should be done at this stage. This may involve the use of additional expertise, providing the added value of this section is justified.
19. **Preliminary analyses.** A preliminary analysis consists of the integration of the modelling results and the results of the cross-correlations and analysis tasks. This should be done by the core team, but a brain storming session with the extended team may help to initiate the work.
20. **Users and agents feedback.** The preliminary analyses should be presented to the users to pulse their first impressions. Modifications may be needed at this stage if the work does not match with the requirements of the patrons, if new ideas arise in this phase, or if more intensity is to be put in at any particular point. A preliminary, agreed document can then, and only then, be presented to the public and agents if deemed as necessary. Further modifications may also arise in this stage.
21. **Proposals and scenario development.** As will be discussed in the last part of this book, it involves defining likely scenarios and developing proposals to improve the rural area. Nonetheless, a rural analysis study does not necessarily need to have a proposals part. The natural clients for rural analyses are the local corporations themselves or regional or national authorities. However, private clients exists and they may want the study of the rural area for marketing or selling purposes, or to characterize the area in order to plan future initiatives, such as a new dam, a motorway, etc.
22. **Third field work visit.** A new, as-fresh-as-possible look to the area should now follow. One aim is to make sure that no relevant element has been left out. The other aim is to test the actual feasibility of the proposals and scenarios.
23. **Draft report and tools for public participation.** A draft version of the final report and the tools devised for allowing public participation in the proposals debate should now be provided.
24. **Final meeting with the client.** The draft should be presented to the client to ensure that the specs have been met and the objectives achieved.
25. **Public presentation and participation.** The public presentation of the report is not only a social act, but an opportunity to engage people into future projects: if the agents can see the usefulness and logic of the project many future obstacles can be easily overcome.
26. **Integration of results from public participation.** This usually happens when the previous phase generates new input and ideas that would benefit the work. If so, they should be integrated into the final report, either as an appendix or in the relevant part.
27. **Delivery of the final report.** The final report is the centerpiece of a rural analysis. It represents the sum of the analytical efforts and is a permanent record of the data gathered and the interpretation done. It is advisable to divide the report in a brief (one page) summary with the major points, a cartographical appendix, the main text and as many appendixes as required. In terms of language, the main text should be as clear as possible. It should be kept in mind that a rural

analysis report is not a literary work. It does not have to be interesting or catching, nor even visually appealing. The language should be as plain as possible, not necessarily in the line of what is known as "plain English" but in terms of avoiding unnecessary rhetoric and jargon. Technical terms should be kept to the appendixes, if possible.

Most of the usefulness of a good report lies, paradoxically, in its aridness. The ability to extract all and nothing but what is worth to be said of a rural area will be reflected in the painstakingly gathered information, in the summary and treatment of measurements, and in the clear formulation of the observed relationships. Comprehensive databases of linked information are also a part of the memory and should be provided. One of the real problems that may arise in a rural analysis is not taking a wrong decision or providing a spurious interpretation, but to forget a key element or not to explore a relevant relationship. If each and every element is explored and the methodological decisions are written down, other analyses can be discrepant but the memory will be consistent.

The structure of the memory should be as standardized as possible. The scheme provided in this handbook is only one possible option; perhaps one with a minimum set of requirements in a real, not only academic, rural analysis and can be adapted to the needs and experiences of the practitioners.

References and Bibliography Used

C. J. Andrews, 'Analyzing quality-of-place', Environment and Planning B-Planning & Design 28 (2001), 201–217.

C. Blackorby, W. Bossert and D. Donaldson, 'Intertemporal Population Ethics - Critical-Level Utilitarian Principles', Econometrica 63 (1995), 1303–1320.

A. Blowers, 'Critical Rationalism and Planning Methodology - Faludi, A', Regional Studies 21 (1987), 279–280.

R. S. Bolan, 'Planning ethics: A reader in planning theory, practice and education - Hendler, S', Journal of Planning Education and Research 16 (1996), 66–67.

I. Bonn and J. Fisher, 'Corporate governance and business ethics: insights from the strategic planning experience', Corporate Governance-an International Review 13 (2005), 730–738.

W. Bor, 'Toward a New Planning Methodology', Journal of Town Planning Institute 52 (1966), 405–413.

R. Brownhill and L. Merricks, 'Ethics and science: Educating the public', Science and Engineering Ethics 8 (2002), 43–57.

E. K. Burns, 'Planning ethics: A reader in planning theory, practice and education', Urban Geography 18 (1997), 469–470.

B. Burrows, 'Essay Review - the Greening of Business and Its Relationship to Business Ethics', Long Range Planning 26 (1993), 130–139.

H. Campbell and R. Marshall, 'Moral obligations, planning, and the public interest: a commentary on current British practice', Environment and Planning B-Planning & Design 27 (2000), 297–312.

H. Campbell and R. Marshall, 'Towards justice in planning: A reappraisal', European Planning Studies 14 (2006), 239–252.

A. J. Catanese, 'Ethics in Planning - Wachs, M', Growth and Change 17 (1986), 74–75.

A. Cogan, 'Planning ethics: A reader in planning theory, practice, and education - Hendler, S', Journal of the American Planning Association 62 (1996), 264–265.

M. Collins, 'Dilemmas of Planning Practice - Ethics, Legitimacy and the Validation of Knowledge - Thomas, H, Healey, P', Regional Studies 26 (1992), 215–216.
C. Cormio, M. Dicorato, A. Minoia and M. Trovato, 'A regional energy planning methodology including renewable energy sources and environmental constraints', Renewable & Sustainable Energy Reviews 7 (2003), 99–130.
W. Cragg, D. Pearson and J. Cooney, 'Ethics, surface mining and the environment', Resources Policy 21 (1995), 229–235.
R. Darke, 'Dilemmas of Planning Practice - Ethics, Legitimacy, and the Validation of Knowledge - Thomas, H, Healey, P', Environment and Planning A 24 (1992), 452–453.
R. Darke, 'Planning ethics: A reader in planning theory, practice and education - Hendler, S', Environment and Planning A 29 (1997), 373–375.
S. Donnelley, 'Nature, polis, ethics - Chicago regional planning', Hastings Center Report 28 (1998), S1–S1.
J. Eriksson, A. Ek and G. Johansson, 'Design and evaluation of a software prototype for participatory planning of environmental adaptations', Ieee Transactions on Rehabilitation Engineering 8 (2000), 94–106.
A. Faludi, 'Critical Rationalism and Planning Methodology', Urban Studies 20 (1983), 265–278.
A. Faludi, 'Why in planning The Myth of the Framework is anything but that', Philosophy of the Social Sciences 28 (1998), 381–399.
K. A. Fox, 'Describing and measuring socioeconomic systems: Prerequisites to planning', Social Indicators Research 39 (1996), 281–305.
B. Gleeson and R. Kearns, 'Remoralising landscapes of care', Environment and Planning D-Society & Space 19 (2001), 61–80.
D. Goulet, 'Ethics of Development-Planning', Studies in Comparative International Development 11 (1976), 25–43.
J. Gross and J. Talavage, 'Multiple-Objective Planning Methodology for Information-Service Managers', Information Processing & Management 15 (1979), 155–167.
L. Holecek, 'System Elements in Long-Term Planning Methodology', Politicka Ekonomie 30 (1982), 1071–1084.
R. Home, 'Measuring Trends in Town-Planning Control through Decision Statistics', Local Government Studies 13 (1987), 51–62.
A. Hooper, 'Critical Rationalism and Planning Methodology - Faludi, A', Town Planning Review 57 (1986), 334–335.
R. C. Hoover, 'A View of Ethics and Planning', Journal of the American Institute of Planners 27 (1961), 293–304.
E. Howe and J. Kaufman, 'Ethics and Professional Practice in Planning and Related Policy Professions', Policy Studies Journal 9 (1981), 585–595.
B. B. Johnson, 'Ethical issues in risk communication: Continuing the discussion', Risk Analysis 19 (1999), 335–348.
M. Kihl, 'Planning ethics: A reader in planning theory, practice, and education - Hendler, S', Policy Studies Journal 23 (1995), 551–554.
Y. G. Kim, S. H. Yu and J. H. Lee, 'Knowledge strategy planning: methodology and case', Expert Systems with Applications 24 (2003), 295–307.
R. W. Lake, 'Planning and Applied Geography - Positivism, Ethics, and Geographic Information-Systems', Progress in Human Geography 17 (1993), 404–413.
J. Macbeth, 'Towards an ethics platform for tourism', Annals of Tourism Research 32 (2005), 962–984.
S. J. Mandelbaum, 'Ethics in Planning - Wachs, M', Journal of Policy Analysis and Management 5 (1986), 830–832.
S. J. Mandelbaum, 'Acting on Ethics in City-Planning - Howe, E', Journal of Urban Affairs 17 (1995), 328–330.
P. Marcuse, 'Professional Ethics and Beyond - Values in Planning', Journal of the American Institute of Planners 42 (1976), 264–274.
P. Marcuse, 'On Ethics, History, and Planning Practice', Journal of the American Planning Association 55 (1989), 500–500.

B. Mitchell, 'Ethics in Planning – Wachs, M', Professional Geographer 38 (1986), 456–457.
D. C. Perry, 'Dilemmas of Planning Practice - Ethics, Legitimacy and the Validation of Knowledge - Thomas, H, Healey, P', Journal of Planning Education and Research 14 (1994), 67–71.
J. Richmond, 'Ethics in Planning - Wachs, M', Journal of the American Planning Association 52 (1986), 370–371.
J. M. Rubenstein, 'Acting on Ethics in City-Planning - Howe, E', Annals of the Association of American Geographers 85 (1995), 612–614.
P. E. Salkin, 'Legal ethics and land-use planning', Urban Lawyer 30 (1998), 383–404.
H. Thomas, 'Planning ethics: A reader in planning theory, practices and education - Hendler, S', Urban Studies 33 (1996), 1233–1234.
B. Townley, 'Managerial technologies, ethics and managing', Journal of Management Studies 41 (2004), 424–445.
J. Tyrwhitt, 'Idea Method - New Development in Regional Planning Methodology', Ekistics 26 (1968), 185–195.
N. Verma and T. Clapp, 'Planning ethics: A render in planning theory, practice and education', Journal of Architectural and Planning Research 15 (1998), 173–174.
M. Wachs, 'Ethics in Planning'. New Brunswick: Center for Urban Policy Research, Rutgers University, 1985.
P. Whitehead, 'Addressing Planning Methodology - Unanswered Questions - a Review Article', Town Planning Review 59 (1988), 325–328.

Chapter 2
The Measuring Process in Rural Analysis

Science is made of measurements. If something can be quantified, either as pure numbers or as variables, then a scientific explanation is possible. If not, the problem is not suitable for scientific explanation. But even if a process can be quantified, it is necessary to make sure that the results conform to a coherent epistemological theory.

Several procedures have been devised to test the degree of scientificity of a given methodology. One of the most productive is Popper's criteria: a given statement can be said as scientific if it is possible to find out a fact that negates the statement. An even harder view of what is science and what is not is the Archimedean (after Archimedes) approach, in which with a set of self-consistent axioms generates a hypothesis that is tested against measurements of our world. Modern physics for example follows an Archimedean method. Nonetheless, no Archimedean social or management science has (yet) been devised. This would represent a formidable step for human knowledge, but several factors hinder the progress in that direction, such as the difficulty of collecting the large amount of required data, or the scarcity of suited methods to specifically filtering out spurious social information. Inversion constraints also affect the progress of this area: if more resources would be devoted to better estimate uncertainties in official questionnaires, social sciences would be able to give more precise measurements and thus improve our knowledge on the relationships between the environment, our society and us.

Nonetheless, all we can do at the moment in rural analysis is to devise a method to pave the way towards that objective. The basic requirements of any method to analyze the rural areas can be summarized in several broad points:

- The analysis should be based on available data. One of the ends of rural analysis is rural management, and managers need evidence and not just an educated guess to proceed.
- The analysis should be able to be checked. A person having the same tools should arrive to the same conclusion using the same data. This is a basic requirement for repeatable experiments, and also for many aspects of social research.
- The analysis should be comprehensive. All the relevant elements must be recorded, and the effects and feedbacks should be analyzed.
- The method of analysis should be able to be taught. If the method followed depends on an expert appraisal acquired after years of experience, it would be useless to attempt any rural analysis.

The method described in this handbook fulfils these four criteria. It is not a closed methodology, however, and can be adapted to different situations and available means if needed. What is advisable is to make sure that the methodological proposal is consistent and based upon objective estimations.

2.1 Variables and Errors

Variables can be either numerical or categorical. Most of the categorical variables can be set as numerical: logical values (yes/no) can be set as (1/0), degrees of public health warnings can be scaled from 0 (insignificant) to 10 (extremely high). The degree of error in the measurement can be expressed by using error bars: if a global change scenario predicts a $4°C \pm 3$ increase in surface temperature for the next ten years that means that the actual value provided by the model is 4 Celsius and the uncertainty 3 Celsius. It should be noted that the error interval can be greater than the actual estimate. The difference between accidental and systematic errors has also to be considered in establishing the error bars.

It is sometimes useful to express data in the form of indexes. This enables comparison between otherwise difficult to grasp values. Normalizing the values to a 100-base, or between 0 and 1 can help to study evolutions or to compare dissimilar data. Zero usually corresponds with the lowest value and 1 with the highest, but the scale can be inverted in some cases to ease interpretation.

The procedures to define and calculate suitable indexes and to normalize scales can be found in most statistical textbooks. One particularly central index in the social sciences is Gini's index, used to analyze inequality. This is so widely used that a working knowledge of it would be of use for the analyst. The actual procedure to do so can easily be found in the economic literature, and most statistical software packages include this function, including SPSS and Matlab. Other less used estimates include Atkinson's measure and Theil's index.

2.2 Basic Indicators

At the same time that the relevant elements of the analyses are described, the means to quantify those elements has to be found. If the nutritional status of children is deemed to be a relevant social element to be considered, a procedure to calculate or estimate this variable has to be described. The same applies for environmental elements, where the same variable can be known under different names depending on the discipline.

Most of the analysis depends on a suitable selection of the variables. As an example of variable selection, Tables 2.1 to 2.3 show some Agenda 21 sustainable development indicators. The measuring process in a rural analysis project goes beyond the requirements of sustainability programs but it is helpful to notice how variables are selected for this particular problem. As can be seen, the indicators

Table 2.1 Environmental indicators of sustainable development, UN Commission on Sustainable Development

ENVIRONMENTAL INDICATORS		
Theme	Sub-theme	Indicator
Atmosphere	Climate Change	Emissions of Greenhouse Gases
	Ozone Layer Depletion	Consumption of Ozone Depleting Substances
	Air Quality	Ambient Concentration of Air Pollutants in Urban Areas
Land	Agriculture	Arable and Permanent Crop Land Area
		Use of Fertilizers
		Use of Agricultural Pesticides
	Forests	Forest Area as a Percent of Land Area
		Wood Harvesting Intensity
	Desertification	Land Affected by Desertification
	Urbanization	Area of Urban Formal and Informal Settlements
Oceans, Seas and Coasts	Coastal Zone	Algae Concentration in Coastal Waters
		Percent of Total Population Living in Coastal Areas
	Fisheries	Annual Catch by Major Species
Fresh Water	Water Quantity	Annual Withdrawal of Ground and Surface Water as a Percent of Total Available Water
	Water Quality	BOD in Water Bodies
		Concentration of Faecal Coliform in Freshwater
Biodiversity	Ecosystem	Area of Selected Key Ecosystems
		Protected Area as a % of Total Area
	Species	Abundance of Selected Key Species

Table 2.2 Social indicators of sustainable development, UN Commission on Sustainable Development

SOCIAL INDICATORS		
Theme	Sub-theme	Indicator
Equity	Poverty	Percent of Population Living below Poverty Line
		Gini Index of Income Inequality
		Unemployment Rate
	Gender Equality	Ratio of Average Female Wage to Male Wage
Health	Nutritional Status	Nutritional Status of Children
	Mortality	Mortality Rate Under 5 Years Old
		Life Expectancy at Birth
	Sanitation	Percent of Population with Adequate Sewage Disposal Facilities
	Drinking Water	Population with Access to Safe Drinking Water
	Healthcare Delivery	Percent of Population with Access to Primary Health Care Facilities
		Immunization Against Infectious Childhood Diseases
		Contraceptive Prevalence Rate
Education	Education Level	Children Reaching Grade 5 of Primary Education

(continued)

Table 2.2 (continued)

SOCIAL INDICATORS		
Theme	Sub-theme	Indicator
		Adult Secondary Education Achievement Level
	Literacy	Adult Literacy Rate
Housing	Living Conditions	Floor Area per Person
Security	Crime	Number of Recorded Crimes per 100,000 Population
Population	Population Change	Population Growth Rate
		Population of Urban Formal and Informal Settlements

Table 2.3 Economic and institutional indicators of sustainable development, UN Commission on Sustainable Development

ECONOMIC and INSTITUTIONAL INDICATORS		
Theme	Sub-theme	Indicator
Economic Structure	Economic Performance	GDP per Capita
		Investment Share in GDP
	Trade	Balance of Trade in Goods and Services
	Financial Status	Debt to GNP Ratio
		Total ODA Given or Received as a Percent of GNP
Consumption and Production Patterns	Material Consumption	Intensity of Material Use
	Energy Use	Annual Energy Consumption per Capita
		Share of Consumption of Renewable Energy Resources
		Intensity of Energy Use
	Waste Generation and Management	Generation of Industrial and Municipal Solid Waste
		Generation of Hazardous Waste
		Management of Radioactive Waste
		Waste Recycling and Reuse
	Transportation	Distance Traveled per Capita by Mode of Transport
Institutional Framework	Strategic Implementation of SD	National Sustainable Development Strategy
	International Cooperation	Implementation of Ratified Global Agreements
Institutional Capacity	Information Access	Number of Internet Subscribers per 1000 Inhabitants
	Communication Infrastructure	Main Telephone Lines per 1000 Inhabitants
	Science and Technology	Expenditure on Research and Development as a Percent of GDP
	Disaster Preparedness and Response	Economic and Human Loss Due to Natural Disasters

are as univocally described as possible, making likely an objective assessment of sustainability.

This example also raises a methodological point. Here, the variable selection actually helps to define the subject, removing any ambiguity. Sustainable development can be a controversial term to define by itself, but if we can describe how it is estimated, the methodology acts in an axiomatic way. The official definition of sustainable development (cf. Chap. 14) is 'a development that meets the needs of the present without compromising the ability of future generations to meet their own needs'. However, it is not clear how such compromise can be assessed: as we do not know what we still do not know, it is logically impossible to derive the consequences of our current actions. Nonetheless, by explicitly stating how we measure something called sustainable development we are providing an objective definition: sustainable development is a weighted combination of such variables whose values are measured using an established protocol. While one can discuss the suitability of the definition, no one would argue that the procedure is an objective and checkable analysis method, whatever the name of the combined estimate. From the epistemological point of view, that is even more relevant than a consensus on the definition regardless of the usefulness or application of the concept.

The set of indicators in Tables 2.1 to 2.3 can be used for the preliminary part of the rural analysis by adapting the indicators to the available information and to the rural specificities (cf. appendix).

References and Bibliography Used

D. Alard, J. F. Bance and P. N. Frileux, 'Grassland Vegetation As An Indicator Of The Main Agroecological Factors In A Rural Landscape - Consequences For Biodiversity And Wildlife Conservation In Central Normandy (France)', Journal of Environmental Management 42 (1994), 91–109.

A. Akkerman, 'Fuzzy Targeting of Population Niches in Urban-Planning and the Fractal Dimension of Demographic-Change', Urban Studies 29 (1992), 1093–1114.

E. R. Alexander, 'A mile or a millimeter? Measuring the 'planning theory-practice gap', Environment and Planning B-Planning & Design 24 (1997), 3–6.

M. A. Altaf and J. A. Hughes, 'Measuring the Demand for Improved Urban Sanitation Services - Results of a Contingent Valuation Study in Ouagadougou, Burkina-Faso', Urban Studies 31 (1994), 1763–1776.

I. Askew, B. Mensch and A. Adewuyi, 'Indicators for Measuring the Quality of Family-Planning-Services in Nigeria', Studies in Family Planning 25 (1994), 268–283.

N. C.Brady and R. R. Weil. 'The Nature and Properties of Soils'. New Jersey. Prentice Hall. (2002)

G. Bramley and C. Leishman, 'Planning and housing supply in two-speed Britain: Modelling local market outcomes', Urban Studies 42 (2005), 2213–2244.

S. D. Brody, 'Measuring the effects of stakeholder participation on the quality of local plans based on the principles of collaborative ecosystem management', Journal of Planning Education and Research 22 (2003), 407–419.

S. D. Brody, V. Carrasco and W. E. Highfield, 'Measuring the adoption of local sprawl - Reduction planning policies in Florida', Journal of Planning Education and Research 25 (2006), 294–310.

S. D. Brody, W. Highfield and L. Alston, 'Does location matter? Measuring environmental perceptions of creeks in two San Antonio watersheds', Environment and Behavior 36 (2004), 229–250.

S. D. Brody, W. Highfield and V. Carrasco, 'Measuring the collective planning capabilities of local jurisdictions to manage ecological systems in southern Florida', Landscape and Urban Planning 69 (2004), 33–50.

S. D. Brody and W. E. Highfield, 'Does planning work? Testing the implementation of local environmental planning in Florida', Journal of the American Planning Association 71 (2005), 159–175.

S. D. Brody, W. E. Highfield and S. Thornton, 'Planning at the urban fringe: an examination of the factors influencing nonconforming development patterns in southern Florida', Environment and Planning B-Planning & Design 33 (2006), 75–96.

I. Brotherton, 'A New Approach to Measuring Trends in Town-Planning Control', Local Government Studies 19 (1993), 156–162.

R. J. Chaskin, R. M. Goerge, A. Skyles and S. Guiltinan, 'Measuring social capital: An exploration in community-research partnership', Journal of Community Psychology 34 (2006), 489–514.

A. Cicerchia, 'Indicators for the measurement of the quality of urban life - What is the appropriate territorial dimension?' Social Indicators Research 39 (1996), 321–358.

M. W. Downton, 'Measuring tropical deforestation: Development of the methods', Environmental Conservation 22 (1995), 229–240.

R. Ewing, R. Pendall and D. Chen, 'Measuring sprawl and its transportation impacts', In Travel Demand and Land Use 2003 (2003), pp. 175–183.

E. Fein and I. Staff, 'Measuring the Use of Time', Administration in Social Work 15 (1991), 81–93.

K. Field, 'Measuring the need for primary health care: an index of relative disadvantage', Applied Geography 20 (2000), 305–332.

P. C. Fishburn, 'Measuring and Planning Quality of Life - Drewnowski, J', Social Indicators Research 2 (1976), 489–495.

C. Kremen, 'Managing ecosystem services: what do we need to know about their ecology?' Ecology Letters 8 (2005), 468–479.

E. V. Maas and S. R. Grattan, 'Crop yields as affected by salinity'. In R. W. Skaggs and J. van Schilfgaarde (Eds) Agricultural Drainage. Agron. Monograph 38. ASA, CSSA, SSA, Madison. (1999)

S. Pavoine, S. Ollier and A. B. Dufour, 'Is the originality of a species measurable?' Ecology Letters 8 (2005), 579–586.

R. L. Pressey, T. C. Hager, K. M. Ryan, J. Schwarz, S. Wall, S. Ferrier and P. M. Creaser, 'Using abiotic data for conservation assessments over extensive regions: quantitative methods applied across New South Wales, Australia', Biological Conservation 96 (2000), 55–82.

J. Triantafilis, A. I. Huckel and I. O. A. Odeh, 'Field-scale assessment of deep drainage risk', Irrigation Science 21 (2003), 183–192.

O. Vigiak, G. Sterk, R. J. Romanowicz and K. J. Beven, 'A semi-empirical model to assess uncertainty of spatial patterns of erosion', Catena 66 (2006), 198–210.

G. C. Zalidis, M. A. Tsiafouli, V. Takavakoglou, G. Bilas and N. Misopolinos, 'Selecting agri-environmental indicators to facilitate monitoring and assessment of EU agri-environmental measures effectiveness', Journal of Environmental Management 70 (2004), 315–321.

Part I
Elements of the Analysis

A rural analysis project comprises of two major tasks: analysis and synthesis. In the first phase, the natural (environmental) and social (anthropic) elements in a rural area are identified, and then individually analyzed. This provides a first approximation of the rural area being studied. The next step is to analyze the cross-relationship between the natural elements on one side, and the social elements on the other side. This will help in finding the prevalent elements, that is, those that have a greater influence on the area and therefore requiring more attention in a second, more detailed phase.

From a human perspective, the Earth has a natural environment in which we live (and that affects us) and has a set of elements created by us in order to survive and prosper. Also, there are a number of relationships between both worlds. These *invisible* networks also affect the rural areas and may even characterize a rural area.

Another reason why it is convenient to study environmental and anthropic elements separately is that traditionally much of human knowledge has been organized in this way. Rural analysis can benefit from previous work in ecology, hydrology and sociology, to name only three relevant fields.

The rationale behind differentiating between environmental and anthropic elements is due to the anthropic perspective of rural analysis. Nonetheless, humans are part of the natural world, and as such no differentiation between natural and artificial (or man made) can be done. The only difference with other animals is the magnitude and technical capabilities of the humans.

While it can be argued that humankind does not have any ethical responsibility to the natural world, the key to survive is a wise management of the natural resources and a careful monitoring of the environmental conditions in which we live. In the end, the ultimate goal of biodiversity is to avoid running out of the resources the natural world offers to us. We might need of this variety to improve our lives, albeit we do not necessarily know how or when. It is thus advisable to understand the processes involved, and to that end the first step is to study the elements individually.

Chapter 3
Environmental Elements in Rural Analysis

Environmental elements define the natural framework of a rural area of study. These must be taken into account in any agrarian analysis, even if the area is highly anthropized. The emphasis to be put on each of these elements should be decided by performing an initial reading of the environment in a first visit to the field. In this visit, the analysis would decide if, for example, geology deserves a deeper analysis than just a few notes or if soils are likely to play a crucial role in rural improvement perspectives. For instance, a flat sedimentary rural area could require far less geological analysis than a rural area near ores or in a very anfractuous region. Conversely, agricultural areas affected by soil degradation require different methods of analysis than soil degradation in forest or recreational areas.

Several tools can help to prioritize the importance of each element in the analysis:

- Regional geographical monographs are a privileged source of expert knowledge on many areas, but they are not always available.
- Meetings with the local people can also help to provide a first overview of the relevant natural and anthropic elements.
- Research on information systems, literary reviews and trip's taken to the area to carry out field analyses can also help to provide a first idea on what must be investigated and to what extent.

A deeper study would require specialized (technical) analyses. The work of the rural analyst is to not only cover all aspects of the investigation, but to be able to select and coordinate several experts. They must provide an integrated approach using their findings and propose recommendations to the 'decision-makers'. In order to achieve this, one must possess a deep understanding of several fields and have a clear idea of what is needed to construct a synthesis. It would be worthless, for example, asking a Pedologist for a complete soil characterization if the analyst does not know what a 'fluvisol' is and its relevant importance in agriculture. A multidisciplinary talent is required; a skill which can, with practice, be developed. The analyst does not need to be an expert in every field, but should be able to co-ordinate and integrate all factors involved.

It is advisable to be systematic in the analysis of environmental factors. This chapter provides a minimum set of elements to be considered by the analyst. This means that it is possible that some other elements appear in the analysis of a given area, but the elements described here have to always be considered. Also, cross-relations between ought to be examined. This task may seem tiresome and repetitive, but it is essential for a comprehensive analysis.

This section provides several items that the analyst should use to construct his own checklists. An initial (careful) consideration of all of them and their cross-relation can take hours or days, and the actual analysis of the most relevant of them can take weeks of hard work, but this procedure presents two major advantages compared with other approaches. First, it provides the analyst with a guide that helps to avoid forgetting crucial aspects that can be hidden at the first sight. Secondly, this task generates objective documents that can be used to make the case for the final proposals. An example of the first situation was a case in which a rural area had a promising large extension of fertile riparian land to be used as new agricultural area. The edaphological analysis showed good prospects for the required diversification of the local economy by cultivating fresh vegetables for the regional market. The hydrological analysis showed that irrigation was guaranteed and sustainable, but two other reports shown that a) the economy of the area would be heavily dependent on these crops and b) there was a return period for floods of about 25 years for the area, with the potential of devastating effects including personal losses. The analysts balanced this (and other) information and presented the community with the opportunities and risks involved in such a decision. Mitigation strategies were thus developed in case of a positive decision on the diversification option being made, including insurance cover and a primitive (but effective) early warning system. It is doubtful that a disorganized appraisal of the situation would have revealed all these elements, considering that the flood hazard was almost unknown to the local people. A careless approach would have resulted in severe losses for the newly-developed local diversified economy beyond its recovering point, and even in human casualties. The professional obligation of thinking in each and every topic can avoid terrible mistakes. Besides, this protocol provides a learning curve for the analyst. It is true that many factors can be overseen, but if a guide is followed, the probability of these mistakes greatly decreases.

The environmental elements checklists are not only a binary, yes/no matrix to be fulfilled. Each element to be considered develops itself so the analyst should also considerer the scheme he needs to provide to the professionals in order to acquire further information from them. Specific questions such as the persistence of snowfall or hail would not be obvious topics to tackle for the climatologist, so they have to be specifically asked for. If the analyst is concern about potential economic impacts of changes in the neighboring areas, the economist will have to provide a report on this kind of sensitivity for the rural community. While most of these issues would naturally appear in the professional reports, it would save time and money to provide the consultants with a guide on what is expected to be learnt with their expertise: a set of tables, forms and documents suitable to be integrated into presentations and reports would help in the advanced stages of the project.

3.1 Geology

The first element of the analysis looks at the characteristics of the rural substrate. Geology can be a key factor to considerer in the analysis, or might only be a secondary one, depending on factors such as geological stability, the availability of minerals or the productive orientation of the area.

Geology conditions rural areas in many ways. Firstly, geology determines the geomorphology of the area. Climate and geology generates different geomorphological landscapes (Sect. 3.3 below), and these landscapes present a different degree of historical human occupation depending on the geomorphological characters. For example, human activities can be completely affected by the existence of slopes and their orientations with respect to the Sun. Northward slopes are suitable for different crops than southward ones, and geology conditions the soil. Also, some natural risks are related to geology: earthquakes are unlikely in large sedimentary areas while they are a serious issue in the convergence areas of the crust. Plutonic areas can be prone to radon emissions and thus to the habitat characteristics and the provisions made in daily life.

In the past, geology determined the availability of building materials. The character and aesthetics of landscapes in some rural areas are linked to the use of some particular building rocks such as slate. The replacement of ancestral building methods and fabrics by modern materials is not always justified by the properties provided by them such as durability, price or energy efficiency. Bioclimatic architecture is now well aware of the fact that some old-fashioned fabrics and materials are sometimes more suitable for solving modern problems than modern ones. Nonetheless, this truth should be analyzed in terms of the modern usage of the 'living space'. What was ideal for a farmer's house in the past may not be appropriate for commuters today. In any case, the preservation of the visual character of some rural areas usually justifies compulsory planning measures and can help to develop local industries and services. An example of good practice can be seen in the Tuscan countryside, where tough regulations have helped to preserve the natural beauty of the area using native materials and building techniques.

The geological characterization of rural areas can be done in terms of the geology of the area; which is the substrate, the rocks, minerals and stones found and the geological features that appear.

A good knowledge of geological maps is instrumental for this part of the rural analysis task. The next paragraph does not pretend to be a substitution to proper training in geological map reading but conveys the basic elements to be considered when the analyst is presented with one.

Geological ages and periods (Table 3.1) are represented in these maps as different colors. Several nomenclatures exist in the colorization of the maps depending on the predominant age: while in America primary age dominates the landscape, younger ages are more common in Europe. Using the same nomenclature would provide too homogeneous maps on either side. Therefore, several systems are used. Nevertheless, each map comes with its own legend and these normalization efforts are only indicative.

Table 3.1 Geological time series 2004, from F.M. Gradstein, et al. 2004: A Geologic Time Scale. 2004, Cambridge University Press

Geological Era	Geological Period	Time before present (Start, Millions of Years)
Cenozoic	Neogene	23
	Paleogene	65.5
Mesozoic	Cretaceous	145.5
	Jurassic	199.6
	Triassic	251
Paleozoic	Permian	299
	Carboniferous	359.2
	Devonian	416
	Silurian	443.7
	Ordovician	488.3
	Cambrian	542

Geological ages give the type of rock found beneath the surface and many of the geological characteristics of the area, such as the tectonic stability, the possibility of ores or petrol and natural gas resources, and of course the landforms. Geology is also central in finding out about the availability of water resources, and hydrogeology plays an key role in analyzing environmental impacts of human settlements and activities.

The legend of these maps provides a fieldwork indication of notable geological features. A basic element to look at is the contacts between different rock units. On the maps, a fine line shows contacts unless the contact is a fault, which is a sharp discontinuity. If the type of fault is not clear, it is depicted by a heavy line (Fig. 3.1). While most faults are inactive they are of primary interest in determinating possible geological risks. All faults were active at some time in the past, but a fault in a map today is usually just forensic evidence.

Other symbols aim to provide a 3D view of the terrain. The strike-and-dip symbols show the averaged direction of the rock layers such as bedding planes

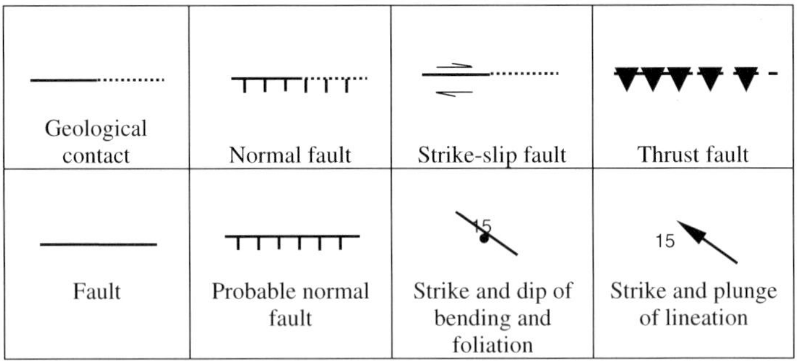

Fig. 3.1 A small sample of some usual symbols for faults found in geological maps. Dotted or dashed lines indicate an inferred element. A comprehensive list can be found in specialized publications such as the Mawer, C.H. 2002. Cartographic standard geological symbol index. Version 3. *British Geological Survey Research Report*, RR/01/01. 49pp

in sedimentary rocks. Other symbols indicate the lineation or the more relevant landforms. Geological maps, however, are less objective than other kind of maps. They are heavily dependent on fieldwork and the expertise of the geologist, so discrepancies can arise either when compared with reality or with other geologist's work. Nonetheless, the differences are usually too small to be of the interest in rural analysis and the only reason to mention this is to avoid confusion when confronted with different versions of the same map.

Rocks are classified into igneous, sedimentary, metamorphic and human-made rocks. Tables that identify a given rock were once useful, but expert systems, simple databases or web applets can help to classify rocks in a short time, depending on the requirements. In some studies, the geology is a valuable element of the landscape by itself, and rare varieties of rock, minerals or landforms justify a major interest in this issue and a professional approach would be required. In that case, it would be desirable to know the particularities of the area in greater detail to convey this and other information to the decision makers in a wider framework.

3.1.1 Mineral Resources

Some rural economies are or have been heavily linked with natural resources to the point that element is the main factor to considerer. If the area of interest has (or had) a prominent mining or quarrying activity it is worthy to analyze the dependence of the rural economy to these activities. Planning optimization proposals can target many issues such as an improvement of the extracting activities, the restoration of mining areas once the extractive activities have ceased or the integration of this activity into the regional economy.

The ability to match their needs and requirements with rural life, natural values, other economic activities and quality of life can be achieved by negotiated procedures between local agents and promoters or companies. Impacts can be mitigated by previous impact assessment studies based upon realistic views: a misguided conservationist frontal opposition can be more damaging for the environment than a clever and well enforced development plan following a holistic logic. In some cases, such as massive quarries, the balance between economic growth, rural life and natural and scenic values can be difficult to achieve, but a mutually-beneficial consensus can be sought. From the point of view of urban tourists or second residence owners, the process can be perceived as catastrophic but for rural people, the agreement can greatly improve their quality of life and the rural development.

The sudden discovery of new mineral resources or the exploitation of known resources due to new economical interest can yield a radical change in rural communities. It is unrealistic though to think that rules and regulation can stop these activities by themselves if they have a major impact on the environment. Only when vital natural values have to be preserved that tough measures can be enforced, and even in these cases, additional monitoring may be necessary. The economical logic of the markets would always press on these areas and the agents involved usually find a way to exploit the resources. In any case, mineral resources are needed for the economy and their exploitation has to be promoted. Whether this would benefit both the rural population and the region as a whole, or if it would generate tensions and diseconomies depends on sensible rural management. It may seem to be a past process, when concerns about rural welfare were subordinated to rapid national economic growth. Nonetheless, it is unlikely that compelling strategic interest would stop the exploitation of mineral resources even if the most beautiful landscape were to be affected. It has happened before with the gold mines in Roman times, uranium at the end of WW2, and petrol during the cold-war period. The effect on remote communities could be devastating if incorrectly managed: the recent discovery of economically profitable diamond mines in Botswana has provoked tensions between the San people and the central government/companies binomial. Illegal practices such as deportations and evictions can only be avoided by firstly enforcing the law and secondly, by increasing the knowledge of the locals in their rights and by improving their ability to manage the environment. A similar problem occurs with the tantalum (Coltan mining) in other parts of Africa, where the high revenues brought by the need of tantalum for electronic components have fueled several civil wars.

In the case of the Wolfram exploitation in the deprived areas of the west of Spain during WW2 is also enlightening. Several centuries ago wolfram enriched stones were used for cattle fencing by the locals. At WW2 this new strategic resource was in the interest of one of the contenders, but the fences were required to sustain a precarious local economy. The solution found to avoid a calamitous effect on the rural life and circumvent local opposition was to buy not only the fences, but the whole exploitations, giving the people enough resources to start new activities. In a happy ending, the fences were used to extract wolfram but the land was given back as a gift to the previous owners. The example is quite extreme, since the price of metal then highly exceeds the land price, but it shows that sometimes compromise is possible when both parties find a beneficial conclusion.

Today, if a new mineral resource is discovered in a rural area the rural expert should consider the kind of development this can generate. The rural manager should propose a strategic plan to maximize the benefits of this economic resource and at the same time avoid the problems and imbalances this could cause. Amongst others, the effects of continuum or punctual exploitation might need to be considered. How this can be achieved depends on the maturity of the civil rural society and the socio-economic conditions of the region.

3.2 Soils

What is worth knowing about edaphology in terms of rural analysis and management is related with how soils affect rural activities. While only in some cases a specific study will be required, a good knowledge of how soils affect rural affairs is desirable. It is not expected that the rural analyst will become an expert in soils with only the information that follows. Rather, the idea is to provide the analyst with enough background to allow a fruitful interaction with the professional soil scientist. This does not guarantee that the analyst will understand every aspect of a soil report, but it will provide guidance.

Soils are the interest of two sciences: pedology (from the greek pedon, meaning soil or earth) and edaphology (from the greek edaphos, soil or ground). From a pedological point of view soil is a natural entity produced by nature, and so it is studied in terms of physical analysis. Edaphology, on the other hand, considers soil as the base for plant production, thus having an immediate practical interest for humankind. Rural analysis and of course rural management, is more interested in the edaphological point of view than in the pedological one, in spite of that when considering civil engineering actuations a pedological approach would be needed. The interest is twofold. Firstly, soils are the foundation of agriculture rural life and thus have to be precisely defined to characterize a particular rural area and its circumstances. Secondly, soils can be managed for example, to be capable of sustaining agriculture in areas where natural conditions are not suitable for certain crops. Also, clever management can help to maintain this basic need: soils can be easily destroyed by bad practices, but with proper management and restoration soil is a renewable resource that can continue to serve mankind's needs in the future.

A soil is the unconsolidated mineral and/or organic material on the immediate surface of the earth that serves as a natural medium for the growth of land plants. Soils also operate as a regulator of water supplies, by storing and filtering water, and also as a recycler of organic waste acting as a regulator of global climate by the accumulation of carbon and nitrogen. A third function of soil is as the habitat for soil organisms, which is relevant in terms of one of the most valuable of Earth's resources: biodiversity. Soils are the habitat of many of the living forms of our planet, most of them poorly studied. Finally, soils are the foundation for human structures such as buildings or railways.

Soils are usually structured into layers called horizons. They represent physiological units named by starting at the beginning of the alphabet, thus horizon 'A' being the one in contact with the atmosphere or the ice/water. These horizons can be further distinguished into sub-horizons, such A00, A0, B3, and so on. (see below). The surface layer is denoted as topsoil and what lies immediately beneath is the subsoil. Another name is the *furrow slice,* referring to the portion of the soil turned by the plough. While the topsoil can be easily modified by human action, the subsoil represents a more stable layer that is seldom altered. It is the subsoil that gives the soils their more defined characteristics in terms of long-term decisions. Beneath the soil stands the rest of the *regolith* (a layer of loose material created by weather and biological processes) that rests on the bedrock.

A soil profile with its horizons characterizes a given soil. The rules for naming and numbering horizons are quite complex and only interesting in terms of rural analysis and management in that they can be used to compare different areas.

The basic tools for analyzing soils are the maps. In some cases, ancillary small-scale soil maps may be available, but if they are not and soil is considered a crucial part of the analysis expert advice is required and a thematic expert should be called in.

Soils are also interesting since they act as a canvas in which meteorology, climatic and anthropic practices are reflected. This is why we should devote some attention to this element, apart from its immediate anthropic use or its effects in human life. Ancient Babylon is a classic example of how soils can affect human activities. It is thought that the decline of this civilization around the tenth century BC, was largely due to a severe degradation in soil fertility due to over-irrigation. Vast amounts of water from the Tigris and Euphrates rivers were constantly diverted to crops using canals. After decades of irrigation, crops surplus began to reduce eventually creating a subsistence crisis.

3.2.1 Soil Composition

Soils are made up of solid, liquid and gas in different proportions and the relationships between these three phases determine the soils properties. For example, a well-balanced soil for plant growth contains about a half in volume of minerals and organic matter and two quarters of air and water.

3.2.1.1 Minerals and Rocks

Most of the composition of the soil consists of minerals, both as aggregates of several types of minerals or as a single mineral. This inorganic part of the soil is made up of silica, oxides of several metals including manganese, iron or aluminium, etc. Minerals in the soil have been generated by the weathering of rock fragments and they are responsible for most of the physical properties of soils, since very small size signifies that the effective surface (the ratio between the surface area and the volume) are very high. The greater the effective surface, the greater the ability of mineral particles to be involved in chemical reactions or physical processes such as the adsorption, which is the preferential partitioning of substances from the gaseous or liquid phase onto the surface of a solid substrate. That is, the smaller the particle the greater its ability to retain gaseous or liquid molecules stuck to its surface, and thus to preserve them in the soil. Particles that have been adsorpted by the minerals are then absorbed by the plants and their roots.

Minerals form macro and micro nutrients (Table 3.2). Macronutrients are elements used by plants in relatively larger amounts (orders of magnitude) than micronutrients. These nutrients are not usually directly available to plants but as cations (positively charged ions) of minerals such as feldspar, micas or calcite, embedded into a watery soil solution. The concentration of ions at a given level in the soil is far less than is needed by a crop. As they are absorbed by the plants new ions are released in a nutrient-supplying natural cycle that can be modified by human action by supplying fertilizers.

3.2.1.2 Organic Matter

Organic matter and its distribution (organic matter profile) are important not only for agriculture but also for physical properties of the soil and global environmental feedbacks. The amount of water a soil can hold increases with more organic matter,

Table 3.2 Major macro and micronutrients of the soil

Macronutrients		Micronutrients (Oligoelements)	
Mostly from air and water		From minerals	
Carbon	Nitrogen	Iron	Sodium
Oxygen	Potassium	Boron	Strontium
Hydrogen	Calcium	Manganese	Silicon
	Phosphorus	Molybdenum	Iodine
	Sulfur	Zinc	Barium
		Copper	
		Chlorine	
		Cobalt	

and most of the carbon dioxide on earth is not in the atmosphere or the oceans, but is stored in the soils: microbial respiration produces CO_2 one part of which is released into the atmosphere and the other stored in the soil.

Organic matter represents a tiny fraction of soils total mass, ranging from 1 to 5% in soils suitable for agricultural practice. Most of a plants food comes from the organic matter, including nitrogen and carbon. A soil with a good proportion of organic matter is less compact and thus has better drainage than a poorer one, allowing for plants to grow. Soil regeneration practices usually involve some kind of balancing between the organic matter and other constituents.

Generation of organic matter is a rural economic activity. Some areas have a surplus of organic matter than can be exported to other less favored areas. Organic matter can also be produced using earthworms and other soil harvesting methods. The earthworms digest raw material and generate a fertile top soil containing a rich proportion of water soluble forms: all the elements required of the earth for plant nutrition. The shipping cost makes this economic activity more suitable for local or regional markets. Today, the household horticultural market is a possible selling outlet for this production. Nonetheless, some sub-arid or sub-desert areas can become a profitable market for big scale productions.

3.2.1.3 Water

The holding capacity of soils determines the physical properties of soils. Also, water is vital for the biological functions of soils. The soil solution -the organic mater- is where macro and micronutrients are dissolved in water and can thus be distributed. Water contributes to soil formation and regeneration and is fundamental for the climate. Soil moisture is a primary source of water vapor in generating clouds, and thus precipitation in sometimes far-away areas. The role of soil moisture in the hydrological cycle is critial, and therefore is a key part in meteorological modeling.

The status of soil water can be measured using several techniques such as the gravimetric method, neutron scattering, TDR (time-domain reflectrometry) or the ERB (electrical resistance block) method. These techniques directly measure the water content of soils but they are relatively expensive. The so-called 'soil water potential' -or the difference between the potential energy of pure water and the potential energy in a given soil- can also be measured using tensiometers, thermocouple psychrometers or pressure membrane instruments. The techniques used to estimate both magnitudes depend on the scale of the work and the need of an accurate estimate of these two magnitudes. Both magnitudes give the soil water status. The hydraulic conductivity of soils, on the other hand, illustrates how water flows in the soil, which is relevant in terms of hydrology and pollutant transport.

More qualitative (and thus cheaper, subjective and less accurate) estimates of soil wetness exist. Terms such as 'saturation', 'field capacity' and 'wilting point' are used to provide an estimate of water status in soil samples.

The design of drainage solutions for runoff water is also a possible outcome of rural management for soils with a great economic value, but only for fragile or large-scale landscape design for the private sector. In these cases, a soil scientist would be required to perform a careful analysis and provide a solution, if the cost-benefit analysis of the action is deemed as appropriate.

3.2.1.4 Gases

Soils contain variable proportions of nitrogen, sulfur, oxygen and many other gases. Amongst them, nitrogen plays a crucial role in life. Nitrogen is the raw material of amino acids and thus proteins and enzymes, nucleic acids, and of course chlorophyll (green chlorophyll: $C_{55}H_{72}O_5N_4Mg$). Nitrogen deficiency in plants, known as chlorosis, destroys crops and tree productions. However, the oversupply of nitrogen is dangerous for livestock and babies when accumulated in the form of nitrates. When natural nitrogen is not enough for plants to grow, fertilizers are used. A correct and timely proportion of nitrogen fertilizers are therefore crucial for the environment and for agricultural production. Slow-release fertilizers (or nitrification inhibitors) can help to avoid undesirable accumulations of nitrogen in the soil, but attention must be paid to agricultural practices in the area of study to optimize costs and avoid diseconomies.

The nitrogen cycle refers to the recycling of nitrogen in a closed cycle; in the same way that the water cycle tells us that the same water appears in several forms without changing its nature. The analysis of the nitrogen cycle is central for numerical modeling environmental and agricultural processes. Figure 3.2 illustrates a conceptual diagram of this cycle.

3.2.2 Physical Properties of the Soils

3.2.2.1 Density

The density of soil can be expressed either as *particle density* or *bulk density*. Particle density is the mass of a unit volume of soil solids, which depends on the chemical

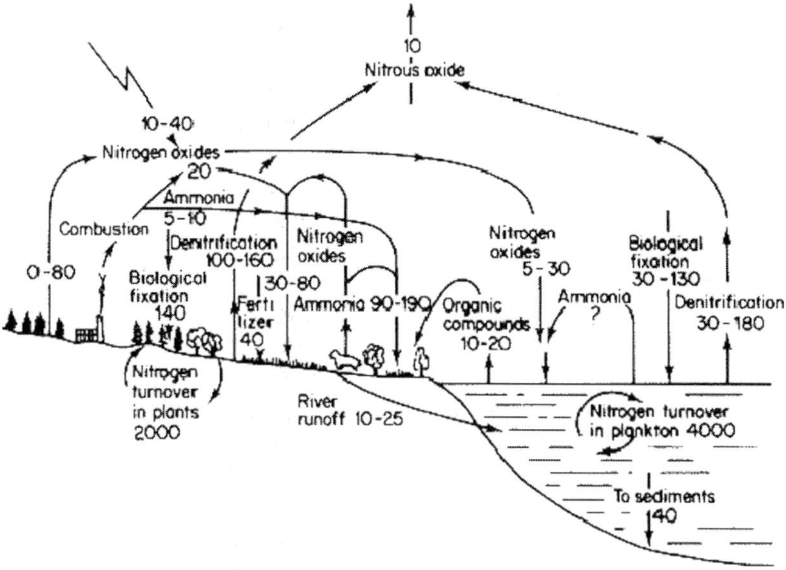

Fig. 3.2 The global nitrogen cycle. Units are in Tg(10^{12}g) N yr^{-1}. From Söderlund & Rosswall (1982) based on Söderlund & Svensson (1976)

composition and crystal structure of the mineral particles. The bulk density is defined as the mass of a unit volume of dry soil, which includes both solids and pores. That is, bulk density takes into account the compactness of the soil. Typical values of particle density are 2650 Kg/m^3. Soils containing most mineral matter can have particle densities as large as 2750 Kg/m^3, while a soil composed by only organic matter can have a particle density of about 1200 Kg/m^3. Typical bulk densities for cultivated soils range from 980 to 1450 Kg/m^3. Notice for comparison, that pure water would have a particle density of 1000 Kg/m^3.

3.2.2.2 Porosity

Porosity is one of the key elements for plant growth. If no pore spaces are present, roots cannot penetrate. Compact soils have around 25% porosity whereas 60% porosity is typical of well-granulated conditions. The value of 50% is typical in soils suitable for plants. Porosity is roughly estimated using the following formulae:

$$\% \text{ pore-space} = 100 - (100 \times \text{bulk density/particle density})$$

This coarse estimate does not take into account the pore size distribution, and it is only indicative. Direct measurements, however, are too complicated and only justifiable in some cases.

Table 3.3 Traditional soil particle size classification

Name	Size
Clay	$< 2\,\mu m$
Silt	$0.02\,mm - 2\,\mu m$
Fine Sand	$0.2\,mm - 0.02\,mm$
Coarse Sand	$2.0\,mm - 0.2\,mm$
Gravel	$> 2.0\,mm$

3.2.2.3 Particle Size and Texture

The texture of the soil is in the proportion to particles of different size ranges. The size of the particles ranges from the familiar sand particles (2.0 to 0.05 mm) to the colloidal particles ($<1\,\mu m$ [$= 0.001$ mm]), that can only be seen by using a scanning electron microscope (SEM). Particles with size between 0.05 mm and $2\,\mu m$ are referred to as 'silt'. Silt can be easily differentiated from clays because wet clay forms a sticky mass, whereas wet silt is a smooth but non-sticky material.

The International Society of Soil Science classifies soil particles according to their size in the classes in the Table 3.3. The size distribution of the soil is a valuable parameter to characterize inhomogeneous soils and can be obtained by calculating the histogram (frequency diagram) for several bin sizes.

The traditional terminology of soil texture differs from country to country, but some international agreement is possible thanks to quantifiable magnitudes such as porosity or hydraulic conductivity. Table 3.4 shows some typical soil textures in terms of these physical quantities. When available, rural analysis should use these quantifiable values instead of more subjective estimates.

3.2.3 Chemical Properties

3.2.3.1 pH (Acidity)

The soil solution presents a given hydrogen ion (H^+) and hydroxide (OH^-) anion concentration that is used to measure the acidity or alkalinity of the soil. By defi-

Table 3.4 Soil textures in terms of physical measurements such as porosity (n, dimensionless), matric potential at saturation (psis, in meters), and saturated hydraulic conductivity Ks (m/s). Data from NASA's DAAC

	Soil texture	N	Psi_S	K_S	Comments
1	Coarse	0.421	0.0363	1.41E-5	Loamy sand values
2	Medium/Coarse	0.434	0.1413	5.23E-6	Sandy loam values
3	Medium	0.439	0.3548	3.38E-6	Loam values
4	Fine/Medium	0.404	0.1349	4.45E-6	Sandy clay loam values
5	Fine	0.465	0.2630	2.45E-6	Clay loam values
6	Ice	n/a	n/a	n/a	-
7	Organic	0.439	0.3548	3.38E-6	Loam values
0	(Ocean)	n/a	n/a	n/a	-

nition, pH (after *hydrogen potential*) is the decimal logarithm of the inverse of the concentration of hydrogen ions:

$$pH = \log_{10} \frac{1}{[H^+]}$$

($[\cdot]$ = concentration). This means that for a soil with 10^{-7} mole/liter the pH is 7, which incidentally, is considered as the neutral pH. Water solutions with a pH greater than 7 are considered as alkaline solutions, while those with a pH less than 7 are labelled acid. The extreme values for most soils range from 4 to 10, but cultivated soils span from only 5 to 9.

3.2.3.2 Salinity and Sodicity

The amount of different salts in the soil affects plants preventing them to absorb water from the soil by osmosis. Salinity can be estimated by measuring the electrical conductivity (EC) of soil samples with different water contents. As a physical procedure, it can be adapted to be used with precision agriculture techniques (Sect. 5.3) providing a means to correct undesired high levels of salt in a field (Table 3.5).

Sodium salts are the most damaging for crop fertility (the sodicity condition of the soil). The sodium state of the soil is characterized in terms of the sodium adsorption ratio (SAR), defined as a percentage:

$$SAR = \frac{[Na^+]}{\sqrt{\frac{1}{2}[Ca^{2+}] + \frac{1}{2}[Mg^{2+}]}}$$

Depending on the pH, the EC and the SAR, soils can be classified into normal, sodic, saline-sodic and acid soils. A precise knowledge of the type of soil in terms of salt relationship is instrumental for assessing irrigation policies. It is estimated that one-fifth of irrigated agriculture is adversely affected by soil salinity. The tolerance of plants to salts is described in specialized literature such as Maas and Grattan (1999) and refers to the EC measure. One of the aims of genetically-modified (GM) crops is to improve the productivity of the plants creating varieties with larger tolerances.

Table 3.5 Classification of salt-affected soils. EC is measured in decisiemens per meter (dS/m)

EC (dS/m)	pH	SAR (%)	Soil type	Soil condition
> 4.0	< 8.5	< 13.0	Normal	Normal
< 4.0	> 8.5	> 13.0	Sodic	Poor
> 4.0	< 8.5	> 13.0	Saline-Sodic	Normal
< 4.0	> 7.8	< 13.0	Acid	Variable

3.2.4 *Soil Characterization*

The characterization of soil can be done in terms of its properties: taking into account the proportion of its constituents and the physical variations in the vertical profile of the soil (the horizons). These are a consequence of the physical and chemical properties we have described above.

3.2.4.1 Soil Texture Classes

Soil texture refers to the relative proportion of sand, silt and clay particles present in the soil. The widely used soil texture triangle (Fig. 3.3) provides an easy way to classify textures in terms of the relative proportions of clay, silt and sand. This is an alternative method to the one presented in Table 3.4.

Soils have other properties that we can easily assess such as thickness or color. Thickness is crucial for agriculture, whilst color gives a first approximation for actual soil characterization. Color is related to the presence of mineral compounds such as ferrous oxide (FeO), ferric oxide (Fe_2O_3) or hydrated ferric oxide ($Fe_2O_3.H_2O$) and to the amount of organic matter in form of humus or decomposed organic matter.

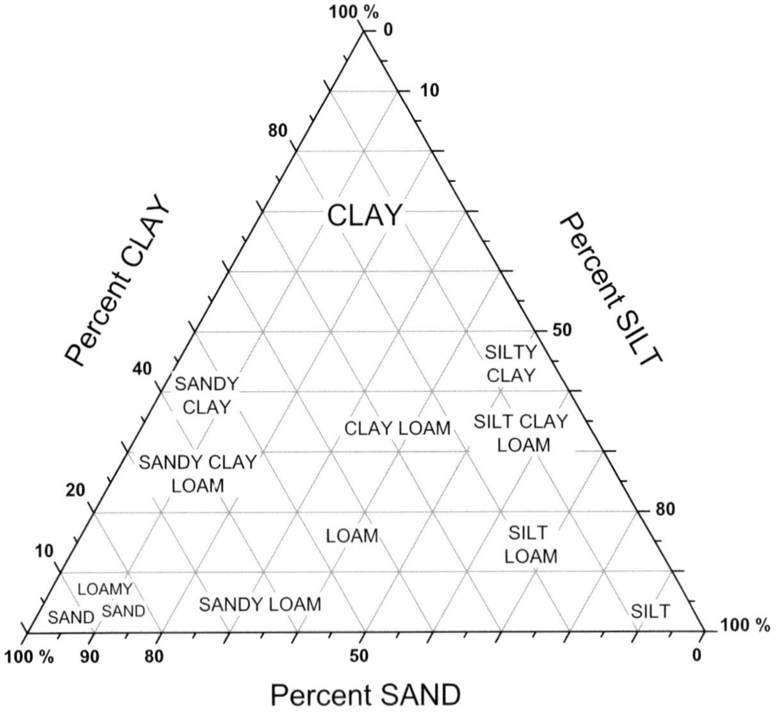

Fig. 3.3 The soil texture triangle. From the soil survey staff, natural resources conservation service, United States Department of Agriculture

Table 3.6 Unified Soil Classification System (from ASTM D2487-00)

Major Divisions			Symbol	Typical Names
Coarse-GrainedSoils More than 50% retained on the No. 200 sieve	Gravels 50% or more of course fraction retained on the No. 4 sieve	Clean Gravels	GW	Well-graded gravels and gravel-sand mixtures, little or no fines
			GP	Poorly graded gravels and gravel-sand mixtures, little or no fines
		Gravels with Fines	GM	Silty gravels, gravel-sand-silt mixtures
			GC	Clayey gravels, gravel-sand-clay mixtures
	Sands >50% coarse fraction passes the No. 4 sieve	Clean Sands	SW	Well-graded sands and gravelly sands, little or no fines
			SP	Poorly graded sands and gravelly sands, little or no fines
		Sands with Fines	SM	Silty sands, sand-silt mixtures
			SC	Clayey sands, sand-clay mixtures
Fine-Grained Soils More than 50% passes the No. 200 sieve	Silts and Clays Liquid Limit 50% or less		ML	Inorganic silts, very fine sands, rock four, silty or clayey fine sands
			CL	Inorganic clays of low to medium plasticity, gravelly/sandy/silty/lean clays
			OL	Organic silts and organic silty clays of low plasticity
	Silts and Clays Liquid Limit greater than 50%		MH	Inorganic silts, micaceous or diatomaceous fine sands or silts, elastic silts
			CH	Inorganic clays or high plasticity, fat clays
			OH	Organic clays of medium to high plasticity
Highly Organic Soils			PT	Peat, muck, and other highly organic soils

G = Gravel, S = Sand, M = Silt, C = Clay, O = Organic, W=Well Graded, P = Poorly Graded, M = Silty, L = Clay, LL<50%, H = Clay, LL>50%

Soils are the foundation of buildings, roads, railways and alike. Soil, both natural and anthropic, can be classified using the US Army Corps of Engineers Unified System of Classification (Table 3.6), which is a classification system for engineering works.

The rural analyst should be familiar with ASTM and ISO documents since they are a comprehensive and comparable source of information.

3.2.4.2 Soil Horizons and Layers

In terms of their vertical characteristics, soils can be structured into layers called 'horizons'. They represent physiological units that can be estimated through the extraction of samples.

The UN Food and Agriculture Organization (FAO) describes seven main horizons and layers. The letters **H, O, A, E, B, C** and **R** represent the so-called master horizons and layers of soils. The capital letters are the base symbols to which other characters are added to complete the designation. Most horizons and layers are given a single capital letter symbol but some require two.

The master horizons and their subdivisions represent layers that show evidence of change, and some layers which have not been changed. Most are genetic soil horizons, reflecting a qualitative judgment about the kinds of changes that have taken place. Genetic horizons are not equivalent to diagnostic horizons although they may be identical in soil profiles. Diagnostic horizons are quantitatively defined features used in classification.

The FAO description of each horizon deserves to be cited to illustrate the degree of detail in the characterization and the properties of each layer. What follows is an excerpt of (FAO, ISRIC and ISSS 1998). The reader is addressed to this publication for a full account:

- **H horizons**: Layers dominated by organic material, formed from accumulations of undecomposed or partially decomposed organic material at the soil surface which may be underwater. All **H** horizons are saturated with water for prolonged periods or were once saturated but are now artificially drained. (. . .)
- **O Horizons**: Layers dominated by organic material, consisting of undecomposed or partially decomposed litter, such as leaves, needles, twigs, moss, and lichens, which has accumulated on the surface; they may be on top of either mineral or organic soils. **O** horizons are not saturated with water for prolonged periods. The mineral fraction of such material is only a small percentage of the volume of the material and generally is much less than half of the weight. (. . .)
- **A horizons**: Mineral horizons which formed at the surface or below an **O** horizon, in which all or much of the original rock structure has been obliterated and which are characterized by one or more of the following:

  - an accumulation of humified organic matter intimately mixed with the mineral fraction and not displaying properties characteristic of **E** or **B** horizons (see below);
  - properties resulting from cultivation, pasturing, or similar kinds of disturbance; or
  - a morphology which is different from the underlying **B** or **C** horizon, resulting from processes related to the surface.

  If a surface horizon has properties of both **A** and **E** horizons but the dominant feature is an accumulation of humified organic matter, it is designated an **A** horizon (. . .).
- **E horizons**: Mineral horizons in which the main feature is loss of silicate clay, iron, aluminum, or some combination of these, leaving a concentration of sand and silt particles, and in which all or much of the original rock structure has been obliterated. An E horizon is usually, but not necessarily, lighter in colour than an underlying B horizon (. . .)

- **B horizons**: Horizons that formed below an **A, E, O or H** horizon, and in which the dominant features are the obliteration of all or much of the original rock structure, together with one or a combination of the following:
  - illuvial concentration, alone or in combination, of silicate clay, iron, aluminum, humus, carbonates, gypsum or silica;
  - evidence of removal of carbonates;
  - residual concentration of sesquioxides;
  - coatings of sesquioxides that make the horizon conspicuously lower in value, higher in chrome, or redder in hue than overlying and underlying horizons without apparent illuviation of iron;
  - alteration that forms silicate clay or liberates oxides or both and that forms a granular, blocky, or prismatic structure if volume changes accompany changes in moisture content; or
  - brittleness. (. . .)
- **C horizons**: Horizons or layers, excluding hard bedrock, that are little affected by pedogenetic processes and lack properties of **H. O. A, E**, or **B** horizons. Most are mineral layers, but some siliceous and calcareous layers such as shells, coral and diatomaceous earth, are included. The material of **C** layers may be either like or unlike that from which the solum presumably formed. A **C** horizon may have been modified even if there is no evidence of pedogenesis. Plant roots can penetrate **C** horizons, which provide an important growing medium. (. . .)
- **R layers**: Hard bedrock underlying the soil. Granite, basalt, quartzite and indurated limestone or sandstone are examples of bedrock that are designated **R**. Air dry or drier chunks of an **R** layer when placed in water will not slake within 24 hours. The **R** layer is sufficiently coherent when moist to make hand digging with a spade impractical, although it may be chipped or scraped. (. . .)

[From the *World Reference Base for Soil Resources*, 1998. Food and Agriculture Organization of the United Nations (FAO)]

3.2.5 Soil Taxonomy

There are many different soil classification systems that depend not only on the criteria used but also on regional preferences. Commonly used in the USA, is the US Department of Agriculture classification, which is one of the most comprehensive as is the FAO soil taxonomy system (Table 3.7), which has the advantage of being consciously international.

3.2.6 The Practical Significance of Soils

The most obvious practical significance of soils is that they are present where most of crops grow. It is technically true that hydroponic agriculture does not require

Table 3.7 FAO soils reference groups and its description

ACRISOLS Soils with subsurface accumulation of low activity clays and low base saturation	ALBELUVISOLS Acid soils with a bleached horizon penetrating into a clay-rich subsurface horizon	ALISOLS Soils with subsurface accumulation of high activity clays, rich in exchangeable aluminium
ANDOSOLS Young soils from volcanic deposits	ANTHROSOLS Soils in which human activities have resulted in profound modification of their properties	ARENOSOLS Sandy soils featuring very weak or no soil development
CALCISOLS Soils with accumulation of secondary calcium carbonates	CAMBISOLS Weakly to moderately developed soils	CHERNOZEMS Soils with a thick, dark topsoil, rich in organic matter with a calcareous subsoil
CRYOSOLS Soils with permafrost within 1 m depth	DURISOLS Soils with accumulation of secondary silica	FERRALSOLS Deep, strongly weathered soils with a chemically poor, but physically stable subsoil
FLUVISOLS Young soils in alluvial deposits	GLEYSOLS Soils with permanent or temporary wetness near the surface	GYPSISOLS Soils with accumulation of secondary gypsum
HISTOSOLS Soils which are composed of organic materials	KASTANOZEMS Soils with a thick, dark brown topsoil, rich in organic matter and a calcareous or gypsum-rich subsoil	LEPTOSOLS Very shallow soils over hard rock or in unconsolidated very gravelly material
LIXISOLS Soils with subsurface accumulation of low activity clays and high base saturation	LUVISOLS Soils with subsurface accumulation of high activity clays and high base saturation	NITISOLS Deep, dark red, brown or yellow clayey soils having a pronounced shiny, nut-shaped structure
PHAEOZEMS Soils with a thick, dark topsoil rich in organic matter and evidence of removal of carbonates	PLANOSOLS Soils with a bleached, temporarily water-saturated topsoil on a slowly permeable subsoil	PLINTHOSOLS Wet soils with an irreversibly hardening mixture of iron, clay and quartz in the subsoil
PODZOLS Acid soils with a subsurface accumulation of iron-aluminium-organic compounds	REGOSOLS Soils with very limited soil development	SOLONCHAKS Strongly saline soils
SOLONETZ Soils with subsurface clay accumulation, rich in sodium	UMBRISOLS Acid soils with a thick, dark topsoil rich in organic matter	VERTISOLS Dark-coloured cracking and swelling clays

a proper soil but a support. However, the economic and social importance of this production system is limited in global terms.

The rural analysis should identify the productive soils that might be desirable to protect. Soils carefully managed over decades or even centuries, are a valuable resource in terms of sustainable development and legacy. It is impossible to fossilize rural areas, but planning must aim to orientate growth and development toward unproductive lands rather than destroying a valuable resource.

Urbanization in rural areas should take into account the effect of runoff in order to avoid soil degradation and even environmental risks. A wide variety of runoff numerical models are available, making it possible to assess the impact of any development plan. Protecting soils from wind or water erosion is another possible factor that can be addressed.

3.2.7 The Geographical Distribution of Soils

Large-scale maps can be useful to understand how soils can affect human activities, how they are related to other elements in the environment, and to assess the dominant soil in which the rural area is situated. While they cannot be used directly in a rural analysis commission, since they lack the required precision, they are illustrative in comparative studies (Fig. 3.4).

As we will see in the third part of this book, a geographical synthesis can help an analyst widen his scope and to understand how larger areas of the territory are structured.

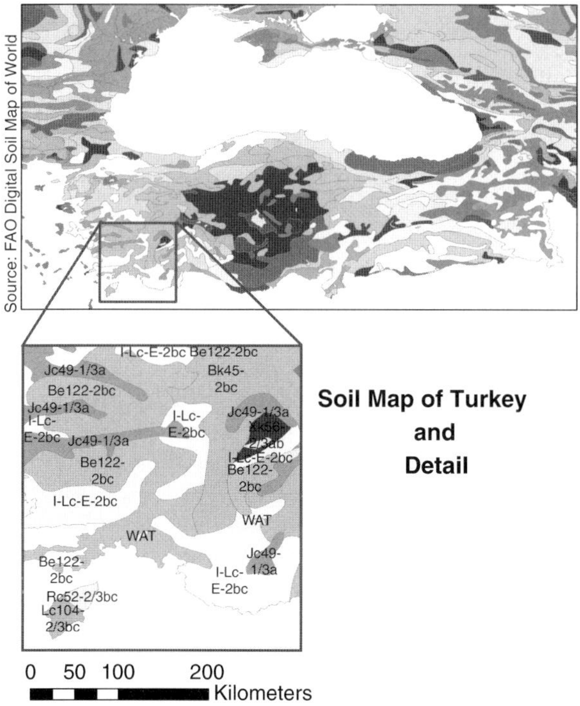

Fig. 3.4 An example of moderately-detailed soil map classification

3.3 Geomorphology

Landforms are the consequence of the combination of weathering, erosion and tectonics. Valleys, mountains and plains are the result of the geological history of the planet. There are purely tectonic landforms, such as recent volcanic forms, but most landforms are the result of hundreds of thousands of years of evolution. Different atmospheric processes act in different climatic zones generating the landforms. Fluvial denudation also generates landforms (and in turn, the climate is affected by landforms). Precipitation and streams create wonderful karsts and pseudokarsts, and glaciers have also produced fluvioglacial and glaciofluvial forms. Wind is also a powerful agent generating the so-called Aeolian landforms and dust deposits such as dunes. In coastal areas, marine landforms appear by abrasion and accumulation forming coastal landforms. Mankind has also generated both anthropogenic and artificial landforms. A useful geomorphologic characterization of an area for rural analysis should comprise of the analysis of all these elements. The rural analyst should ask the expert to highlight any geomorphological singularities that might appear in the area, for instance a particular kind of dune only found in Europe, or extinct volcanoes in a seemingly flat landscape.

Contrary to geology or soil horizons, most landforms are visually available to everyone. Elemental education presented the major landforms to the public, and introducing an appreciation to the more unusual ones is a good cultural resource for quality tourism. Creating educational centers in key areas can reinvigorate associated services and activities (restaurants, coffee-shops, local crafts, direct horticultural products sells, etc.), and some particular landforms (e.g. the Grand Canyon) are an attraction by themselves. Karsts attract speleologists and can hold many geosites (Sect. 3.10).

In some cases, the geomorphology of an area can determine the kind of solutions that can be provided for a rural problem. Firstly, it is advisable to establish whether or not the area belongs to one or several geomorphologic units. If the area is quite homogeneous, a different kind of rural planning is required than if the area is not, since the variety can be used as a resource. Geomorphological maps are expensive to generate, but remote sensing techniques in the visible/infrared wavelengths facilitate this process. With these maps the analyst can identify the possible potential of an area for tourism, the unique or distinct landforms that deserve protection, the natural risks associated to landforms, and so on.

New developments, or new infrastructures, can affect how flash floods are handled by topography: landforms are the result of centuries of weathering and tectonics and their current state represents a minimum energy state (or a maximum entropy state). Modifications can generate new conditions that have to be considered in the light of the current landforms. Changes in the sea level or dams can alter the profile of rivers slowing or fostering the waters with disturbing side effects. Some ancient communities (e.g. Ancient Ephesus) suffered these kinds of problems associated with topography and geomorphology, but it seems that their experience has not been of any use to modern planners. Thus, while it is a gross error to build houses in apparent sleep torrents, some modern rural villages on the coast of Catalonia (Spain) have been built right in the middle of natural drainage channels.

The temporal scale in which extreme events happen is usually larger than one or two generations (one generation is by definition a period of 30 years), and what never happened in living memory can easily occur in the next few years, risking properties and lives. Insurance companies usually consider geomorphological risks for home insurance and raise the prime rates to pay consequentially, but the risk for human lives is unacceptable.

The topography of the area is also of interest as it conditions infrastructures, facilities, transportation, natural risks, etc. Digital Elevation Models (DEM) can be used to derive slope, sun orientation, and visual path maps or can be combined into 3D views that are useful for planners (Fig. 3.5). Several methods can be used to quantify two-dimesional areas such as hill slopes beyond the classical Ruhe classification.

Fig. 3.5 Integrated DEM and satellite images fusion. Royal Site of 'La Granja' in the Central System, Spain. Snow appears in the top of the mountain, whilst the dam and the village are clearly visible in the plain. Forest and roads are also apparent. The landforms of these products can be calculated using automated procedures (see Color Plate on following p. 333)

3.4 Meteorology

Meteorology studies the physical state of the atmosphere at a given moment or during a short period of time (the weather). Climatology studies the effect of the weather during long periods of time in a given location. Meteorological characterization of rural areas can be done using several methodologies that consider the evolution of variables such as temperature, wind speed and direction and precipitation: both in intensity, duration and the presence of extremes. More practical than this classification, are the forecasts (Fig. 3.6). Apart from the more obvious effects such as hail-storms in agriculture or the impact of monsoons, the interest of the meteorologist in rural analysis and management also includes risk assessment, agricultural planning, tourism and fisheries in rural coastal areas.

Wind conditions can be relevant in coastal areas for surfing activities; wind and sea surface temperature are essential in for example sportive fisheries in the Caribbean. Speleologists in places such as Reunion Island really need a rain forecast before a four day trip to the narrow and deep canyons of the island or they may risk drowning. Meteorological real-time systems can save a valuable vintage by triggering fans or heaters in the event of freezing (as the Ribera del Duero vineyards in Castile, Spain) and preventing the spread of fire in the case of a forest fires is directly linked with effective real time information and modeling.

The list does not stop here: many other rural activities not only depend on the climate of the site, but also on the meteorological conditions. It is the analyst's task to carefully evaluate this aspect when analyzing a rural area. Providing accurate meteorological information can be one of the things that can be done from either the public sector or private industry to improve rural economies.

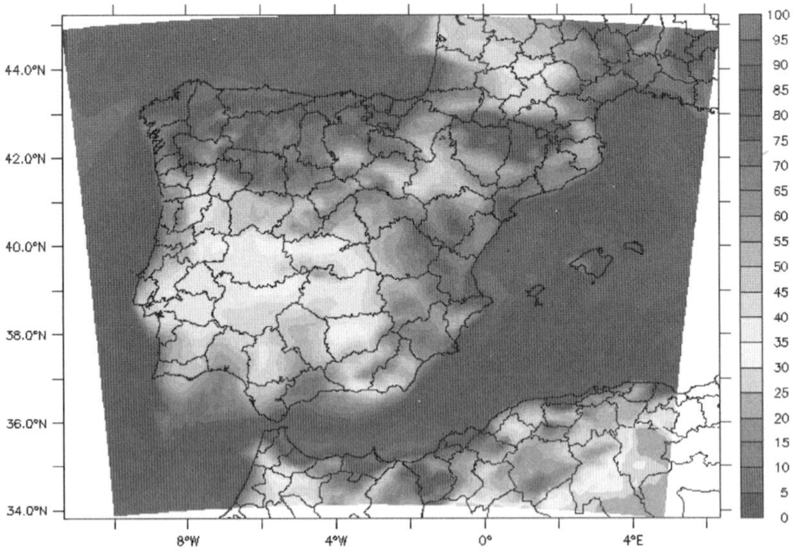

Fig. 3.6 An example of high-resolution weather forecast. Surface relative humidity (%) in Spain for the 15/08/2005 (http://momac.uclm.es)(see Color Plate on following p. 333)

In agriculture, sunlight, humidity and of course rainfall, determine the overall quality of a crop. Soil composition is instrumental for seed germination due to its thermal inertia, but temperature also plays an instrumental role. Temperatures affect the crop during their entire life cycle. Planting below the minimum temperature for germination implies that it will take longer for the seed to germinate resulting in a loss of quality. Some crops even require fluctuating temperatures and/or light conditions to germinate: celery germinates below 10°C in the dark at a constant temperature. With some light, it starts germinating at 20°C and with a 10 degree daily fluctuation will germinate at 30°C. The number of frost-free days also affects crop production and high temperatures can cause not only desiccation, but also pollen sterility due to excessive drying of the pollen, or changes in the enzymes of the plant's cells. In a very few incidents, mitigation strategies can be applied for adverse meteorological cases but in most of these situations the use of this knowledge can be used for planning.

How are the meteorological variables estimated in practice? Except for a very few instances, it is not practical to hire a meteorologist in rural analysis work. National weather services routinely provide up to 96 hours of forecast at no cost. Tailored high-resolution predictions can be expensive, but some research centers are currently provide such forecasts for agrarian and environmental needs. Some companies also provide specific forecasts for medium to long-term strategies. As we will see, one of the tasks of the rural analyst is to make communities aware of these technical possibilities.

Issues related to global change are also of interest for long-term programs, either to avoid crop losses or to mitigate the effects of global changes with intelligent planning. Farmers can control the varieties they use to adapt to predicted anomalies in the seasons or can modify planting dates. Since state of the art numerical models provide reasonably accurate seasonal forecasts, this technique can offer comparative advantages over other areas.

3.5 Climate

Climate classifications are useful in providing an overall picture of a rural area. Historically, there have been two major climatic classifications. While some aspects of these are quite obsolete, they are still widely used. A prior knowledge of these classifications is desirable since we use terms such as 'tropical climate' and 'semi-arid environment' in our common language.

3.5.1 Köppen's Climatic Classification

Vladimir Köppen (1846–1940) devised a climatic system by relating the presence of animals and vegetation to temperature and precipitation (Table 3.8). The reason of selecting these parameters, and no others, was data availability in spite of the clear importance of them for living beings. Köppen distinguishes five major types of climate that he noted with a capital letter:

Table 3.8 Köppen climate classification

Code	Climate
Af	Tropical climate-no dry season
Aw	Tropical climate-winter dry season
BS	Steppe dry climate
BW	Desert dry climate
Cf	Mild climate-no dry season
Cs	Mild climate-dry season
Cw	Mild climate-dry winter
Df	Microclimate-no dry season
Dw	Microclimate-dry winter
ET	Tundra cold climate
EF	Snow-ice climate

A = humid climates in subtropical zones or megathermic plants' climate (all months with a temperature higher than 18°).
B = dry climates or xerofile plants' climate.
C = humid mesoclimates or mesothermic plants' climate (temperature of the coldest month between 18° and −3°)
D = microclimates or microtermic plants' climate (temperature in January lower than −3°, but that of the warmest month higher than 10°);
E = polar climates or polar plants' climate (even the warmest month has a temperature lower than 10°).

He also described transition areas between these rough climatic zones. In order to be more specific, he then added another letter as follows:

f = no dry season climate
s = summer dry climate
w = winter dry climate
S = steppe climate
W = desert climate
T = tundra climate
F = ice climate

Therefore, there are 35 possible combinations between both classifications, some of the results being absurd. The reasonable combinations result in the following table, which is the basic format of the standard form (simplified) of the Köppen classification.

The Köppen system is widely used to define the main climatic types of the world. Figure 3.7 depicts the application of the system to the world using most of the classes. The result is a map of large, homogeneous climatic areas once useful to understand the cross-relationship between plants, animals and climate.

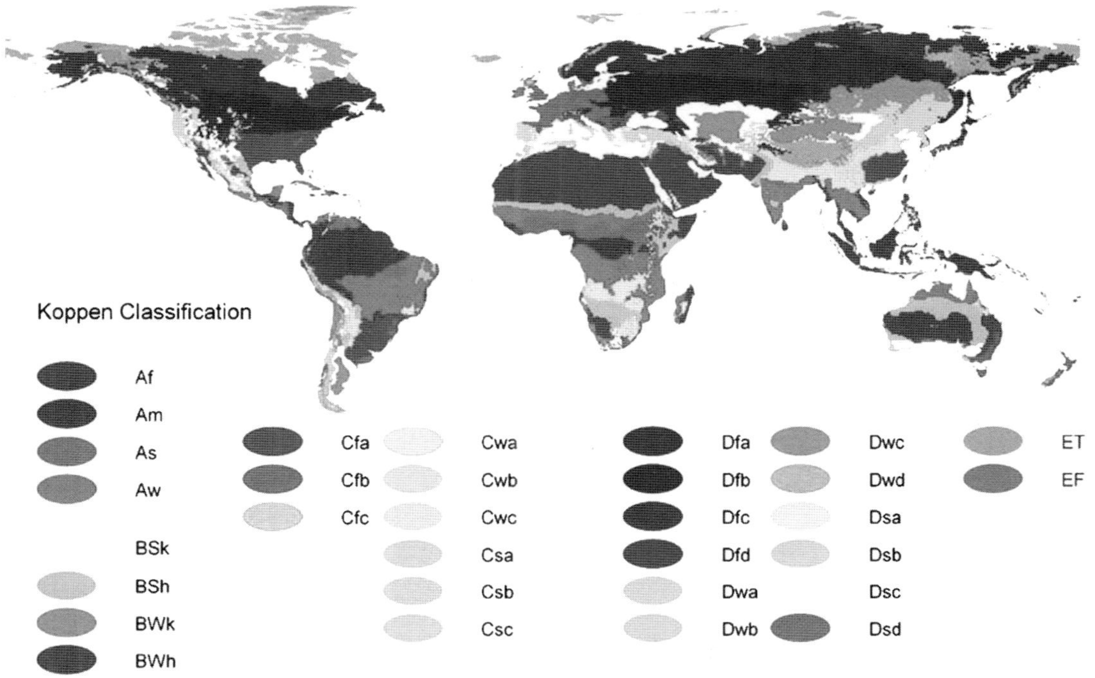

Fig. 3.7 Köppen climatic classification map. Data from *The Agromet Group*, SDRN, FAO

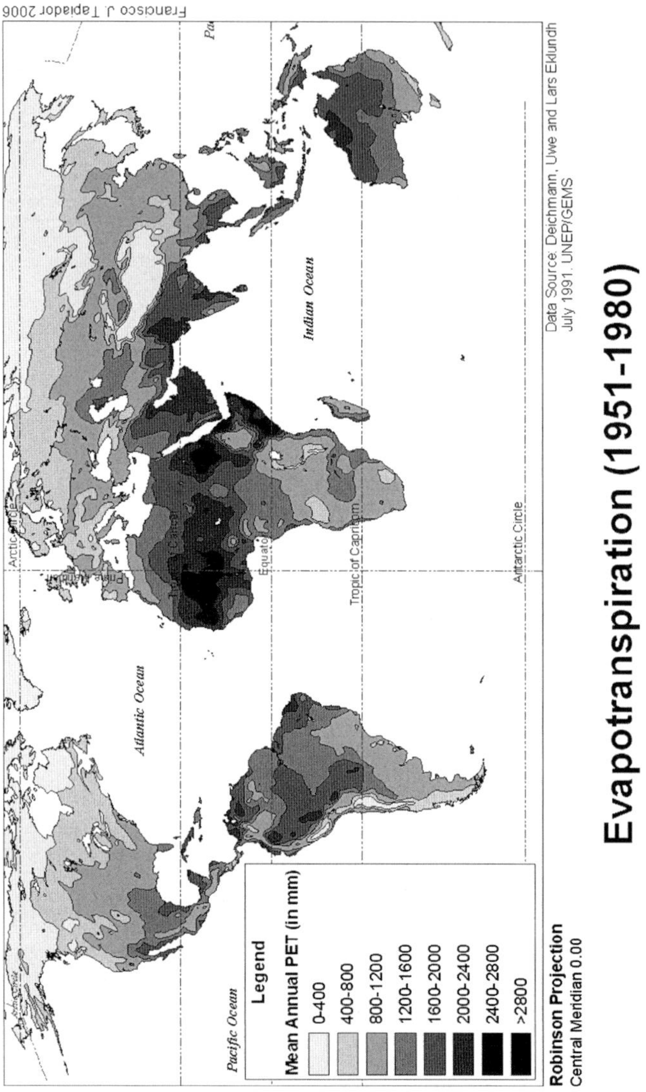

Fig. 3.8 Global-scale PET calculation using Thornthwaite's system

3.5.2 Thornthwaite's Climatic System

Thornwaite created a more quantitative (and therefore preferable) method for the synthetic study of the climate. He combined rainfall and potential evapotranspiration (PET) measurements.

PET (Fig. 3.8) is defined as the water quantity soil would lose due to surface evaporation and plant respiration if continuous soil water storage were available. This variable thus includes the effect of temperature. Thornthwaite devised an index to take into account the balance between precipitation and PET the total moisture index (MI) calculated as:

$$MI = 100 \cdot \frac{(s-d)}{PET}$$

Where s is the surplus and d the deficit of water. The formulae only means that if precipitation is lower than PET (s = 0 and d > 0), MI is negative and, and if precipitation is higher than PET (s > 0 and d=0), MI is positive. MI negative values correspond with a dry climate, and positive values with a humid one (Table 3.9). Thornthwaite eventually went to a more complex system including thermal efficiencies and a monthly computation of indexes, but the nucleus of his system is the MI estimate.

Other indexes that can be found in literature are the index of dryness (DI) and the index of humidity (HI). They are calculated using the number of days (d) with water surplus (or deficit) and the relative amount of water deficit (d) and surplus (s), respectively:

$$DI = 100 \cdot \frac{d}{PET} \qquad HI = 100 \cdot \frac{s}{PET}$$

Both the Köppen and Thornthwaite systems aim to provide a short and synthetic nomenclature for the whole planet. They are useful in giving a crude label in global terms, but are generally unusable today when we can easily classify climate using parameters at finer scales. Satellites are able to give hourly estimates at kilometrical resolution, which can be aggregated and fused with ancillary data to produce satellite climatologies at local scale (Fig. 3.9). Today, it is possible to use records

Table 3.9 Elemental climatic types using Thornthwaite's total moisture index (MI)

Symbol	Climatic type	MI
A	Very humid	> 100
B4	Humid	80 – 100
B3	Humid	60 – 80
B2	Humid	40 – 60
B1	Humid	20 – 40
C2	Subhumid	0 – 20
C1	Subdry	−33 – 0
D	Semiarid	−66 – 33
E	*Dry*	−110 – 66

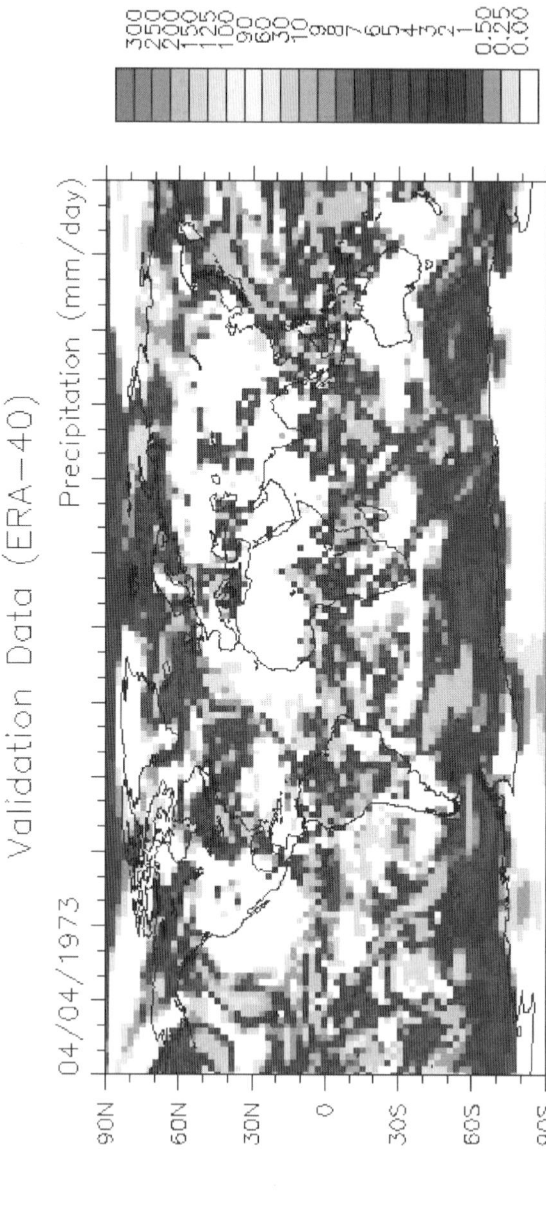

Fig. 3.9 Merged satellite and model global precipitation estimation (see Color Plate on following p. 333)

spanning over three decades, which is the minimum period required for climatological significance.

Far more useful than simply defining that a rural area corresponds with a B2 humid climate in Thornthwaite's system, is to know the hourly frequency of precipitation during the last ten years, daily amounts, maximums and minimums or several parameters (temperature, soil moisture, etc), total sunlight and so forth. The seasonal distributions of various variables permit to build more graphs than the well-known ombrothermic diagrams for temperature and precipitation. Those can be useful in having a complete picture of the climatic element in an area and to analyse the interactions with other environmental and anthropic elements.

3.6 Hydrology

The distribution of fresh water is probably the single most limiting environmental element in rural management. Water is not only a limiting factor for life, but also a major environmental factor for agriculture since crop productions depend on its timely availability. Rural life depends on a constant supply of fresh water for plants, animals, people and human activities. Water bodies also affect land use and microclimates, and global change will probably reduce stream flow due to reduced rainfall and increased evaporation. Not to mention the effects in areas that gravitates around water (such as floodplains). Thus, rice cultivation heavily depends on massive water availability resulting in very specific factors for rural management in large Asian areas.

The examination of this element in a rural analysis should comprise of the identification of water resources; an assessment of the need for agriculture, farming, forestry, economic activities, natural ecosystems, animals and peoples direct consumption; an analysis of the water quality; a rainfall-runoff analysis and flood routing alternatives; and a detailed analysis of how any intervention can modify the hydrological system. If it is possible, a numerical model should be used in these tasks. It is also worth studying the effects of hydrological processes in the ecological communities and the movement of soil water and its behavior.

In some deprived areas, hydrology can be the main factor to be addressed. Water may need to be found, and efficient water supplies built. In these cases, the otherwise unavoidable work of hydraulic engineers should be sensibly combined with the rest of the rural management. A key aspect is the management of water resources to ensure that communities will have a sufficient supply of clean water in the future. The concept of 'integrated catchments management' encompasses the analysis of all the hydrological aspects in the rural area and may become a major issue in the rural area at the end of the analysis. How to interlink the management of a particular element with the whole directive process requires a careful and balanced study to avoid converting the rural management into a monothematic enterprise, regardless of the importance of water in the area.

Drought management is also a part of rural management. If this aspect is deemed as relevant for the analysis, a part of the rural management process could be devoted to minimizing the effects of drought periods by creating infrastructures, selecting crops, managing aquifers, and so on.

Runoff should be considered in terms of environmental hazards, soil fertility losses and hydraulic resources. There are many models, such as the Système Hydrologique Européen (SHE) and the Institute of Hydrology Distributed Model (IHDM). Both of these are objective means for calculating runoff. In coastal areas, fresh water may need to be provided by desalinators and the problem of salt intrusion can appear. River restoration can also be an likely part of an analysis and, on the other hand, rural areas can be the passive actors of regional or national projects such as large dams.

3.7 Land Cover

The term 'land cover' refers to the materials that appear in a site such as vegetation, rocks, sand, building and water. It is not to be mistaken with 'land use', which is the pattern of construction and activity land is used for. For example, an area classified as predominantly 'olive tree' in the land cover legend can be either a Mediterranean golf course or an agricultural exploitation in terms of land use.

Land cover on large areas can be identified using satellites. Multi-band data provides a means to identify covers by comparing the spectral response of the cover in several wavelengths. Satellites such as LANDSAT or SPOT are used for land monitoring and observation and their images can be obtained from local or regional authorities. Processing remote sensing data requires specific knowledge: many pitfalls and gross errors can be committed with the powerful software available (ENVI, ERDAS, etc.) and with little or no mathematical or physical background, so land cover maps should be obtained from a trusted source.

Satellites provide data at several spatial resolutions. LANDSAT images can discriminate a feature being more than 30m size, while IKONOS satellites go to 5m. Figure 3.10 shows the visual differences between them: in the IKONOS image, even some large trees can be individualized. The spatial resolution is closer to an aerial image.

The LANDSAT image in Fig. 3.10, being cheaper, can only provide a general picture between irrigated, not-irrigated, fields, rivers and the riparian vegetation. Land use maps usually require ground information since it is not always obvious from a satellite or airplane.

Data fusion procedures exist to combine both types of data. The final result in terms of land cover will be a large or small resolution map. Of course, a fieldwork-based land cover map would almost certainly be better than any remotely sensed map, but even so, practical issues recommend the use of satellite technology (cf. Sect. 5.3).

It is not always required to generate a new land use map or update an old one. Some quality land use datasets already exist for areas such as the EU. Here, the 'CORINE land cover 2000' provides high-quality information suitable for rural analysis (Fig. 3.11). In this inventory the mapping of the changes of land

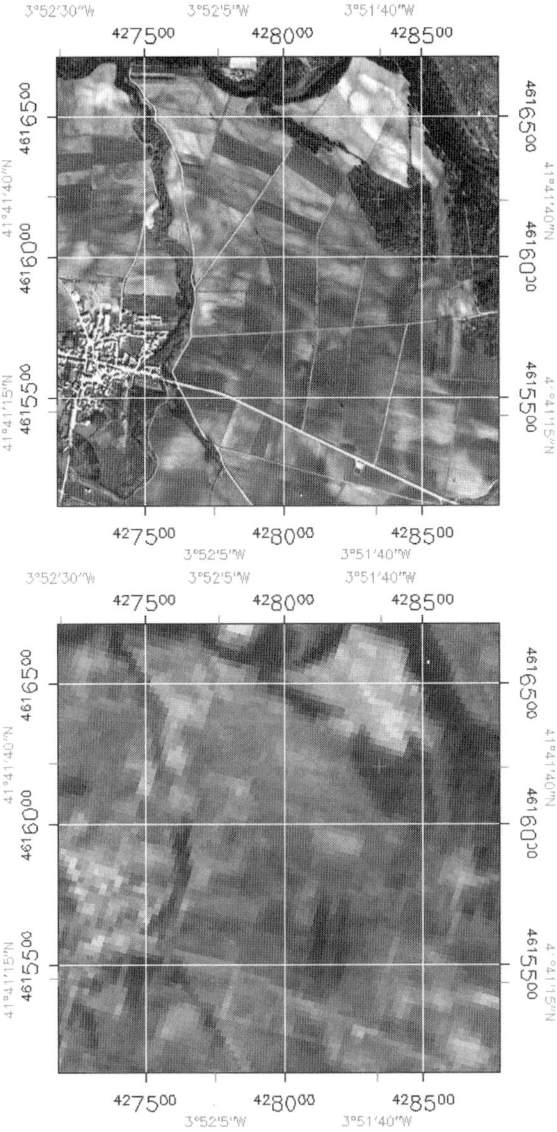

Fig. 3.10 Comparison between IKONOS and SPOT satellite images for the same area. Color combination in the SPOT image has been chosen to enhance irrigated land (see Color Plate on following p. 333)

cover is carried out by means of image comparison using computer assisted image interpretation tools. It also uses ground data and ancillary information to help in the interpretation.

3.8 Ecology

In some cases, a 'land use' or 'land cover' map is not enough for rural planning. This is the case of rural areas with large plant or animal diversity, special natural values, or some rural forest areas such as the Boreal forests in North America. A study of the distribution and abundance of different types of living beings and their relationships (ecology) is therefore crucial.

Ecological maps can help to assess this environmental element. According to Kotelnikova and Kotova (1994) they can be used to map: 1) The natural ecological potential, that is, the estimation of the natural conditions and resources; 2) The unfavorable and perilous natural and anthropogenic natural processes and phenomena; 3) The use of the natural ecological potential of the territory and the anthropogenic influences; 4) The alteration of the state of components of the geographic environment in consequence of the anthropogenic influence; 5) The protection of the nature; 6) Complex ecological relationships; and also to control de ecological situation.

An ecological analysis of a rural environment should comprise –but does not have to be limited to- the identification of species, their interactions, behaviors, population dynamics, food webs, migration corridors, and so on. If the rural area

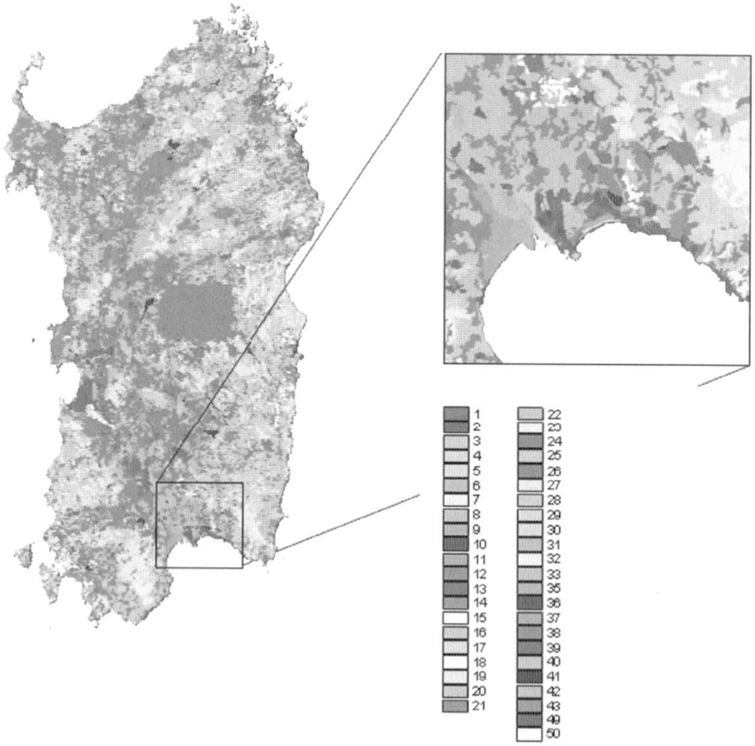

Fig. 3.11 CORINE-2000 land use map example (Sardinia). Key corresponds with the three levels version of CORINE-2000 land cover classification (see Color Plate on following p. 333)

Code	Level 1	Level 2	Level 3
1	Artificial surfaces	Urban fabric	Continuous urban fabric
2			Discontinuous urban fabric
3		Industrial, commercial and transport units	Industrial or commercial units
4			Road and rail networks and associated land
5			Port areas
6			Airports
7		Mine, dump and construction sites	Mineral extraction sites
8			Dump sites
9			Construction sites
10		Artificial, non-agricultural vegetated areas	Green urban areas
11			Sport and leisure facilities
12	Agricultural areas	Arable land	Non-irrigated arable land
13			Permanently irrigated land
14			Rice fields
15		Permanent crops	Vineyards
16			Fruit trees and berry plantations
17			Olive groves
18		Pastures	Pastures
19		Heterogeneous agricultural areas	Annual crops associated with permanent crops
20			Complex cultivation patterns
21			Land mainly occupied by agriculture, with significant areas of natural vegetation
22			Agro-forestry areas
23	Forest and semi natural areas	Forests	Broad-leaved forest
24			Coniferous forest
25			Mixed forest
26		Scrub and/or herbaceous vegetation associations	Natural grasslands
27			Moors and heathland
28			Sclerophyllous vegetation
29			Transitional woodland-shrub
30		Open spaces with little or no vegetation	Beaches, dunes, sands
31			Bare rocks
32			Sparsely vegetated areas
33			Burnt areas
34			Glaciers and perpetual snow
35	Wetlands	Inland wetlands	Inland marshes
36			Peat bogs
37		Maritime wetlands	Salt marshes
38			Salines
39			Intertidal flats
40	Water bodies	Inland waters	Water courses
41			Water bodies
42		Marine waters	Coastal lagoons
43			Estuaries
44			Sea and ocean
49	No data		
50	Sea and ocean	Sea and ocean	Sea and ocean

Fig. 3.11 (continued)

is fortunate enough to host unique ecological niches, then these have to be fully identified and protected.

Ecological unit inventories, if available, may provide a wealth of valuable information. It is up to the rural analyst to recommend carrying out one of the inventories if the area is ecologically valuable. In order to do that, the first step to take into account is that this inventory will be used for rural planning. This may affect the design of map units and guide the style of reports done by the ecologists. The ecological team will then evaluate the personnel and equipment needs and costs, providing a budget and a schedule. To be helpful to planning issues, the ecological inventory should be focused on providing a product for resource management needs, segmented by planning levels. The ecological map unit should contain a methodological section. The delimited units must be fully described in terms of scientifically measurable values, describing the taxonomic components, plant and animal associations and ecological types. The amount of different ecological types in the units would be decided after the field-work, considering the time and money constraints of the work.

It would also be valuable if the ecology team could provide an interpretation for the ecological units in terms of management: including specific ideas for the overall rural project such as vegetation use, wildlife habitat capacity, conservation values, recreation potential, insects, plagues, fire and disease hazards, hydrologic alteration, etc. It is advisable that their final report would be compatible with the formats and specs of the Geographical Information System of the rural project including maps in the same projection, units, nomenclature and digital format. Tables should be in the same Database Management System than the rural GIS (or converted to). Apart from that, the ecological report should include the aim of the work, rationale, methods followed, field reports and notes, costs (including updating costs), samples, the complete records of any experiment done, the interpretation of the units, and any valuable information for the rural planners.

A complementary approach is to use ancillary environmental impact assessment (EIA) information. EIA studies are integrated analyses of the natural and anthropic elements of the environment aiming to predict the influence of a project in a landscape. They are not however, a tool for planning. They have to be used, if required, in the rural planning section (cf. Sects. 14 and 16), once the environmental and anthropic elements have been analyzed. Then, EIA can help to decide between several planning alternatives, ruling out some of them and providing valuable information on the potential effects of the feasible plans. Whether or not a EIA is required is given by law in some countries and depends on the actuation which is being planned.

One part of rural management is to make the interests and needs of the people compatible with the natural communities. The balance between the requirements of a growing population and the natural areas is not easy to achieve, but it is necessary to ensure future resources. Planners have to consider not only how to solve the problem today, but also the consequences of the planned action in the future availability of resources. This is one of the central ideas of the 'sustainable development' concept. An example will clarify this: tourism is one possible strategy for improving rural coastal economies, but depending on the planning strategy, can result in the destruction of natural values in the name of progress. Alternatively, it can result in a careful balanced plan to avoid future diseconomies. A short-term substitute can dramatically improve a damaged rural economy, but the agents involved must know if this is at the cost of destroying rural life and creating dysfunctional blocks of second-residence homes. Nonetheless, a decisive and traumatic action is not always undesirable if it has a place in the regional or national structure of the territory and the agents directly involved agree.

The example of Tellico Dam (Ehrlich and Ehrlich, 1981) in which a small snail darter was threatened by extirpation by the construction of the dam, shows that it could have compelling reasons to proceed. The construction of the dam was stopped by the tribunals when it was well advanced to protect the fish, and legal and political issues arose in terms of balancing human needs with natural ones. The case eventually ended in the amendment of the US Endangered Species Act in 1979, the completion of the dam, and the snails moved to another habitat so they could survive. This case is also illustrative because it was found, subsequently, that the species had other habitats nearby, concluding that the construction of the dam was not as damaging as previously thought.

In order to interpret EIA the analyst has to be familiar with concepts such as 'vulnerable', 'endangered' and 'critically endangered' species. A good knowledge on modern ecological applications is also expected from the rural analysts since ecology is only relevant in areas with one specific ecological value, but in almost any rural area. There is a direct link between agricultural practices and ecological communities. Contaminated land can be reclaimed through restoration ecology techniques, and rivers can be restored to hold native fish as an additional economic resource. Biological or integrated pest control is a classic example of how pure knowledge can be more beneficial than the 'brute-force' approach of using pesticides everywhere. Also, eutrophication due to agricultural practices can be attacked by biomanipulation of the food webs in the habitats.

If it is required, management techniques based on biodiversity principles may have to be applied. It can involve selecting conservation areas, developing guidelines to protect them, and so on. Biodiversity is a value by itself, since it increases the entropy and thus the ability of the earth system as a whole to survive and flourish, but also has an economic importance. The diversity of species can actually be valued in monetary terms. Pharmaceutical companies benefit from natural products; natural areas not only produce food, but have a role in maintaining a stable climate, etc. Under the Kyoto's agreement, carbon dioxide emissions (amongst other products) are a real economic cost to pay and forests act as sinks for it. The concept of 'ecosystem services' includes a proper valuation of all these hidden services. Research indicates that the cost of managing natural areas is well below the actual economic value when properly estimated (Balmford et al. 2002). Contrary to the past, it could be more profitable to manage a mangrove area rather than convert it for agricultural production, since in today's economy services are more profitable than production or manufacturing.

These, and many other cases, show the importance of information and knowledge in our society. A deep study of say, the ecological processes of benthic communities, is not a scholarly luxury but a real economic activity that yields tangible mid-term benefits.

3.9 Natural Resources

All the environmental elements visited so far can be considered as a natural resource. However, apart from these and minerals, crops and timber, rural areas possess several other natural resources such as energy and water.

A rural analysis should identify the possibilities of the area for providing energy, not only for local use (domestic and industrial), but for regional and national markets too. Water resources should also be considered as natural resources and not only as natural elements, in order to analyze the cross-relationships between environmental elements (Chap. 4).

3.9.1 Energy Resources

Energy in rural areas is a twofold issue. Firstly, it is related with the need of providing energy (electricity, heat, combustible) to rural areas, and secondly, it relates with the production of energy in rural areas. In developing countries the second issue is the main concern. In these areas, energy resources should be developed in an integrated way to generate rural development, improve rural economies and to raise rural living standards. It can also be included into environmental policies: organic waste generates methane suitable for domestic use and (as discussed in Sect. 5.15.) contributes to an efficient use of the energy.

The sun provides high entropy radiation that is used by plants and animals to lower their own entropy; it is not the actual amount of solar radiation that makes life possible, but its low entropic form. Solar energy is the ultimate energy source: it is clean, endless and equally distributed on Earth. Solar technologies have been available for a long time now, but only recently have they become efficient enough to be widely implemented.

Solar energy can be used to heat fluids (passive heating) or to directly generate electricity through photovoltaic technologies. There are several types of solar water heaters suitable for rural implementation: the flat-plate type, the vacuum tube type and the stagnation type. All of these are more indicative for areas with good irradiations and can be used for heating houses, greenhouses and even classrooms.

Wind power energy production is another rural resource to be considered. Thousands of square kilometers of bare or unproductive land, either owned by villagers or the municipalities, have found a new use with the improvement in the technology and public support (cfr. Sect. 5.15). Thus, several arid or sub arid areas of Europe have experienced an economical revival by a public impulse of wind turbine installation. The economic importance of this element cannot be dismissed. For instance, a single wind turbine (0.6 Mw) requiring an initial investment of $500,000 USD (2004) can provide a constant annual net income of $70,000USD for a rural community in an otherwise unproductive area. The growth in wind engine installation in some rural areas has been made possible thanks to two strategic policy decisions: the electric companies are obligated to purchase at market price the power generated by any individual and governments have generously subsidized the initial inversions required. In turn, this renewed economic interest in wind power generation has developed rural engineering companies (wind power estimation, wind-farm design, installation and maintenance).

Behind the public impulse of these policies lies not only an economic interest, but also the awareness of the anthropogenic influence on global warming and an unconcealed interest in reducing the strategic dependence on oil producers.

The visual impact of these wind farms is widely debated. For some, they are ugly artifacts that devalue the landscape, while for others they are not only a valuable economic resource, but also a svelte piece of high-tech engineering.

Other energy sources can come from agriculture: some existing crops are suitable for producing biofuels (wheat, sugar beet, potatoes and oil seeds). Some by-products such as cooking oils can also be used as fuel. It is not yet clear whether or not dedicated tree plantations are an economically viable option. The development of forests devoted to fuel production could be desirable in areas subject to severe erosion. The forest can be used to protect the soil with fast-growing highly productive trees. In any case, it is doubtful that these kinds of energy resources will play nothing but a complementary role in the near future.

Fuel cells are also a promising technology, but are still in a very early stage of development. This issue will become influential (or not) in the future, but the curve of technological development tells us that it is too soon to consider it in terms of rural planning.

Geothermal energy depends on suitable geological conditions. If possible and technologically efficient, it should be considered. Central heating systems are currently in use in many rural areas of China. Tidal energy is also a promising source of energy in coastal areas.

Nuclear technology has a potential impact for rural life. While nuclear technology is becoming safer, and accidents such as Chernobyl are quite implausible with the new designs, the problem of nuclear waste remains. The planning issue of 'not in my backyard' has made sparsely populated and depressed rural areas the propitious victims of national policies. In the early sixties, remote areas in the very north of mainland Scotland were used for experimental nuclear fission reactors. The results were not very encouraging. The technology was safe, but the environmental problems associated with nuclear waste were not well understood and radiation contaminated the beaches. The nuclear plant created new job opportunities for the population but ruined parts of the environment, which sadly is a very well known tale in rural history.

On the other hand, hydroelectric power has become less impacting on rural landscapes with the generalization of small-scale generation plants. A recent exception to this rule is the gigantic 'Three Gorges Dam in China', which has modified not only the lives of millions of rural people, but even the climate of the area.

3.9.2 Water Resources

Water does not only act as an environmental element (Sect. 3.6) but also as a resource. Surface water is required for life to flourish and water stored as groundwater is crucial for both domestic and farming use. Analyzing future water availability is relevant for planning recreational areas and resorts. Spas are a possible source of income for some rural communities and canals have played a role in some rural economies.

Water resources need to be fully mapped. The rights of use also need to be considered and special care has to be taken in good practices. Fresh water is sparse; therefore an efficient use must be encouraged. This can be done by using native plants in public and private spaces, offering incentive programs (as rebates or tax credits) to homeowners and businesses to encourage replacement of plumbing fixtures and appliances with water-efficient models, or carrying out water-use audits of homes, businesses and industries. Other ideas include the use of low flow faucet aerators, highly efficient showerheads and leak detection tablets that can also help to save water.

The role of climate change should also be addressed. In some areas forecasts predict higher summer temperatures and less summer rainfall yielding more evaporation. This will increase the need for irrigation, not only for agriculture but also for leisure uses such as golf courses. If the water supply-demand balance is being considered only for the current climate and population, the plan could not be successful and rural life could be seriously affected.

3.10 Scenic Values

The scenic values can be evaluated in terms of visual impacts. First, it is necessary to define the view character, that is, the key elements that define the scenery, and the other characteristics. A next step is to establish the state of the view in terms of what is the state of each of the characters previously defined. The sensitivity of the scenery is then assessed. This is done by assessing the effects in the view of the inclusion of new features.

The next stage is to evaluate the scenery. This can be done in terms of natural aspects, distinctiveness, public acknowledgment of the value of the place and historical factors.

Scenic values are usually assessed taking into account the relief, the presence of water, the visual complexity, visual coherence, visual barriers, visual features, visual detractors, etc. The process is quite subjective, as can be imagined, and critically depends on the educational background of the evaluator. What some consider an ugly post-industrial site, for some it is a site to preserve due to the historical links with the first industrial revolution.

Scenic values are not only related to areas of natural beauty but also areas with a strong individual character. A particularly unimpressive landscape becomes interesting if the site is a battlefield and this fact can be used to explain history. This can attract tourism by the site becoming part of a wider net of historical sites or by creating a center of analysis of the place. Services can appear to serve this new economic activity and local opportunities in terms of crafts or folk music might be developed.

3.10.1 Natural Parks and Estates

The role of rural tourism in terms of development will be explored in Chap. 11.4. Here, we are interested not in the natural parks by themselves, but in the links between rural areas and these areas. Natural parks such as Yellowstone or Yosemite, for example, are only of the interest of rural analysis in that rural economies outside the parks are strongly affected by the presence of the parks, providing work and services for the communities.

Some private estates such as Chatsworth in the UK are perfect examples of the effect of well-managed landscapes and natural areas in a rural economy (and vice-versa). This estate, covering 14,000 hectares is visited by 650,000 people every year. It is estimated that several thousand people depend on it for all or part of their annual income. Regardless of the specificities of the place, this property is an example of how a conscious rural management can create not only beautiful landscapes but a prosperous economy. It is obvious that not all rural areas have the potentialities of Chatsworth to become rural paradises, but this estate is a fine illustration on the relationship between natural resources and rural activities. The estate is now a landscape resource by itself.

3.10.2 Geosites

'Geosite' is a concept created in 1996 by the UNESCO and the International Union of Geological Sciences (IUGS). Both institutions launched the 'Geosites and Geoparks' programmes to improve regional (and thus rural) development. Geosites are defined as small-sized places of geological heritage, whilst Geoparks may be larger areas, which can also include other resources such as archaeological, ecological, historical or cultural values. The idea is to facilitate the preservation of the geological heritage, increase public awareness and support local and regional development plans. They are also educational sites.

One of the requirements for becoming a Geopark is integrating the promotion of geological heritage and regional development. The brand name of a Geosite (or Geopark) is given by UNESCO upon formal application and the fulfillment of a series of requirements.

In Europe, a Geopark is a territory including a particular geological heritage and a sustainable territorial development strategy supported by a European development program. It must have clearly defined boundaries and sufficient surface area for true

Table 3.10 List of natural hazards (after Bryant 1991)

Tropical cyclones	Sea-level rise
Extra-tropical cyclones	Beach erosion
Storms	Droughts
Tornadoes	Flash floods
Dust storms	Thunderstorms
Localized strong winds	Lightning
Oceanic waves	Hail
Sea-ice	Snowstorms
Land instability	Anthropic fires
Blizzards	High magnitude flood
Freezing rain	Natural forest fires

territorial economic development. It is desirable that it will have a direct impact on the territory by influencing living conditions and environment: one of the declared objectives is to allow the inhabitants to re-evaluate the values of their territory and participate in its cultural revitalization.

Apart from proper Geosites, geological heritage can be a resource for rural tourism. Deserts (Tunisia, Canary islands) are attractive to many people for their geological formations. Even a single abandoned mine can be of interest and help to improve rural life, so it is essential not to rule out any possibility.

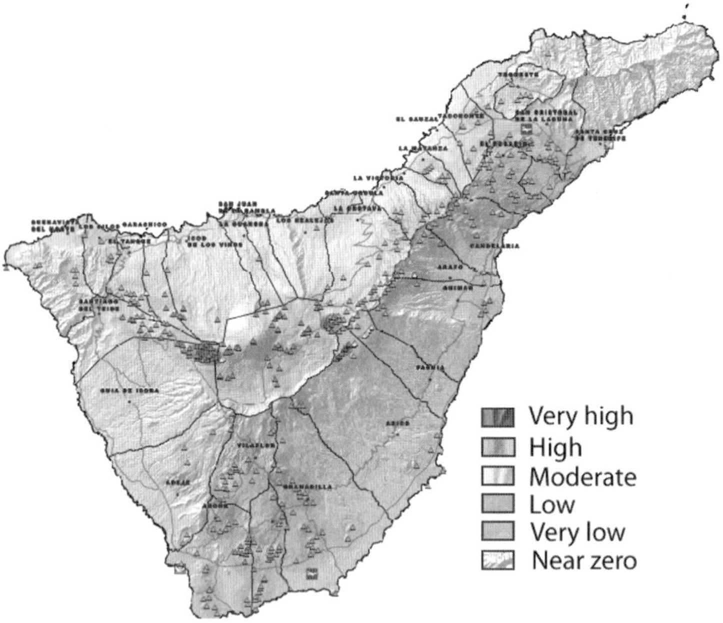

Fig. 3.12 Probability of volcanic events in Tenerife Island (from Tenerife's Cabildo PTEO)

3.11 Natural and Environmental Risks

Hazards and their associated risks have to be considered as part of the environmental inventory. Hazards have to be identified, the risks assessed, and contingency plans prepared if the risk is high. Table 3.10 summarizes most of the natural hazards that can affect a rural area.

National agencies in developed countries provide maps of these risks, usually charging a fee for its professional usage (Fig. 3.12). This makes it difficult for professionals to access this necessary information and may cause a study to be abandoned or only superficially made. In spite of this, the potential losses greatly exceed the cost of providing such a service for free. Sometimes, big scale maps are made freely available and can then be used to evaluate the need for a more detailed study. An example would clarify this point: the publicly-available radon potential map of the UK of Green et al. gives an estimate of radon risk in a 10×10 km mesh. If the rural analysis were referred to as an area with very high radon potential (such as in Cornwall), this would justify the investment for a more detailed study.

References and Bibliography Used

M. Abensperg-Traun, T. Wrbka, G. Bieringer, R. Hobbs, F. Deininger, B. Y. Main, N. Milasowszky, N. Sauberer and K. P. Zulka, 'Ecological restoration in the slipstream of agricultural policy in the old and new world', Agriculture, Ecosystems & Environment 103 (2004), 601–611.

J. Ahmad, B. Goldar and S. Misra, 'Value of arsenic-free drinking water to rural households in Bangladesh', Journal of Environmental Management 74 (2005), 173–185.

R. Aspinall, 'Modelling land use change with generalized linear models–a multi-model analysis of change between 1860 and 2000 in Gallatin Valley, Montana', Journal of Environmental Management 72 (2004), 91–103.

A. Balmford (and 18 co-authors) 'Economic reasons for conserving wild nature'. Science (2002), vol 297, no. 5583, pp. 950–953.

B. P. Bhatt and M. S. Sachan, 'Firewood consumption along an altitudinal gradient in mountain villages of India', Biomass and Bioenergy 27 (2004), 69–75.

I. Blanckaert, R. L. Swennen, M. Paredes Flores, R. Rosas Lopez and R. Lira Saade, 'Floristic composition, plant uses and management practices in homegardens of San Rafael Coxcatlan, Valley of Tehuacan-Cuicatlan, Mexico', Journal of Arid Environments 57 (2004), 179–202.

E. Bryant, 'Natural Hazards', Cambridge University Press (1991)

H. Buller, 'Where the wild things are: the evolving iconography of rural fauna', Journal of Rural Studies 20 (2004), 131–141.

G. Chen and P. A. White, 'The mutagenic hazards of aquatic sediments: a review', Mutation Research/Reviews in Mutation Research 567 (2004), 151–225.

J. Clark, 'Forest policy for sustainable commodity wood production: an examination drawing on the Australian experience', Ecological Economics 50 (2004), 219–232.

M. J. Cohen, K. D. Shepherd and M. G. Walsh, 'Empirical reformulation of the universal soil loss equation for erosion risk assessment in a tropical watershed', Geoderma 124 (2005), 235–252.

O. T. Coomes, B. L. Barham and Y. Takasaki, 'Targeting conservation-development initiatives in tropical forests: insights from analyses of rain forest use and economic reliance among Amazonian peasants', Ecological Economics 51 (2004), 47–64.

S. Dallas, B. Scheffe and G. Ho, 'Reedbeds for greywater treatment–case study in Santa Elena-Monteverde, Costa Rica, Central America', Ecological Engineering 23 (2004), 55–61.

H. Daniel and E. Lecamp, 'Distribution of three indigenous fern species along a rural-urban gradient in the city of Angers, France', Urban Forestry & Urban Greening 3 (2004), 19–27.
B. Deckers, M. Hermy and B. Muys, 'Factors affecting plant species composition of hedgerows: relative importance and hierarchy', Acta Oecologica 26 (2004), 23–37.
A. Deletic, 'Sediment transport in urban runoff over grassed areas', Journal of Hydrology 301 (2005), 108–122.
P.R. Ehrlich and A.H. Ehrlich. 'Extinction: the causes and consequences of the disappearance of species'. Random House (1981), New York
B. H. M. Elands, T. N. O'Leary, H. W. J. Boerwinkel and K. Freerk Wiersum, 'Forests as a mirror of rural conditions; local views on the role of forests across Europe', Forest Policy and Economics 6 (2004), 469–482.
J. Ewan, R. Fish Ewan and J. Burke, 'Building ecology into the planning continuum: case study of desert land preservation in Phoenix, Arizona (USA)', Landscape and Urban Planning 68 (2004), 53–75.
E. Fernandez-Juricic, 'Spatial and temporal analysis of the distribution of forest specialists in an urban-fragmented landscape (Madrid, Spain): Implications for local and regional bird conservation', Landscape and Urban Planning 69 (2004), 17–32.
P. Givone, 'Determination des risques d'inondation, effets de l'amenagement de l'espace', Comptes Rendus Geosciences 337 (2005), 229–241.
M. A. Hoogstra, H. Schanz and K. Freerk Wiersum, 'The future of European forestry–between urbanization and rural development', Forest Policy and Economics 6 (2004), 441–445.
O. Iacob and C. Grecea, 'Public exposure to radon and thoron progeny in Romania', International Congress Series 1276 (2005), 373–374.
R. Jamieson, R. Gordon, D. Joy and H. Lee, 'Assessing microbial pollution of rural surface waters: A review of current watershed scale modeling approaches', Agricultural Water Management 70 (2004), 1–17.
S. Kaloudis, D. Anastopoulos, C. P. Yialouris, N. A. Lorentzos and A. B. Sideridis, 'Insect identification expert system for forest protection', Expert Systems with Applications 28 (2005), 445–452.
N.E. Kotelnikova and T.V.Kotova, 'Ecological maps as one of the directions of the maps for the social development', 60th IFLA General Conference - Conference Proceedings (1994)
L. F. Lumsden and A. F. Bennett, 'Scattered trees in rural landscapes: foraging habitat for insectivorous bats in south-eastern Australia', Biological Conservation 122 (2005), 205–222.
L. Mancini, S. Caimi, S. Ciardullo, M. Zeiner, P. Bottoni, L. Tancioni, S. Cautadella and S. Caroli, 'A pilot study on the contents of selected pollutants in fish from the Tiber River (Rome)', Microchemical Journal 79 (2005), 171–175.
A. Martin-Dominguez, M. T. Alarcon-Herrera, I. R. Martin-Dominguez and A. Gonzalez-Herrera, 'Efficiency in the disinfection of water for human consumption in rural communities using solar radiation', Solar Energy 78 (2005), 31–40.
R. T. Nickson, J. M. McArthur, B. Shrestha, T. O. Kyaw-Myint and D. Lowry, 'Arsenic and other drinking water quality issues, Muzaffargarh District, Pakistan', Applied Geochemistry 20 (2005), 55–68.
D. Oguz, 'Remaining tree species from the indigenous vegetation of Ankara, Turkey', Landscape and Urban Planning 68 (2004), 371–388.
B. K. Paul, 'Arsenic contamination awareness among the rural residents in Bangladesh', Social Science & Medicine 59 (2004), 1741–1755.
K. S.-H. Peh, J. d. Jong, N. S. Sodhi, S. L.-H. Lim and C. A.-M. Yap, 'Lowland rainforest avifauna and human disturbance: persistence of primary forest birds in selectively logged forests and mixed-rural habitats of southern Peninsular Malaysia', Biological Conservation 123 (2005), 489–505.
R. T. Ramessur, 'Statistical comparison and correlation of zinc and lead in estuarine sediments along the western coast of Mauritius', Environment International 30 (2004), 1039–1044.
B. D. Richter, J. V. Baumgartner, J. Powell and D. P. Braun, 'A method for assessing hydrologic alteration within ecosystems', Conservation Biology 10 (1996), 1163–1174.

S. J. Rooklidge, E. R. Burns and J. P. Bolte, 'Modeling antimicrobial contaminant removal in slow sand filtration', Water Research 39 (2005), 331–339.

K. Saarinen, A. Valtonen, J. Jantunen and S. Saarnio, 'Butterflies and diurnal moths along road verges: Does road type affect diversity and abundance?' Biological Conservation 123 (2005), 403–412.

R. Soderlund and T. Rosswall, 'The Natural Environment and the Biogeochemical Cycles' (ed. Hutzinger, O.) 61–81 (1982), Springer, Berlin.

N. S. Sodhi, L. P. Koh, D. M. Prawiradilaga, Darjono, I. Tinulele, D. D. Putra and T. H. Tong Tan, 'Land use and conservation value for forest birds in Central Sulawesi (Indonesia)', Biological Conservation 122 (2005), 547–558.

P. Solis, 'Water as rural heritage: reworking modernity through resource conflict in Edwards County, Kansas', Journal of Rural Studies 21 (2005), 55–66.

M. I. Tariq, S. Afzal and I. Hussain, 'Pesticides in shallow groundwater of Bahawalnagar, Muzafargarh, D.G. Khan and Rajan Pur districts of Punjab, Pakistan', Environment International 30 (2004), 471–479.

L. L. Tieszen, G. G. Tappan and A. Toure, 'Sequestration of carbon in soil organic matter in Senegal: an overview', Journal of Arid Environments 59 (2004), 409–425.

F. T. Wakida and D. N. Lerner, 'Non-agricultural sources of groundwater nitrate: a review and case study', Water Research 39 (2005), 3–16.

D. H. Wang and K. E. Medley, 'Land use model for carbon conservation across a midwestern USA landscape', Landscape and Urban Planning 69 (2004), 451–465.

C. Watanabe, A. Kawata, N. Sudo, M. Sekiyama, T. Inaoka, M. Bae and R. Ohtsuka, 'Water intake in an Asian population living in arsenic-contaminated area', Toxicology and Applied Pharmacology 198 (2004), 272–282.

J. C. Woersching and A. E. Snyder, 'Earthquakes in El Salvador: a descriptive study of health concerns in a rural community and the clinical implications: Part III-mental health and psychosocial effects', Disaster Management & Response 2 (2004), 40–45.

N. H. Wong and C. Yu, 'Study of green areas and urban heat island in a tropical city', Habitat International 29 (2005), 547–558.

M. R. Zak, M. Cabido and J. G. Hodgson, 'Do subtropical seasonal forests in the Gran Chaco, Argentina, have a future?' Biological Conservation 120 (2004), 589–598.

K. Zhu, L. Zhang, W. Hart, M. Liu and H. Chen, 'Quality issues in harvested rainwater in arid and semi-arid Loess Plateau of northern China', Journal of Arid Environments 57 (2004), 487–505.

Chapter 4
Dynamics and Processes of the Environmental Elements

4.1 A First Approach to Complex Analysis

The environmental elements do not have separate effects on rural areas. They interact with one another in very complex ways. Many of the geomorphologic features are consequences of the continuous action of a particular climate during tens of thousands of years, and vegetation clearly depends on soil and atmospheric characters. Some of these relationships are historic, but several others act in very short periods of time being thus relevant for rural planning.

The possible strengths of cross-relationships for an area are illustrated in Fig. 4.1. The matrix represents a host of processes and feedbacks between the environmental elements. It is to be read by selecting first a row on the left and then a column on the top (so the effect of geology in climate is a direct effect and the effect of climate in geology is not important in that given rural area). The idea behind selecting the dominant elements is clear: to keep the number of these sub-relationships to a minimum and avoid unnecessary work, and focus on what is significant for a rural analysis.

Figure 4.1 is the final version of the process, but transient matrices can be created to include a full justification of choice by writing down every bit of information used in each cell. Assigning a weight in terms of the strength or relevance of that relationship for the area is always the last step, but must be backed by explicit evidence.

The matrix does not represent a fixed statement but a particular selection. It can be argued that it is somehow subjective to assign a weight to each link and that for a given area values from different analysts may differ. But what is crucial for rural analysis is to illustrate the principles being used by explicitly stating how the values are given. Furthermore, this matrix will allow for comparison with other analysts' work. An expert discussion based on comparing different matrices is always productive for the rural project and enriches the analysis even if a consensus is not achieved.

The last five rows and columns of the matrix represent the more dynamic and more manageable processes. The effect of geology on soils is well known but little can be done to modify it as it operates over large periods of time. Another example is how geology is related with natural risk such as earthquakes, tsunamis, or in minor extent, radon emissions. On the other hand, how meteorology and climate affect animal distribution must be carefully considered in the light of how climate change can affect a rural area.

ENVIRONMENTAL ELEMENTS	GEOLOGY	MINERAL RESOURCES	SOILS	GEOMORPHOLOGY	METEOROLOGY	CLIMATE	HYDROLOGY	LAND COVER	PLANTS AND ANIMALS DISTRIBUTION	NATURAL RESOURCES	SCENIC VALUES	NATURAL RISKS
GEOLOGY		A	A	B	A	A	A	A	A	A	A	A
MINERAL RESOURCES	-		-	-	-	-	-	-	-	A	A	B
SOILS	-	-		C	-	-	B	A	A	A	A	A
GEOMORPHOLOGY	-	B	A		C	C	A	A	A	A	A	A
METEOROLOGY	-	-	A	A		A	A	A	A	A	-	A
CLIMATE	-	-	A	A	-		A	A	A	A	B	B
HYDROLOGY	-	-	A	A	A	A		A	A	A	A	B
LAND COVER	-	-	A	B	C	C	B		A	B	A	B
PLANTS AND ANIMALS DISTR.	-	-	A	B	B	-	B	A		B	A	B
NATURAL RESOURCES	-	-	-	-	-	-	-	-	-		-	-
SCENIC VALUES	-	-	-	-	-	-	-	-	B	B		-
NATURAL RISKS	-	-	-	-	-	-	-	-	B	B	B	

A DIRECT EFFECT
B VARIABLE EFFECT
C INDIRECT OR WEAK EFFECT
- UNCOMMON OR LONG TERM EFFECT

Fig. 4.1 Cross-relationships between environmental elements. The intensity of the effects may vary from one rural area to another, so a matrix should be build for each rural area

The matrix depicted in Fig. 4.1 contains 132 relationships, some of them not particularly relevant for rural analysis. It is not practical to provide a detailed account of each of the relationships here. The matrix is a guide to help in analyzing the dynamics of the natural elements of the rural areas and devote more attention to the dominant elements. This will lead to extend the analysis to each of the sub-elements in each element with the others.

The practical use of the matrix is as follows: once the environmental elements have been analyzed (and if possible quantified), it is time to gather all that information and start to quantify the cross-relationships. Using the geological maps or the soil maps, for example, the relationships between plants, animals and the soil can be analyzed. How relevant is this factor for the rural area under study? Is the soil a homogenous mass or can several types be found in the area? What kind of relationships can be found between different classes of soil and the natural values? Is the soil a valuable or singular resource here? Thus, the idea is to extend the analysis of each environmental element to the analysis of their mutual relationships.

It is up to the analyst to choose how to use or to fill the matrix: it can be used as a simple checklist to ensure that the possible cross-relationship is being taken into account, or each cell in the matrix can be understood in terms of how relevant the effects are for the analysis; or to directly record the most relevant elements found; or be used for communication with the local communities. In terms of rural analysis as academic discipline, the matrix can be used to characterize a rural area, to compare different areas, and indeed to model processes. What is required is to

Fig. 4.2 The numerical form of the matrix of cross-relationships (named as A)

1	3	3	2	3	3	3	3	3	3	3	3
0	1	0	0	0	0	0	0	0	3	3	2
0	0	1	1	0	0	2	3	3	3	3	3
0	2	3	1	1	1	3	3	3	3	3	3
0	0	3	3	1	3	3	3	3	3	0	3
0	0	3	3	0	1	3	3	3	3	2	2
0	0	3	3	3	3	1	3	3	3	3	2
0	0	3	2	1	1	2	1	3	2	3	2
0	0	3	2	2	0	2	3	1	2	3	2
0	0	0	0	0	0	0	0	0	1	0	0
0	0	0	0	0	0	0	0	2	2	1	0
0	0	0	0	0	0	0	0	2	2	2	1

carefully explore each of the relevant relationships to see if they are relevant for the rural area under study, and to what degree.

The practical utility of this matrix can be enhanced using its numerical form:

This matrix is known in graph theory as the 'adjacency matrix' of the graph. It is convenient to have the main diagonal set to 1s. This matrix form is preferred to the diagram even if the diagram provides a more visual appraisal of the processes, as it helps to forget considering all the possible relations and assign them a value. Algorithms can thus be applied to the numerical matrix to extract all the chains of relationships and closed cycles such as soils-land cover-soils. Concepts such as spanning trees or Hamiltonian cycles are of use here, and environmental models can be developed using this matrix. If this expertise is available, it should always be preferred to the qualitative analysis.

The benefits of modelling and quantification greatly exceed the cost of setting up such technology. It permits a quick update, is easily comparable with the work of other analysts, and allows for rapid corrections if needed. If an error is found in the initial data, one only needs to change the matrix and the model (if it is not misconceived) will provide a new corrected answer. Models also provide a means to analyze the evolution of the process and generate forecasts and scenarios.

Dynamics can be studied by modelling a change in one element and comparing the initial and final states. If changes in land cover are deemed to generate an impact on hydrology or scenic values, then this should be taken into account. The rural analysis can include a dynamic model as a base for the future management of the area. As a toy example, if a simple model states that the environmental elements of the rural area can be characterized by assigning a weight to each one of them, this can be related with the matrix in the following way: first, we build a vector with the 12 weights:

$$x := [0.0\ 0.5\ 1.0\ 0.2\ 0.0\ 0.2\ 0.5\ 0.3\ 0.2\ 0.1\ 0.1\ 0.2]$$

Each value represents the weight given to each one of the environmental elements. If we now multiply this vector (transposed) by the matrix (named A), we get:

$$x \cdot A = [0.0\ 0.9\ 5.2\ 4.3\ 2.4\ 2.2\ 4.7\ 6.6\ 7.4\ 8.9\ 9.0\ 7.2]$$

This is the weight of each element once considered the cross-relationship between them. On the other hand, if we multiply the matrix by the vector we obtain:

$$x^T \cdot A = [9.7\,1.5\,4.9\,8.6\,8.1\,7.7\,7.2\,6.4\,6.4\,0.1\,0.7\,1.0]^T$$

Which represents the effect of applying the vector to the rows. Since A happens to be an invertible matrix (Figs. 4.2, 4.3), we can solve any problem of this type: given the input y, determinate the x such as:

$$A \cdot x = y$$

This is done in matrix algebra by doing:

$$x = A^{-1} \cdot y$$

Which results in the input required to obtain the desired output with the matrix defined as in Fig. 4.2. This is equivalent to apply the combined relationships to another set of variables, for instance the cost involved in managing each environmental factor by itself.

	1	2	3	4	5	6	7	8	9	10	11	12
X	10.0	0.0	10.0	5.0	2.0	7.0	2.0	1.0	0.0	5.0	0.0	1.0
Y	9.6	0.9	5.7	0.0	1.1	5.5	2.8	4.6	2.4	3.7	4.2	2.5

Which indicates to us that, given the relationships between elements and our original allocation of resources, we should not be expending anything in factors 2, 9 and 11. Instead, we can minimize the expenses for factor 4.

The plots in Fig. 4.4 show how different the allocation is if the cross-relationship is either considered or not. The example is an extremely simple one, but serves to show how matrix algebra can be applied to numerically treat the cross-relationship table providing non-intuitive results.

1	-21/5	-21/10	3/5	-29/20	27/10	1/20	-21/10	-1/10	53/5	-223/20	263/20
0	3/5	-6/5	1/5	3/5	2/5	-2/5	-6/5	4/5	-19/5	6/5	4/5
0	-11/5	-37/20	11/10	-23/40	29/20	7/40	-27/20	3/20	51/10	-241/40	301/40
0	6/5	-9/10	-3/5	19/20	3/10	-11/20	-9/10	11/10	-28/5	53/20	-13/20
0	6/5	3/5	-3/5	7/10	-6/5	-3/10	3/5	3/5	-18/5	39/10	-39/10
0	0	3/2	0	-3/4	-1/2	3/4	3/2	-3/2	3	-1/4	-11/4
0	2	3	-1	0	-2	0	3	-1	-2	6	-10
0	-4/5	-3/20	2/5	-17/40	11/20	13/40	-13/20	-3/20	19/10	-79/40	79/40
0	-1/5	-3/5	1/10	3/10	1/5	-1/5	-3/5	2/5	-7/5	1/10	7/5
0	0	0	0	0	0	0	0	0	1	0	0
0	2/5	6/5	-1/5	-3/5	-2/5	2/5	6/5	-4/5	4/5	4/5	-14/5
0	-2/5	-6/5	1/5	3/5	2/5	-2/5	-6/5	4/5	-4/5	-9/5	19/5

Fig. 4.3 The inverted matrix (A^{-1}) of the cross-relationships between environmental elements

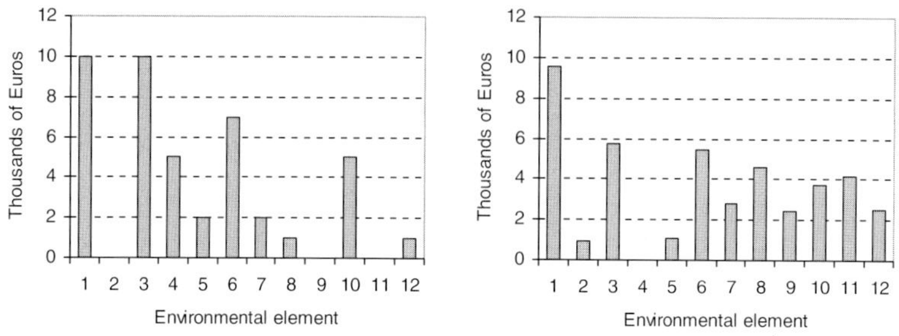

Fig. 4.4 A simple example of the original (X, left), and adjusted (Y, right) allocation of resources considering the cross-relationship between the twelve environmental elements

References and Bibliography Used

L. An, M. Linderman, J. Qi, A. Shortridge and J. Liu, 'Exploring complexity in a human-environment system: An agent- based spatial model for multidisciplinary and multiscale integration', Annals of the Association of American Geographers 95 (2005), 54–79.

M. Antrop, 'Landscape change and the urbanization process in Europe', Landscape and Urban Planning 67 (2004), 9–26.

M. Antrop, 'Why landscapes of the past are important for the future', Landscape and Urban Planning 70 (2005), 21–34.

A. M. Fisher and S. J. Harris, 'The dynamics of tree cover change in a rural Australian landscape', Landscape and Urban Planning 45 (1999), 193–207.

F. Hochtl, S. Lehringer and W. Konold, ' "Wilderness": what it means when it becomes a reality–a case study from the southwestern Alps', Landscape and Urban Planning 70 (2005), 85–95.

J. J. Kennedy and N. E. Koch, 'Viewing and managing natural resources as human-ecosystem relationships', Forest Policy and Economics 6 (2004), 497–504.

M. Yokohari, Y. Morimoto and N. Nakagoshi, 'Ecological dynamics of urban and rural landscapes in East Asia', Landscape and Urban Planning 70 (2005), 193–194.

Chapter 5
Anthropic Elements in Rural Analysis

The second major group of elements to analyze is the anthropic elements. They represent humankinds action on the environment and therefore on rural areas. The quantification of anthropic elements is more difficult than the quantification of the natural elements, and special care must be taken to try and be objective in the information gathering process, in developing protocols and methods when confronted with a new problem, and in explicitly writing down the methodological decisions to be taken.

The chapter is divided into three parts. Firstly, the spatial elements are described. They include legal and ownership issues, the location of services and utilities amongst others, and represent what can be mapped. All of these can be located in a space and should be integrated into a Geographical Information System (GIS) to store the data and be able to perform spatial analysis in a complex, non-trivial way.

The non-spatial elements refer to those elements that are present in rural areas but no individual location can be established. Issues such as culture, folklore and public health cannot be assigned to a single physical element in a space but to the area as a whole. They are however important, since they may be either opportunities or constraints for further development.

The last part of this chapter is devoted to the technological elements of agrarian activity. Since most of the rural areas to be managed are agrarian, the impact of the technical advances is instrumental to an integrated management approach. The role of irrigation policies, biotechnology, remote sensing and other technological issues have to be investigated and used if deemed as suitable for the rural area. Some areas can improve their quality of life simply by accessing available technology and using the local knowledge. It depends on the skills of the analyst to correctly identify the needs and potentialities of a rural area, and to propose guidelines and actions to improve the life of its inhabitants. To do so, as in the environmental elements analysis, the analyst should resort to specialists in diverse fields, but only after a first approximation of the area has been carried out from a multidisciplinary, geographical perspective. To re-iterate: the need to quantify and justify every aspect of the research is critical for the future success of the plan.

5.1 Spatial Elements

The spatial elements are those that can be assigned a precise geographical co-ordinate or set of co-ordinates. That includes agrarian plots, population, buildings, leisure areas, utilities, services, utilities or transportation networks amongst others. The term 'georeferencing' refers to assigning a co-ordinate to these elements so we can map them. Only when mapped can the spatial relationships be analyzed to try to establish relationships between them and to understand the role these elements play in the territory.

Spatial structure can be quantified beyond the crude qualitative estimate offered by terms such as *openfield* or *bocage*. Numerical measurements can be used to parameterize the agrarian plots, the habitat or the commercial activities. Thus, to quantify the morphology of a given rural area, we can calculate the following estimates:

- Density: the average number of units per area.
- Continuity: the degree of connection between areas within the same class.
- Proximity: the degree to which elements of the same class are close to each other.
- Clustering: the degree to which the anthropic elements cover a relatively small portion of the area.
- Centrality: weighed distance to the centroid.
- Entropy. In this context, entropy (S) is defined as

$$S = -\sum_i p_i \cdot \log(p_i)$$

The p_i are the measured frequencies for each class. From this formulae result, the maximum entropy appears when all the frequencies have the same value, that is, when the area has different classes homogenously distributed in the space. The entropy decreases if one element clearly stands out from the rest, since this increase the uncertainty of the choice. Figure 5.1 shows the result of applying this estimate to the land use map of Ireland. High entropy areas correspond with highly diversified land uses whereas low entropy appears in more homogenous areas. Thus, entropy gives us a sensible and a numerical estimate of spatial homogeneity. At its turn, spatial homogeneity is a key indicator for analyzing scale economies or tourism potential. These relationships are well-known issues for a geographer, who can also provide other objective spatial measurements adapted for the area. These objective measures can then be integrated in models to forecast tendencies and to test hypotheses.

Spatial metrics can be applied to both environmental and anthropic elements. For instance, one of the objectives when quantifying patterns in settlements is being able to quantify sprawl. In many developed countries, agricultural areas are increasingly being transformed into residential properties within a process called 'counter-urbanization' (Sect. 6.3). Providing a numerical estimate for the evolution of counter-urbanization is crucial to assess the rhythm of the processes, and this can only be done properly done by using morphological measures.

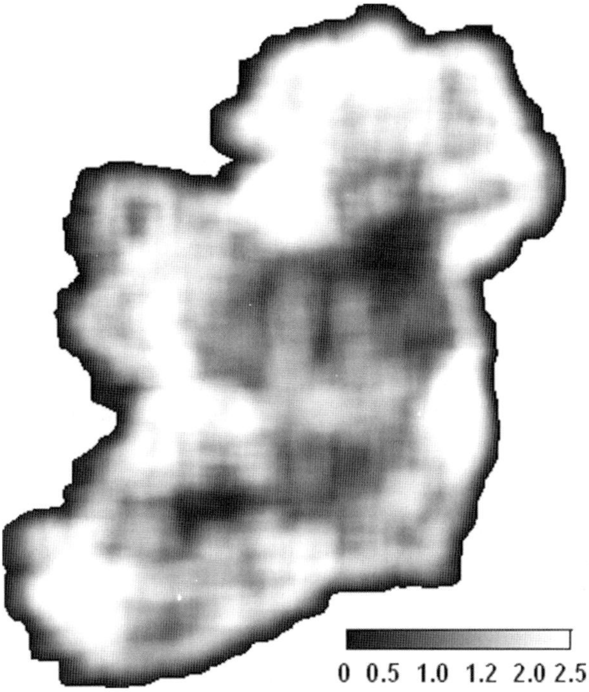

Fig. 5.1 Dark tones correspond with low entropy whereas light tones are high-entropy, inhomogeneous areas

5.1.1 The Rural Habitat

How dwellings are built and how are they distributed in a territory are two aspects to analyze. The variety of dwellings is greater in rural areas than in their urban counterparts, including semi-permanent villages and cave dwellings. Many buildings have architectural interest in that they represent a singular dwelling solution, adapted to local environmental conditions (Fig. 5.2). Buildings such as mills, schools, and civil building provide a distinctive character to some areas, singularizing them. Rural habitat tourism routes exist (i.e. the Pyrenees), following the link between agriculture, habitat and culture. Nonetheless, houses are after all a necesity, and as such should be considered in most rural areas of the world where a place to live is not always granted.

The analysis should contain a description of the building techniques and materials used in the construction of the villages, a list of singular buildings, and a distribution map of the habitat with all its peculiarities. The spatial distribution of the buildings is relevant: concentrated and disperse forms of occupation in the rural space have historical roots, and have different functions in the local economy and in the way of life.

Fig. 5.2 A patio in a rural manor house. Patios play an important role in Mediterranean villages

5.1.2 Agrarian Land

Characterizing the agrarian land can be done by mapping using a sensible classification system, either a general, comprehensive one or another judiciously adapted for the area of study. It is necessary to record not only the actual location of the spatial elements but also additional information such as the size distribution of farms, the economic activities and the number of people involved in a holding. In terms of GIS jargon, both the cartography and the database should be completed once the data model has been agreed.

There is one key economic parameter that characterizes agricultural land: the Standard Gross Margin (SGM). The SGM is the difference of the standard value of the total production, corresponding to the average situation of the agricultural holdings in a given region (less any losses and including subsidies linked to products, to area and/or to livestock) at basic prices, value tax not included; and the standard value of certain specific costs in that region, which can be directly attributed to the production. The data used for calculating the SGMs correspond to a reference period of 12 months, that is, either to a calendar year or to an agricultural production year. The SGMs used in the EU for farm structure surveys are averages of three successive reference periods, that is, 36 months. In practice, SGMs are estimated using average basic data from farm accounts or specific surveys as the average of three successive calendar years or agricultural production years for each type of crop (per ha) and each livestock characteristic (per animal). This gives an estimate such as 'SGMs for one hectare of barley', or 'for one dairy cow'.

In other words, the SGM is the average value of gross margin corresponding to the average situation in a given region for each agricultural category. In plain English, the SGM is the 'mean unit of productivity' for one agricultural enterprise in a given area.

The economic size of the holding is derived from the SGM. This value is calculated by multiplying the quantity (in ha and/or number of animals) of each enterprise in the holding by the SGM value in the region. In the EU, the economic size is quantified as European Size Units (ESU), which is calculated considering that 1,200 € of SGMs corresponds (by definition) to one ESU.

5.1.2.1 Typology of Agricultural Holdings

The type of farming carried out in a holding can be determined by the economic size itself and/or the relative contribution of different enterprises to the total economic size. A step-by-step description of the administrative procedure to classify the holding by economic size is:

- Step 1: Identify the enterprises (types of crops and types of animalcategories) belonging to the farm concerned.
- Step 2: Determine the area of each crop (in ha) and the number of animals per animal category.
- Step 3: Calculate the total gross margin of each enterprise by multiplying the scale of each enterprise with the corresponding SGM value.
- Step 4: Calculate the total gross margin of the farm concerned by summing up the total gross margins of the various enterprises.
- Step 5: Determine the economic size of the farm concerned by dividing the total gross margin of the farm by the reference number (1,200€ in the EU)

The resulting value is then used to generate a first typology of an agricultural holding by economic size. In the European case, Table 5.1 depicts the thresholds currently in use.

A more complete classification is to assess the type of farming, which includes the relative contributions of each production to the total economic size. The procedure to assess the type of holding is as follows:

- Step 1: Calculate the total economic size of the farm concerned.
- Step 2: Calculate the relative contribution (in %) of the various enterprises of the farm concerned to the total SGM of that farm.

Table 5.1 Typology of the European agricultural holding by economic size

ESU	<2	2–4	4–6	6–8	8–12	12–16	16–40	40–100	100–250	>250
Type	*I*	*II*	*III*	*IV*	*V*	*VI*	*VII*	*VIII*	*IX*	*X*

- Step 3: Apply the formulas defining the various types of farming. The formulas are based on 1/3 and 2/3 thresholds defined in the Commission Decision 377/85/EEC.

This generates a classification for agricultural holdings. In the EU, the typology is established in the legislation (L127/48, 23/05/2003). The system (Table 5.2) is adapted to the productions in the EU, but can be used in other areas of the world, or adapted to local needs.

The EU typology contains four hierarchical levels: general types of farming (one digit in Table 5.2), principal types of farming (two digit), particular types of farming (three digit), and subdivisions of certain particular types of farming (four digit).

Table 5.2 Typology of the European agricultural holdings (European Journal of the European Communities L220, 17/08/1985 and L127, 23/05/2003)

1				**Specialist field crops**
	11			Specialist cereals
		111		Specialist cereals (other than rice)
		112		Specialist rice
		113		Cereals and rice combined
from 1985 to 2003	12			General field cropping
		121		Specialist root crops
		122		Cereals and root crops combined
		123		Specialist field vegetables
		124		Various field crops
			1241	Specialist tobacco
			1242	Specialist cotton
			1243	Specialist oilseed and textile crops
			1244	Various field crops combined
	13			Specialist cereals, oilseeds and protein crops
		131		Specialist cereals (other than rice), oilseeds and protein crops
		132		Specialist rice
		133		Cereals, oilseeds, protein crops and rice combined
	14			General field cropping
from 2003 ahead		141		Specialist root crops
		142		Cereals and root crops combined
		143		Specialist field vegetables
		144		Various field crops
			1441	Specialist tobacco
			1442	Specialist cotton
			1443	Various field crops combined
2	**(20)**			**Specialist horticulture**
		201		Specialist market garden vegetables
			2011	Specialist market garden vegetables - outdoor
			2012	Specialist market garden vegetables - under glass
			2013	Specialist market garden vegetables, outdoor and under glass combined
		202		Specialist flowers and ornamentals
			2021	Specialist flowers and ornamentals - outdoors
			2022	Specialist flowers and ornamentals - under glass
			2033	Specialist flowers and ornamentals - outdoors and under glass combined

(continued)

Table 5.2 (continued)

	203	General market garden cropping	
		2031	General market garden cropping - outdoor
		2032	General market garden cropping – under glass
		2033	Specialist mushroom
		2034	Various general market garden cropping combined
3	**Specialist permanent crops**		
31	Specialist vineyards		
	311	Specialist quality wine	
	312	Specialist wine other than quality wine	
	313	Quality and other wine combined	
	314	Vineyards for various types of production	
		3141	Specialist table grapes
		3142	Specialist raisins
		3143	Mixed vineyards
32	Specialist fruit and citrus fruit		
	321	Specialist fruit (other than citrus)	
		3211	Specialist fresh fruit (other than citrus)
		3212	Specialist nuts
		3213	Fresh fruit (other than citrus) and nuts combined
	322	Specialist citrus fruit	
	323	Fruit and citrus fruit combined	
33	Specialist olives		
	330	Specialist olives	
34	Various permanent crops combined		
	340	Various permanent crops combined	
4	**Specialist grazing livestock**		
41	Specialist dairying		
	411	Specialist milk production	
	412	Specialist milk production with cattle rearing	
42	Specialist cattle - rearing and fattening		
	421	Specialist cattle - mainly rearing	
	422	Specialist cattle - mainly fattening	
43	Cattle - dairying, rearing and fattening combined		
	431	Cattle - dairying with rearing and fattening	
	432	Cattle - rearing and fattening with dairying	
44	Sheep, goats and other grazing livestock		
	441	Specialist sheep	
	442	Sheep and cattle combined	
	443	Specialist goats	
	444	Various grazing livestock - no dominant enterprise	
5	**(50) Specialist granivores**		
	501	Specialist pigs	
		5011	Specialist pigs rearing
		5012	Specialist pigs fattening
		5013	Pig rearing and fattening combined
	502	Specialist poultry	
		5021	Specialist layers
		5022	Specialist poultry-meat
		5023	Layers and poultry-meat combined
	503	Various granivores combined	
		5031	Pigs and poultry combined
		5032	Pig, poultry and other granivores combined

(continued)

Table 5.2 (continued)

6 (60)	**Mixed cropping**
601	Market gardening and permanent crops combined
602	Field crops and market gardening combined
603	Field crops and vineyards combined
604	Field crops and permanent crops combined
605	Mixed cropping, mainly field crops
606	Mixed cropping, mainly permanent crops
6061	Mixed cropping, mainly market gardening
6062	Mixed cropping, mainly permanent crops
7	**Mixed livestock holdings**
71	Mixed livestock, mainly grazing livestock
711	Mixed livestock, mainly dairying
712	Mixed livestock, mainly grazing livestock other than dairying
72	Mixed livestock, mainly granivores
721	Mixed livestock; granivores and dairying combined
722	Mixed livestock; granivores and grazing livestock other than dairying combined
723	Mixed livestock; granivores with various livestock
8	**Mixed crops - livestock**
81	Field crops - grazing livestock combined
811	Field crops combined with dairying
812	Dairying combined with field crops
813	Field crops combined with grazing livestock other than dairying
814	Grazing livestock other than dairying combined with field crops
82	Various crops and livestock combined
821	Field crops and granivores combined
822	Permanents crops and grazing livestock combined
823	Various mixed crops and livestock
8231	Apiculture
8232	Various mixed holdings
9	**Non-classifiable holdings**

5.1.2.2 Land Ownership and Tenure

The concepts of 'land ownership' and 'tenure' refer to who owns the land and who has an interest in it. Ownership is usually secured by law and can be conferred to individuals or legal entities (such as villages). The status of the ownership varies from owning the entire property, to more limited ownerships such as the fee tail, which consists in a hereditary but non-transferable ownership for life, or the life estate that is an arrangement for the life of the owner.

The owner/s can work on its land or can agree with someone else to exploit the land; if the ownership status confers that right. The agreement can be done in the form of rental or leasing for a given period of time. Another possibility is 'sharecropping', in which the land is exploited by a second party, but the production is shared between the owner/s and the people who work on it.

When the property is collective (several people and/or legal entities share the ownership) or public, other forms of tenancy appear such as the right to use the commons. The UN defined 'customary land tenure' as 'the rights to use or dispose of use-rights over land which rest neither on the exercise of brute force, nor on evidence of rights guaranteed by government statute, but on the fact that they are recognized as legitimate by the community, the rules governing the acquisition and transmission of these rights being usually explicit and generally known though not normally recorded in writing'.

If there is a public interest in using the land, an easement can be established to permit the use of the land regardless of ownership. This can be in the case of mines: the minerals belong to whoever exploits the mine and the ownership remains in the hands of the original owner/s but they must allow the exploitation to take place, usually through a fee. Another classical example is the easing for public utilities or the right of pass to neighbouring properties.

The relationship between ownership and tenancy provide an indicator of the rural economy and helps to characterize (and therefore compare) a given area. Thus, over the same plot several combinations of ownership and tenancy can coexist: a plot can be privately owned, but collectively exploited, and the product shared with the owner. This kind of arrangement is related to cultural and historical reasons, including productivity, political and social issues and obviously local legislation. Every combination has its own advantages and disadvantages.

The role of cultural factors in land tenure cannot be diminished. As an example of the complex relationships that might appear, a case study in Fiji (Table 5.3) illustrates the differences between a western approach and the customary land tenure of that area.

5.1.2.3 Concentrated and Fragmented Agrarian Land

Properties can be concentrated or fragmented. If owners have several plots or the tenants exploit several plots, the property is said to be fragmented. Fragmentation usually comes from historical reasons – mainly heritage along generations – and is not as economically profitable as a concentrated exploitation can be. Land consolidation, which is one of the processes that can transform fragmented to concentrated agrarian landscapes, is desirable for improving local economies.

Historically, the process of land consolidation has been important in European rural economies and is considered crucial to improve regional productivity in those areas in which property is very fragmented.

Technically, land consolidation consists of rearranging the plots to provide the maximum of topologically connected exploitations. That helps in reducing fuel and working time costs, and allows for cost-effective mechanization. Figure 5.3 shows a very simplistic example of this process (valid for illustrative purposes only). The idea is to make the number of plots owned by owner the lowest as possible, maintaining at the same time the average quality of the land that is owned and, when possible, the average distance to the plots to the village.

In the case of Fig. 5.3, the area belonging to each owner has been conserved, but this is not always the case: some owners can be compensated a loss in surface by an increase in the quality of the land and vice-versa. Other elements, such as the proximity to facilities or to the village, water availability, and topography are

Table 5.3 Comparative attributes of land tenure systems in Fiji, from Cultural Issues in Land Information Systems, Chukwudozie Ezigbalike, M.T. Rakai and I.P. Williamson, Land Reform Bulletin, FAO 1996

	Customary land tenure		Western land tenure
	Informal	Formal	
Rightslegitimized by:	Recognition and acceptance by community (and reinforced by occupation and amicable relationships within community)	Statutorily by legislation	Statutorily by legislation
Land rights transmitted by:	For landowning members: inheritance qualified by place of residence, need, gifts, number of children, etc. For immigrants: transmitted in response to their needs or other circumstances, e.g. gift, reward	Statutorily, only by inheritance or legally documented gift By sale, for registered leases All by statute	Inheritance By sale All by statute
Land used:	Individually	Individually	Individually
Land owned:	By group Groups vary in size and type, according Expiry of temporary user rights	By group Group size and type permanently defined, legally (e.g. commonly *mataqali*)	By individual or corporation, but held allodially by state/crown
Conditions for reversion of land rights from individual back to group	Abandoned land Extinction of individual user subgroup Renouncement of allegiance to group	Appointed term, e.g. two years of absenteeism from village, but rarely, if ever, enforced	Not applicable, since rights held either in perpetuity for freeholds or for term of lease for leaseholds
Social, mystical or religious attributes attached to land	Sanctity of spiritual sites Ancestral sites	Not legally acknowledged or recognized	None, apart from reserves set aside for cemeteries or religious sites
Degree of rights held	Varies according to political/social status May not be distinct or clear-cut	Varies according to political/social status Varies according to size of land allocation	Equally by individuals

usually also considered making land consolidation a complex resource-optimization problem that requires a vast amount of negotiation and technical work.

By definition, land consolidation represents a change in ownership for a given plot, but not a global change in ownership. The variation in absolute terms of the

Owner	Number of connected plots	Agrologic class sum	Score (1/class)	Plot distribution scheme				Plot Quality			
Before consolidation											
A	2	8	0.13	A	B	D	E	1	2	1	2
B	2	3	0.33	A	E	D	E	2	2	2	3
C	1	7	0.14	A	C	C	C	3	3	2	2
D	2	5	0.20	B	D	A	A	1	2	1	1
After consolidation											
A	1	9	0.11	A	A	B	B				
B	1	3	0.33	A	E	E	E				
C	1	7	0.14	A	C	C	C				
D	1	4	0.25	A	D	D	D				

Fig. 5.3 A simple illustration of the idea behind land consolidation

amount of land owned by each owner is usually small: it is a classical example of zero sum redistribution: in which – at least in theory – no one is damaged and everyone is benefited. Land consolidation is seldom a mandatory process. The owners soon realize the advantages of having all their properties being connected as possible and even if the changes may represent a problem for some tenants, the process benefits the rural area as a whole.

The technicalities of land consolidation have long been solved using GIS technology integrating simple algorithms to minimize an objective function subject to a set of constraints. Methods include the simplex method, neural networks or simulated annealing.

5.1.3 Leisure Places

The analysis of leisure activities is relevant not only for developed societies, as could be first thought, but also for the less developed areas. Several studies have concluded that leisure is of primary importance to fix the young population to rural areas. Even if job opportunities are still the first factor, the existence of adequate leisure means can seduce young couples into not migrating to cities. The apparently subjective perception of how lively a village is depends on how leisure needs can be satisfied. The analysis can help to harmonize these needs with economic development in a global development plan. As always, the first step is to assess the current state.

It has to be noticed that leisure could be the main component of rural regeneration for some areas. The economic relevance of leisure in modern societies can provide a way of life for some rural areas without hindering their rural character. It is important to analyze how people enjoy leisure facilities, but also to help in leisure awareness in rural areas, which is also a task to accomplish in rural development.

Leisure activities also comprise of the effects of mass-sports in some rural areas. Villages near stadiums can have an economic input as direct jobs or sport-related services such accommodation, catering, etc. Extreme cases such as the Olympic Games modify extensive areas, and a careful rural planning should be done to take advantage of this impulse and avoid potential pitfalls and diseconomies.

More common cases involve the analysis of sport facilities for local and possibly even for regional areas. Swimming pools are expensive facilities that can be integrated into villages with a natural potential and provide service to the neighbouring villages if only a limited budget is available.

The number of potential leisure activities even permits specialization at regional level. Sports facilities such as golf, tennis, soccer and other ball games, or other more regional sports such as curling or cricket can act as focal points for more comprehensive leisure activities. Gyms, health clubs, yoga, tai-chi and other martial arts can be attractive for all ages. Crafts workshops can benefit from rural ancestral knowledge providing a link between these activities and the present, thus conserving (and improving) the rural traditions. Pottery, jewelry, sculpture and painting are attractive choices for many people, providing not only leisure and relaxation services, but also skills and education. Community life can be enhanced thorough cultural and art centers. A good case study is the Bridport Arts Center (Dorset, UK). This village shows how creative activities help rural areas development. Arts and musical festivals can help to define and put rural communities on the map, with associated advantages in terms of economic growth.

In most areas, activities such as cycling, diving, walking, horse riding and hiking can be possible leisure options. Sometimes, it is hard for local people to see the potentialities of their own area for such activities. Climbing can be an obvious choice for some mountain areas, but day-trips to natural or agrarian landscapes are also activities in demand, not only for urban populations but also for rural populations. Other unusual activities include paintball camps, organized shopping trips, kart racing, extreme sports shows, plane modelling, or fairs (antiques, second-hand, etc.). It is up to the analyst to identify the potentialities of the area. Hunting, fishing or other more fauna-friendly activities such as bird watching are also potential choices. Synergisms between several fields can be particularly interesting: school farms can both help farmers' economies and provide a valuable educational service for the local children.

Local TV and radio stations are also useful means to enhance cultural life and leisure in rural areas. They can involve people in the actual making or be used to communicate and entertain people with local topics. A sort of cultural planning can be established to help individual initiatives in several areas. More obvious leisure activities such as bars and discos are in less need of planning since they obey to the economic market logic and need very little public impulse.

5.1.4 Utilities and Services

Utilities have long been associated with electrification. There is however other services such as gas, telephone, water and sewer utilities. More sophisticated utilities are centralized heating systems (Iceland, Canada). Basic utilities are central for the well being of people and for rural development. Actually, the degree of development of rural areas can be measured by the extension and state of utilities: areas with poor or inexistent water utilities cannot be considered as developed. If this is the case, one of the first tasks to accomplish is to provide an appropriate service in drinking water, sewerage, electricity and secondarily on gas and telephone.

Utility assessment should be based on data. Sources of data are official reports from supply companies and direct inspection. The map of the utilities is the first step to building management systems to optimize resources. Integrating the utilities map with other maps will help to understand the needs of the area and the priorities to be addressed.

To quantify the quality of these services, a questionnaire such as the one provided by the Local Utilities and Services Enquiry (EIEL in the Spanish acronym) in Spain can help (Table 5.4).

Table 5.4 Comprehensive list of the variables required to fully characterize a rural area in terms of utilities and services (slightly adapted from JCyL 1999)

Category	Variable	Code
Population and housing	Censed population	A1
	Area	A2
	Population density	A3
	Average height of the place	A4
	Residents	A5
	Maximum seasonal population (MSP)	A6
	Total households	A7
	Occupied households	A8
	Secondary residences	A9
	Unoccupied households	A10
Surface distribution	Total residential use area	B1
	Total industrial use area	B2
	Total developed residential area	B3
	Total developed industrial area	B4
	Not for development area	B5
	Total (B1+B2+B3+B4+B5) [not necessarily equal to A2]	B6
Pavement	Not paved surface	C1
	Paved surface	C2
	. . . in good condition	C3
	. . . in regular condition	C4
	. . . in bad condition	C5
	. . . in execution	C6
Roads	Total length of the roads in the area	D1
	. . . in good condition	D2
	. . . in regular condition	D3
	. . . in bad condition	D4
	. . . in execution	D5
Public lights	Total power	E1
	Number of lights	E2
	Households without public lights	E3
	Length of the streets without public lights	E4

(continued)

Table 5.4 (continued)

Category	Variable	Code
Water pipes and tanks	MSP [A6] with sufficient water availability	F1
	MSP with insufficient water availability	F2
	MSP with drinkable (purified) water	F3
	Total length of the water pipes	F4
	... in good condition	F5
	... in regular condition	F6
	... in bad condition	F7
	... in execution	F7
	Water tanks, total volume	F8
	Total added volume of the water tanks in good condition	F9
	... in regular condition	F10
	... in bad condition	F11
	... in execution	F12
Water distribution	Households with public water supply	G1
	Households with water supply	G2
	Households without water supply	G3
	Pipe length required to cover the water supply deficit	G4
	Total length of the water distribution net	G5
	... in good condition	G6
	... in regular condition	G7
	... in bad condition	G8
	... in execution	G9
Sewerage	Households with sewerage	H1
	Households without sewerage	H2
	Total length of sewers required to service all the households	H3
	MSP without sewerage	H4
	Households with independent sewerage systems	H5
	MSP with independent sewerage systems	H6
Sewers & sewage treatment plants	Flux of the combined sewers	I1
	Volume treated by the sewers	I2
	Sewers total length	I3
	Sewage treatment plants capacity	I4
Waste collection services	Solid waste production	J1
	MSP without refuse collection service	J2
	Households without refuse collection service	J3
	Dumps capacity	J4
Equipments	Covered sporting facilities, total area	K1
	Outdoors sporting facilities, total area	K2
	Covered leisure and culture centers, total area	K3
	Outdoors leisure and culture centers, total area	K4
	Covered gardens and parks, total area	K5
	Outdoors gardens and parks, total area	K6
	Covered markets and fairs, total area	K7
	Outdoors markets and fairs, total area	K8
	Slaughterhouses, total area	K9
	Slaughterhouses, total capacity	K10
	Cemeteries, total area	K11
	Saturation degree of the cemeteries	K12
Health centers	Number of beds available	L1
	Total area	L2
	... in good condition	L3

(continued)

Table 5.4 (continued)

Category	Variable	Code
	... in regular condition	L4
	... in bad condition	L5
	... in execution	L6
Care centers	Number of places available	M1
	Total area	M2
	... in good condition	M3
	... in regular condition	M4
	... in bad condition	M5
	... in execution	M6
Learning centers	Number	N1
	Number of places available	N2
	Total area covered	N3
	Area outdoors	N4

Gathering this data is mandatory in Spain for municipalities below 50,000 inhabitants in order to help planning. The database contains easily measurable values that must be assessed by a panel of experts or technicians. This task usually involves the use of a GIS, which can be subsequently used to create spatial models. These variables can be combined into synthetic indexes (Table 5.5) to facilitate interpretation and to develop strategies. The indexes are designed to provide a 0 value to the lower values and a 1 (or 100) to the higher value.

By mapping them, municipalities can be easily compared (Fig. 5.4): thus, when the rural area under analysis comprises of several municipalities, mapping the results can help in prioritizing needs. On the other hand, the indexes by themselves are illustrative even for a single village and can help to identify hidden problems and needs.

A key factor in assessing the quality of life in rural areas is an adequate number of services for the population. The availability of health centres, elderly centres, kindergartens and key facilities such as a diversified commerce, post offices, petrol stations and primary schools are instrumental for high standards of living. Some elements may require more attention than others, depending on the degree of development in the area. For example, telephone utilities have experienced a profound evolution in the last few years. Wireless technologies have provided telephonic facilities to areas where wiring was too expensive. Telephone antennas have become part of the rural landscape and mobile technology is now part of rural life. The internet has arrived to areas where three decades ago were difficult to access by car. This represents an opportunity for development, providing new services such as tele-medicine, tele-assistance and remote learning.

5.1.5 Transport and Accessibility

Rural travel patterns are different from urban travel patterns in most countries. Table 5.6 shows some of these differences for the UK, which are representative

Table 5.5 Some derived indexes using the variables in Table 5.4

Index	Formulae	Interval and measurement unit
Water availability index	$\frac{F1}{A6}$	[0,1] dimensionless 1: Population well served
Tanks capacity index	$\frac{F8}{A6}$	m^3/inhabitant Depends widely
Tanks condition index	$\frac{F11}{F8}$	[0,1] dimensionless 1: Good condition
Water distribution coverage index	$\frac{1}{2}\left(\frac{G3}{A7}+\frac{F2}{A6}\right)$	[0,1] dimensionless 1: Good coverage
Water distribution net condition index	$\frac{F7}{F4}$	[0,1] dimensionless 1: Good condition
Sewerage index	$\frac{1}{2}\left(\frac{H2}{A7}+\frac{F2}{A6}\right)$	[0,1] dimensionless 1: Good service
Refuse collection index	$\frac{J2}{A6}$	[0,1] dimensionless 1: Good service
Pavement extension index	$\frac{C1}{B6}$	[0,1] dimensionless 1: All the streets are paved
Pavement condition index	$\frac{C5}{B2}$	[0,1] dimensionless 1: All the streets are in good condition
Public lights index	$\frac{E2}{C1+C2}$	Lights/m, one light each 10 m is considered suitable
Leisure and cultural centers index	$300\frac{K3}{A5}$	Centers/inhabitant, one center by 300 inh. is considered suitable
Sport centers index	$100\frac{K1}{A5}$	Facilities/inhabitant, one facility by 100 inh. is considered suitable

of the EU. It can be seen that the annual distance travelled in rural areas is greater than in urban areas. Also, the proportion of households in rural areas within thirteen minutes walk of at least an hourly bus service is lower than in the rest of the country, which shows the rural dependence on a private car.

The UK Countryside Agency acknowledges that restricted rural services, longer journey distances and remoteness mean that rural people can be denied full access to, for example, medical care, shopping, jobs and further education. There may be particular difficulties for poorer households, disabled people and those left at home while the car user is at work. In the case of developing countries, the problems are exacerbated by poor infrastructures and a more or less limited access to motor vehicles.

Accessibility is a key element to measure. It is a more comprehensive indicator than isochrones, since the time required to reach a place is irrelevant if there is not a population that can benefit from it. The definition of accessibility varies, but in geographical terms, accessibility is a locational characteristic that measures how well a place can be reached from populated places. It can be applied to a single mode of transportation such as rail or to several combined modes (intermodal transport). The actual calculation of the accessibility depends on these factors being considered:

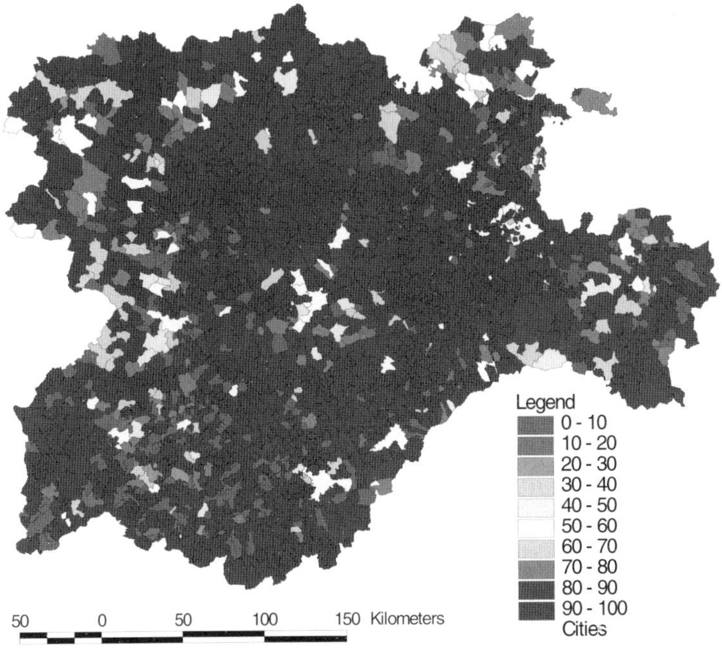

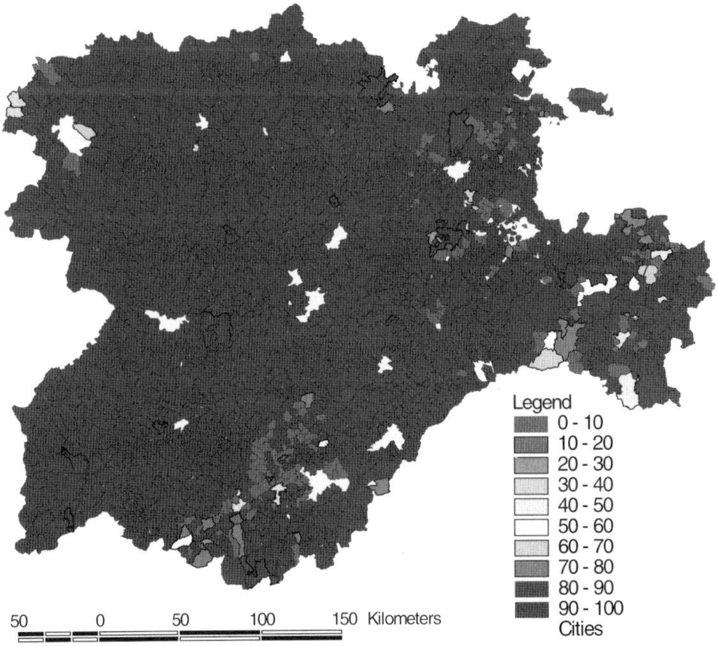

Fig. 5.4 Water availability (top) and refuse collection indexes (bottom) in Castilla y León region, Spain. Data has been scaled to percentages by multiplying by 100 the actual values of the indexes

Table 5.6 Rural travelling facts in the UK. [source: Countryside Agency, 2005]

Variable	Rural areas	Whole of UK
Households with access to a car	85%	72%
Households with two or more cars	38%	26%
% of those over 17 with car driver access	81%	71%
Annual mileage per 4 wheeled car	10,400 miles	9,200 miles
Households with 13 minutes walk of an hourly or better bus service	48%	89%
Households with 13 minutes walk of a food store	60%	82%
% of all journeys by car	73%	63%
% of all journeys made on foot	20%	26%
Total distance traveled per year (all modes)	*8,900 miles*	*6,800 miles*

topography, the state of the roads and congestion charges can be included in the calculation.

The result is a population-weighted accessibility map such as those in Fig. 5.5. Desktop software packages such as the International Center for Tropical Agriculture (CIAT) accessibility tool make easier the actual calculation of accessibility, providing the required input data is available.

A rural transport strategy should comprise of the analysis of the accessibility, road safety, congestion and environmental effects (pollution, noise) of the traffic. It should also analyze the forecasted tendencies in rural traffic and the intermodality of the solutions provided. National and regional planning should also be taken into account to provide services to local areas and to facilitate global cohesion. Specific policies such as graduated fuel pricing should be considered and the views of interest groups carefully balanced with the interests and needs of several scales.

For new transport services, several criteria can be used for planning or to assess proposals. For instance, the VIRGIL project provided 14 criteria to consider:

1. Operational viability: Is this service the only means of public transport in the area? Is it a supplementary service to regular services? Does the service substitute a previous service?
2. Use of telematics: Has telematics been utilized in: Reservation of service? Ticketing system? Information system? Traffic planning, e.g. routing?
3. Potential for integrated passenger / freight transport: Freight in passenger traffic? Passengers in freight traffic?
4. Applicability: Is a similar service concept in use in the same region / in the same country / in other countries?
5. Economic viability: Can economic viability be determined based on the following indicators: Total annual cost of operation, Ticket (fares) revenues (passengers and freight), Annual surplus / deficit, Financing of deficit, Continuation of financing?
6. Geographical coverage: Has the service brought new areas within service which have not had previous public transport services?
7. Social viability: Has the service particularly improved / decreased the mobility for: Elderly people / Disabled people / School children?

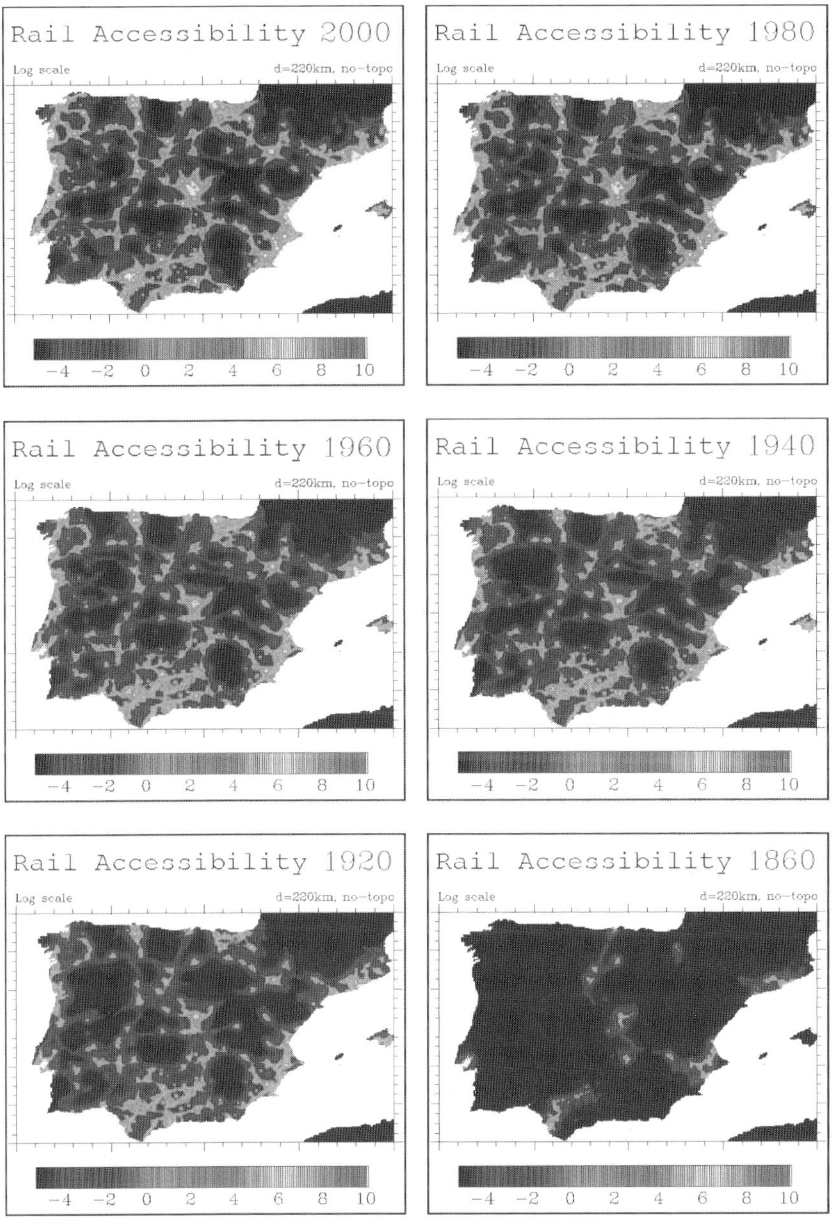

Fig. 5.5 Changes in the Iberian rail accessibility. Large rural areas still present a low accessibility by train (see Color Plate on following p. 333)

8. Flexibility of service: Is there any degree of flexibility of the service with regard to: Route /Stops / Checkpoints / Timetable?
9. Technical suitability of vehicle fleet: What is the standard passenger capacity? Are there any special comfort features? / Is there wheelchair access? /Is there low-floor access? / Is there separate access to a freight space?

10. Training of personnel: Is there special training for vehicle drivers and assistants / Reservation personnel?
11. Efficiency of reservation and data collection system: Have the following systems been considered: Reservation via Travel Dispatch Centre or in-vehicle mobile phone or radio link /Automatic, semi-automatic or manual data collection?
12. Reallocation of labor: Has the service resulted in re-allocation of resources between different public sectors (measurable cross-sector benefits)?
13. Users' opinion: Is the users' opinion positive about the service?
14. Marketing: Is there sufficient marketing of the services to users?

This criterion is sensible and general enough to be applied in developed countries. Nonetheless, evaluating the rural transport scheme of the area may require developing a new set of estimates, since the technical constraints of the area (or the travel behaviour) may differ.

5.1.6 Energy

Energy issues are relevant in two aspects: first, from the point of view of energy generation, rural areas host most of the energy production installations. Thermoelectric, nuclear, and hydroelectric plants are mostly located in rural areas and have an impact on rural life. On the other hand, rural areas need energy and that should be provided for a reasonable cost. It is estimated that more than 2,000 million people worldwide lack access to modern energy resources and most of them live in rural areas.

One of the emergent sources of energy worldwide is biomass, which is naturally suited for production in rural areas since their raw materials are agrarian. It represents today (2006) 15% of global primary energy consumption. In developing countries, biomass is about one third of their total primary consumption. Biomass can be a profitable source of income using otherwise useless material, contributing to limiting methane emissions and the use of fossil fuels. Also, it can be understood as an industrial activity, with plantations designed for production of fuel fed stock.

Wind energy is also a growing form of energy production. It is clean, easy to install and maintain, renewable and can be easily dismantled if not longer desired. The effects on the environment are negligible, but the visual impact can represent an issue when dozens of turbines scatter the landscape. Some rural areas have benefited from the installation of wind farms in terms of council taxes devoted to public services, and individuals renting land for turbine installation. The profitability of such initiatives depends on the natural conditions and on the current energy price. In coastal areas offshore wind farms can be installed. This location minimizes the visual impact and provides a potential income source for the locals.

Nuclear energy on the other hand remains a conflictive issue. In spite of its inarguable advantages for the environment, no one desires a new nuclear plant in his backyard. Even when the new methods and designs used in new installations make it almost impossible for a new Chernobyl, the opposition of communities makes it difficult to develop new plants in western countries. Compensations are a possible way of compromise for deprived areas. The installation of new plants in rural areas is a trade-off between the possibility of new jobs and economic possibilities for the area and the almost negligible risk of radiation or accidents. It is remarkable that most of the opposition to nuclear plants comes from people alien to the affected rural communities. It is however imperative for the rural community to press for a strict and independent security control of a nuclear plant, and for the installation of a well-calibrated environmental measurement network operating 24/7 in real time. It is also advisable that the authorities should provide contingency plans, insurance policies, and regular health checks for the population. In some cases, the best way to tackle fears and disinformation is by providing exceptional warranties and services, even if they seem to be unnecessary, redundant or disproportionate. Nonetheless, the safety of the population is always the first priority. The very few nuclear accidents that have happened in the past were due to incorrect management, obsolete installations and budget cuts in safety provisions. While nuclear technology has long advanced from the Chernobyl accident, there are lessons to be learnt on how human failure propagates into an organization. A local, expert-based watchdog on how the Corporate Social Responsibility (CSR) guidelines of the nuclear plants are to be put into practice is also advisable, and local communities should press for establishing independent bodies formed by experts.

Another aspect of energy in rural areas is the rural energetic needs. Development brings new needs and more energy is required. As in the urban case, a key aspect is energy conservation. Domestic fuel saving is the key for energy conservation in large areas such as China (and therefore at a planetary scale, since China represents one fifth of the World's population). Following with the Chinese example, in 1995 almost 90 percent of domestic energy consumption in rural China was devoted to heating and cooking. It is estimated that the use of subsidized efficient stoves in 120 million households has saved at least 22 million tons of coal-equivalent each year. Just by increasing the number of households using methane, the average combustion efficiency of domestic fuels was increased to 20%. Even when methane stoves are eight times more expensive than coal ones, the savings in coal compensate the initial inversion.

In the case of developing countries increasing rural mechanization, the savings in replacing diesel with electric power is about 20% in energy consumption and about four times in operating costs. This replacement reduces the economic dependence on oil and is arguably beneficial for the environment.

5.1.7 Industrial and Other Economic Activities

There are many ways to measure the economic activity of a rural area. The basic information for a rural analysis is the location of the activities, an estimate of the productivity and the economic sector of the business (primary sector: agrarian activities; secondary sector: industrial activities; tertiary sector; services). Sometimes an 'advanced tertiary sector' item is used to differentiate between classic services and R&D activities.

Geo-referencing the business part in a GIS and characterizing their activities is always required. Depending on the available information, this task can be more or less exhaustive. A minimum corresponds with the location, size, number of employees and activity of the factories, workshops, shops and other business. Functional relationships between businesses are also interesting to record if this information is available.

The number of people engaged in each economic activity is a valuable indicator of the economic orientation of a rural area and can help to devise development strategies. Also, the analysis of the evolution of the occupational structure could be significant to track the economic evolution of the area.

5.2 Non-spatial Elements

Not everything of importance is visually apparent on the territory. Some of the characterizing elements in some rural landscapes are related with non-spatial elements such as gender relationships, capital fluxes, public health, politics or traditions. Cultural issues may play a determinant role in, for example, preferring apparently unprofitable solutions to problems in favour of keeping a way of life. Also, laws are something that we cannot assign to a particular place, and laws differ from one place to another. Elements such as social equity or salaries are also within this category.

Some of these elements are not easily quantifiable in economic terms. They are, however, quantifiable in the sense that we can assign a number based upon objective (that is, shared) values. Vote intention is immaterial, but can be integrated into a GIS as a joint or conditional probability, helping the analysis and the policy-making. To develop means to integrate this wealth of information into an information system is a science in itself: Geography deals with environmental, human, spatial and non-spatial elements in a unified mathematical framework, and the need to accomplish this task to provide models.

5.2.1 Social Structure: Demography, Occupations, Education and Training

The distribution of population in the rural areas of the world is quite different from one area to another. Some areas suffer the problems of overpopulation while others present problems associated with depopulation. However, to quantify what represents over and depopulation depends not only on the actual number of people (and their age and gender distribution) but also on economic factors.

The density of population (ρ, the Greek letter called rho) measures the number of people by the area they live in. Calculating ρ for small areas (French *communes*, Spanish *municipios*, etc.) can give an initial idea of the population pressure, but comparison with bigger units is sometimes problematic. The use of administrative boundaries to calculate ρ must lead to gross errors since these units have not been developed to this end. The population density map of Sweden at NUT-2 administrative units is a typical example. Swedish counties *(län)* such as Norrbotten or Västerbotten are extremely large compared with the equivalent units in France or Italy, but have the population concentrated in a tiny part of the county. A population density map using these units can be misleading, and do not provide useful information on the population distribution of the area. An alternative approach is to avoid using administrative units and calculate ρ in geometric grids. This requires fairly disaggregated data which is not always available, but that can be obtained for the typical sizes in which rural analysis takes place.

Another quantitative measure is the 'demographic capacity index', which estimates the maximum number of people that a given area can sustain. This depends on the degree of industrialization, the regional and national integration of the area, the natural resources and the actual population. The calculation of this index is far more complicated than the density of population, but is crucial for estimating the needs and potentialities of the area under scrutiny.

The existence of a population optimum (PO) has been discussed in literature. It is considered that equilibrium should exist between the number of people and the rural resources. If the PO can be parameterized as a function of several factors, sensitivity analyses can be performed.

Both the population density and the DCI provide a single value for a given area. The next level of analysis is the distribution of the population within the area being studied. People have historically settled down following several patterns, depending on natural or historical factors. Hilly areas are more prone to disperse settlements while settlements in the planes have tended to concentrate structures, mainly for defensive purposes. How today's population is distributed affects services, social relations, economy and infrastructures, and this variable should be considered in the analysis. Nevertheless, how it should be considered will depend on the cross-correlations with other factors, elements and dynamics.

The internal dynamics of the population is another area to be considered. It can be estimated by plotting the structure by age and gender (Fig. 5.6). This graph, also known as population pyramid, gives valuable insight into the demography of the rural area. A symmetrical graph (approximately the same number of males and females by age) is the natural form: any dissymmetry reflects traumatic effects suffered by the population, such as wars or emigration.

The pyramid indicates an expansive population if it has a large base and a small top. This means a large proportion of children and a small proportion of the older population. A stable growth is reflected in the pyramid with a structure with indentations that even out and reflect slow growth over time. Stationary pyramids have a narrow base and roughly equal numbers in each age group, tapering off at the older ages. Declining pyramids represent a high proportion of aged persons and declining numbers. Pyramids such as Gómara's or Torremocha's in Fig. 5.6 are typical of these declining rural populations, with a large number of elderly population and a very limited number of young people.

The beauty of this graph is that it provides a combined and visual appraisal of several actual indicators such as fertility rate and life expectancy and it should always be constructed and analyzed for the rural community under study.

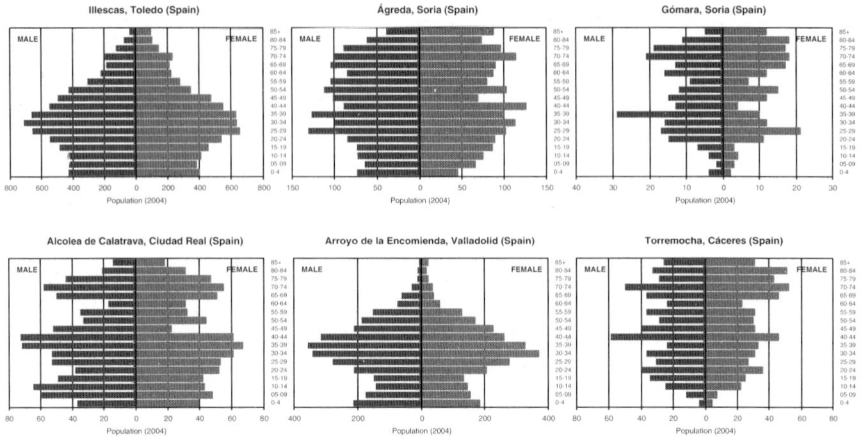

Fig. 5.6 Population pyramids for several Spanish municipalities in 2004 [Data source: INE 2005]

When immigration is a relevant issue, the pyramid can be disaggregated to show the age and gender structure of a foreign population. This can provide useful sociological information and help in planning social services, transportation and education.

Another indicator of the dynamics of the rural population is the life expectancy at birth. The actual number is easily calculated through a mortality table, which can be readily implemented in desktop spreadsheets.

Life expectancy is one of the easiest indicators to misinterpret. It is common knowledge that life expectancy in the middles ages was 30 years old and that that allegedly means that people of 30 years old were regarded as elderly people. This is not always the case. Life expectancy heavily depends on infant mortality. When infant mortality is very high, life expectancy falls since the survival rate of the population is lower. Life expectancy is high in developed societies because most children survive (and is very high when these children grow and have a healthy life). A life expectancy of 30 years old in Roman times does not mean that people over 30 years old were the elderly population as we consider it now, but rather that not many children survived infancy and those who survived tended to die younger than today.

The distribution of the population has historically depended on migration and colonization policies. These are effects that should be considered when analyzing a rural society, since it can introduce a bias in other elements. Massive socially-engineered deportations have generated devastating effects in rural societies, some of which can still be affecting the area today. Colonization policies have, by definition, presented a certain degree of planning whose aims and strategies can be worth considering. When documentation is available it can provide information on not only past but also current issues regarding the social evolution and the potentialities of the area.

The degree of social homogeneity in terms of ethnicity should be considered when it represents an issue. It could be argued that in a well integrated society, ethnicity should only represent a lateral element of the analysis but sadly this is not the case in most societies. Also, the role of multiculturalism in modern societies is an active research topic and no definitive conclusions can be drawn, but it should be considered in the analysis of the rural society.

Finally, it is required to characterize the rural population in terms of classical sociological indicators such as level of education, job market, training, rent, and family size. These variables have all to be presented in aggregated form, either as statistical figures or as anonymized records, in order to maintain privacy. Knowing if the area has enough qualified people to support advanced economic activities is crucial when devising development strategies. Alternatively, improving the qualifications of the people can be a first step to development.

5.2.2 Gender Relationships

Rural gender issues are a topic usually neglected in rural geographies. Actually, it is somehow a shame to have to devote a chapter to this issue, which demonstrates the persistence of one of the inequalities of our societies even in developed countries. Hopefully, in a few years this topic will become obsolete and people reading this section will wonder on the fact that in the early twenty first century gender was still an issue to be separately considered in rural analysis.

Education of equal gender roles and contributions to the society is of paramount importance to the material and cultural advances of a society. The multicultural concept, namely, the alleged cultural differences that justify social and political differences also in gender, is sometimes used as an alibi to inadmissible practices. It is true that some cultures have a strong gender component but patriarch or (matriarch) is one the trademarks of poorly developed societies. Historically, progress has been linked to gender equality. The challenge for the analyst is to make compatible this axiom with local customs, involving people in the process.

In some close communities, a hard approach to eradicate *soft* discriminations – that is, discriminations not directly against the life or personal safety- are not always advisable since it can generate powerful distortions and unwanted effects in the community. The key factor is education. Women, who are the main subject of discrimination, have to have the same educational opportunities that men do. When applicable, there is nothing unethical about linking external aid to compliance with basic human principles such as gender equality, personal freedom and child rights. Erasing any gender difference in the decision-making process of rural societies is a must, and start by promoting equitable gender participation. Rural management performed under unacceptable customs can only generate perverse effects. Thus, any rural management initiative requires a suitable social structure that, if not available, requires the need of a previous social and political management. No benefit can be obtained by applying the principles contained in this volume in areas where, for instance, women suffer from political discrimination or where racial segregation exists.

Education also plays a central role in contra conception policies. Women are the major actors in these policies, and specific attention should be paid to ensure they have the best possible information in order to make a decision. The same can be said about HIV/AIDS awareness and the importance of preventive measures. Also, the costs and benefits men and women experience from rural innovations often differ.

Pregnancy services are also to be considered in the analysis, as are the different nutritional needs required by pregnant women in areas with a limited food supply. The relative workloads of men and women should also be quantified in order to avoid unbalances even when in some policies such as Conservation Agriculture, women may play a more preeminent role.

In terms of sexual tendencies, it is an acknowledged fact that rural areas in developed and developing countries are less socially permissive to the sexual tendency of their members than urban communities. It can be sociologically argued that small communities are less prone to tolerance than others since the contact with different ways of life have been limited. Thankfully, most countries have laws against any form of harassment, and as the information society advances, effective education of difference and tolerance also advances. Aiming to be a free, rational and open rural society in which every person can develop oneself is one of the pillars of any sensible policy, and thus of rural management policies.

5.2.3 Public Health Issues

One of the major concerns of public authorities is to provide access to quality health care to rural people. This can be done by providing a rural perspective to national departmental efforts to help improve the services. These rural health stakeholders include provincial and territorial governments, rural health care providers, rural communities, NGOs, etc. Providing innovative solutions to the identified problems and supporting the development of common strategies to these problems are two major tasks to accomplish in rural management. In order to do this, these problems first need to be defined and characterized.

Many factors may obstruct a quality rural health service. The accessibility of the area, the degree of remoteness, the unavoidable need to concentrate expensive medical equipment in major cities or towns, the ageing of the rural population, a limited number of rural medical doctors, and rural specificities such as the existence of endemic illnesses or harsh environmental conditions are all factors to take into account. Innovative solutions such as telehealth initiatives can help to optimize resources and to solve problems once difficult to tackle. The possibility of having remote diagnoses to common but worrying symptoms can benefit many populations where IT services are available.

Less critical, albeit still pertinent issues, are related with the existence of noise and odours. This problem seems to be worse in peri-urban areas, where counter-urbanization has attracted people not so used to secular rural life to the local population. What is acceptable for some, can be a nuisance for newcomers, and conversely, rural peace can be perturbed by newcomers and generate distress to the original population. The degree of medical attention that these problems require is a debatable issue. The smell from intensive pig farming can be hard to endure even if one is used to it, but the limits of what is acceptable today or what is a cause of mental illness (depression, etc.) is not as clear as expected. Cultural issues are applicable here, and no recipe can be provided. The balance between development, rural life and modern standards also applies to health issues.

As an example of rural health issues, the US Agency for Health Care Policy and Research (AHCPR) has conducted and funded projects whose findings include points such as:

- Almost one in three adults living in rural America is in poor to fair health. Nearly half have at least one major chronic illness. Yet, rural residents average fewer physician contacts per year than those in urban communities.
- Traumatic injuries are more common in rural areas, and residents face worse outcomes and higher risks of death than urban patients, partly because of transportation problems and lack of advanced life support training for emergency medical personnel. Identifying these factors may help shape solutions.
- Rural hospitals show a greater shift toward outpatient services, and greater declines in admissions and lengths of stay than urban hospitals. Economic pressures have driven rural hospitals to shift rapidly to outpatient care.
- Alcoholism and drug abuse are growing problems in rural areas. With a scarcity of mental health professionals in rural areas, fewer than one in five rural hospitals has treatment services for these conditions.
- Rural and urban residents are equally likely to lack health insurance. Underinsurance is as much of a problem for rural residents as being uninsured.
- Analytic geographic mapping techniques are valuable for rural health policy and health services research. A guide to using these decision making tools now is available as an innovative resource in examining rural health care delivery.
- Different solutions are required to *keep* primary care providers in rural areas than are needed to *attract* them there initially.

Improving Health Care for Rural Populations. Research in Action Fact Sheet. AHCPR Publication No. 96-P040, March 1996. Agency for Health Care Policy and Research, Rockville, MD.
http://www.ahrq.gov/research/rural.htm

All of the above applies to developed and developing countries, but neglects the problems of the majority of the world population. In 1996, the 38 countries in the world with the highest under-5 mortality rates also have the highest proportion of rural populations (73%). In 2005, the combined rural populations of developing countries exceeded 550 million. Using the UN indicators for health care, the situations for large areas of the world are quite different and the magnitude of the problems involved make some of the previous discussion laughable. Many epidemic diseases such as malaria or dengue fever are endemic to many rural areas of the planet. People living near marshes or everglades and with precarious medical attention can half their life expectancy compared with countries such as the UK or the USA.

A common public-health concern in those countries is the shortage of doctors and medical personnel (nurses, paramedics, etc). Medical schools exist in most countries, but many highly qualified people tend to leave undeveloped countries once they have finished their specialized training. This brain-drain appears in several areas across the occupational spectrum, but it is particularly damaging when it appears in the health system. The rural areas of these countries suffer this shortage in greater extension than urban areas where public administration tends to be concentrated. Cities create a demand for medical care that the private sector can fulfil, but the countryside cannot exert enough demand. Moreover, a young population structure and a high proportion of rural population characterize undeveloped and developing countries. Neonatal care, pediatrics and women's health are usually poorly covered in these areas. This shortage is complicating the fight against AIDS and other diseases. In this situation, clinical officers are the backbone of health systems, even in developed countries, and are instrumental for the health of rural populations in deprived areas where medical doctors are scarce.

A rural analysis should identify to what extent these factors – and possibly many more- affect the area under study. Estimates such as infant or neonatal mortality, life expectancy, or objective health screenings in a sample of the population could help to gauge the current state and needs. Particularities in the counting should be taken into account: in Korea for example, age counting starts at one year-old for newborn children, and every January a year is added. That means that a child born the 14 of November 2010 would be counted as a two-year old person in February 2011. If any comparison with other areas is to be made being unaware of this fact could lead to a result being biased.

Unexpected connections with other human and environmental elements can be found. Energy needs and available resources are linked with rural health in remote and deprived areas, where developing a reliable and cheap source of energy for primary health care services is still an open challenge, both for direct energy supply and for vaccine refrigeration. As photovoltaic units are becoming cheaper, their use is a potential solution, if a continuous energy supply can be assured. Besides, it can provide complementary services.

A recent episode -the avian flu outbreak in 2005-, is indicative of the health challenges faced by many rural areas of the planet. The staggering absence of elemental hygienic standards in some rural areas favors the spread of viruses and their mutation from domestic animals to people. The causes of such a lack of hygiene is clearly related with a lack of education, which, at in turn, is less-obviously related with economic underdevelopment due to several factors, such as environmental diseconomies, anthropic pressure, democratic shortage or other complex processes, one being the quick and traumatic penetration of market economies into traditional societies. To tackle this serious problem mixed strategies can be devised, but the bottom line of the problem is to improve living conditions as a whole through economic development.

5.2.4 Political Issues

An initial aspect to consider in this part of the analysis is the existence of basic political freedom in the area. If basic liberties such as freedom of speech, association, assembly, movement, religion, press or thought are not granted, the first conclusion of the analysis should be that these basic rights must be respected. It is not possible to conduct any rural planning without democracy, because rural planning is about the will and interest of the rural people.

In advanced societies where freedom exists and decisions are democratically taken, rural politics are an integral part of the rural life. For the rural analyst, the political factor is also relevant for management, both in the decision-making and in the monitoring processes.

Politics in the analysis has to be understood in terms of constrains, resources and as a mean to understand the priorities and the will of the population. The degree of political polarization is of interest as are the political priorities of the ruling party. Local politicians are not only the usual commissioners of the rural analysis, but the legal interlocutors in the whole process. The analyst cannot pretend, in any case, to impose his own views on how a rural landscape should be, or which one should be the main economic activity in the area. The priorities of the people living in the area should be put in first place, although the best possible advice must always be provided.

A fact that should be acknowledged at this stage is that, in democratic societies, politicians are the only actual representatives of people's will. NGOs can be beneficial but represent the views of lobbies of particular interests and as such should be considered in the rural analysis and management. If politicians fail to deliver welfare, democracy has mechanisms to balance the situation. Relying on local lobbies for the managing process is sometimes unavoidable and can be very beneficial: these groups are usually composed of educated people, who are very concerned about rural life and strongly involved with their communities. They can provide first-hand information for the analyst and help in putting into practice the proposal. They are not, however, the people's representatives.

Mainstream political sensibilities should be identified and represented in the whole process. It is usually worthless to impose a partisan rural development program to the communities. It is likely that the most controversial issues will be scrapped in the next election, unbalancing the whole rural development process. Thus, in order to avoid unwanted situations, a compromise should be pursued. Whether or not this is possible, it is a scholarly debate.

A rural project is never apolitical. Firstly, democracy is a must for the development of rural societies. Without democracy, a rural analysis becomes a social engineering project for the ruling power. In any case, within this situation rural management cannot be successful, since it involves a great deal of social involvement. Secondly, within a democracy co-exists several tendencies and many ways to try to improve people's life. The analyst can propose several technical solutions to solve a problem, but it is up to the politicians to choose to apply them or not, depending on their ideology, interests and plans.

In terms of the analysis of anthropic elements the rural power structure should be identified. The links between property, paternalism and power, how the rural area is governed, the kind of rural policies applied in the past, how politics are interlaced with agriculture and with regional, national, and supranational realities.

5.2.5 Rural Culture, Traditions and Folklore

For many people, rural culture is a basic part of the rural appeal. Most of the old traditions (forgotten in major cities) remain, if not pristine, at least in a preserved state in the countryside. Also, the sense of place of some communities is linked with a perceived common culture or customs, regardless of the fact that what is acceptable in some areas can be unacceptable in others. Sometimes, some irrational or highly questionable practices define a rural area. A typical example is the recently-abolished fox-hunting practice in the UK. The statement that fox hunting was a way of life was not any less true that it is for twenty-first century people finding the practice highly inhumane, not only by the actual suffering of the animal but also because the paraphernalia that surrounds what is sometimes presented as an act to restore the ecological equilibrium. This case also exemplifies the conflict between rural and exogenous values. The pace in which ideas evolve in rural areas are arguably slower than in cities and some social conquests take some time to permeate to the whole society. The diffusion of ideas is a well-studied process and it seems clear that even in the Internet age, the conditions for the spread of innovation are less favourable in rural areas. This is a particularity of those areas that should be considered, since conflict can easily occur when new ideas and practices are brought in at an inappropriate time. This fact is not an excuse to fossilize questionable practices. Traditions cannot oppose the law of the country or human rights. Remote rural areas are the most inertial areas and special care must be taken in ensuring that the law is obeyed.

Another aspect to be considered in the analysis is whether these traditions limit scientific, technical and cultural development. Recently, a former UN officer acknowledged astrology as one important cultural component of the Egyptian culture. When one philosopher noted that this does not make astrology less false in Cairo than in Helsinki, the officer replied that this was a too Eurocentric view. One may wonder with the philosopher if the Egyptians are to be congratulated for having astrology while other societies advanced in the Eurocentric fields of nanotechnology and genomics.

It can be argued that arts, traditions and folklore should not be just a mere complement of one strong and common global society vertebrate around the UN Declaration of Human Rights and the scientific revolution as principal components. Following this line of thought, any attempt to limit human freedom based on cultural specificities should be, by principle, banned. It is however less clear whether cultural specificities can be a catalyst for societal improvement. So, it is debatable if initiatives such as promoting dance groups in the deprived areas of Brazil help children to avoid crime, or it would be better to directly devote resources to help them to develop less colourful, albeit more sensible skills, such as basic arithmetic. The risk of limiting true development by well-intentioned, but misguided initiatives should be considered. Even when the initiatives are intelligently-planned the danger is thinking that instruments such as *samba* or *capoeira* schools as pillars for long-standing development. These initiatives can be very effective in integrating societies as a first step for further education, but the danger of *folklorizating* the society instead of developing it is real if they are considered as ends by themselves or deemed as a part of one alleged *cultural identity* of the country. It can however be presented as such to attract tourists, if this policy is considered as suitable.

In reality, it is perfectly compatible to preserve and celebrate traditions and beliefs without magnifying their actual importance for societal advance. Santa Klaus is a colourful component of the Artic circle, providing some money to remote rural communities from tourism, but not many Scandinavians would identify themselves with a Santa-based culture. Instead, they would probably see their society as one based on constant scientific innovation and a strong welfare state. Defending the cultural specificities of developing countries has a strong colonial component and may have the effect of impeding development. It is not clear to what extent cultural clichés are conscious strategies in generating competitive disadvantages to other countries, but many developing projects from Western countries might have that effect. It is up to the locals to ensure that foreign initiatives truly benefit them and suit the projects to their needs.

5.2.6 External Fluxes. Rural Tourism

External fluxes affect rural areas in many ways. It can affect planning policies, the environment, the jobs and economic prospect of the inhabitants of the area and even the agriculture by providing additional income making farming a viable option. Advantages include an improvement in the sustainability of local business, leisure and cultural facilities through the 'added' spenditure that visitors bring; added incentive for conserving and enhancing our environment; the creation of new job opportunities for local people; enhance desirability of an area for potential inward investors; and can even instil a sense of local pride (North Dorset Tourism Strategy 2004–2007).

As far as the analysis of the human factors are concerned, external fluxes refers to those elements on the territory having a direct connection with this economic activity. That includes from theme parks to private lodgings (B&Bs), tourist pathways to closed circuits. The analyst should identify not only the existing elements, but potential ones, taking into account that tourism can be either beneficial or a heavy mortgage for rural areas. Some coastal areas of the Mediterranean clearly show the effect of unbounded and unplanned development that has seriously damaged not only the natural beauty of the landscapes, but also the future tourist potential of entire villages in Spain. Also, the analysis should reflect the institutional framework involved in the development and management of rural tourism. Even considering that rural tourism is still one of the less regulated economical activities in developed countries involving a non-negligible amount of informal economy the legal framework should be known and considered. Policy dimensions of rural tourism are relevant both in the sense of being capable of offering realistic solutions, and in orienting the policies to tourist development if required.

The benefits of this economic activity are clear. Tax revenues based on lodging, food and recreational activities increase, as do capital investments in real property. Also, amenities and infrastructure development increase and in general, a whole tourism-related employment niche is created. If a more diversified rural economy is targeted, rural tourism can provide a complementary resource, but it is advisable to assess the place of tourism in the framework of the current social and economic development, or in the plan if known.

If an area can be identified as a tourist area, the number and type of jobs created, the position of the locals to the new situation, and the new resources of additional income should also be identified in the analysis. If it is clear from the beginning the work that a plan for tourism will be proposed, it would be helpful to assess the training and education of the locals to tourism.

In the analysis, it is crucial to understand the variable demand for tourism and leisure activities from other parts of the country and from abroad. Fashionable areas might soon become saturated leaving costly facilities unused. The links with environmental elements can be crucial. Meteorology and climate can be relevant in some cases: intermittent water flows can enhance the tourist potential of an area in some seasons or during a favorable climatic cycle, but can be very unattractive in the drought periods. Other environmental elements, such as the geology of the area should also be considered in terms of tourism potential, as was discussed in the first part of this book.

Tourism potential in rural areas can be evaluated following several procedures. Heneghan (2002) proposed a number of critical factors which rural communities should consider when examining the possibilities for rural tourism in their areas:

1. A Critical Mass to Provide the Holiday Experience
2. Co-operation and Group Development
3. A Strategic Plan
4. Focused Marketing
5. Education and Training

Critical mass implies having enough tourist interest to provide a destination rather than a place to 'stop-off'. For areas in which the potential of providing a *holiday experience*, the main orientation in rural development could easily be tourism rather than agriculture or forestry. The critical mass factor can be evaluated considering other elements such as accommodation, accessibility, services availability, scenic values and activities. It is clear, however, that not every rural area has such capabilities. For these cases, tourism can also be beneficial in so far that it is possible for a rural economy and society to take advantage of short stays.

Co-operation is vital for any rural development initiative. A specific plan is also fundamental for the efficient use of resources and funds, and should be either integrated in the rural analysis plan or be a consequence of that plan (that is, the rural analysis may conclude that an in-depth tourism plan is needed). Pooling resources for effective marketing at a local scale is also a must in terms of attracting, and more importantly, for sustaining tourism. Finally, regarding education and training, courses on hospitality, dealing with conflict, health and safety, local geography, business planning and management can be offered to local entrepreneurs. Depending on the scale of the area, improved results can be achieved by taking a regional approach, in which education and training is offered in a few central places. This may help to cut costs substantially, also creating synergism between rural communities in the same region: critical mass can be achieved by an intelligent design of itineraries alluring tourists to stay in the same region for more than one day. A single village or municipality can be unable to attract visitors for more than a few hours, but a floating tourist population can be created at a regional level if villages specialize in a particular factor of attractiveness. Tourists may expend money quite evenly over the whole area, redistributing wealth amongst the rural communities acting as one unit. Too-local approaches can miss many opportunities. Budget airlines in Europe have created a new trend in tourism that can reinvigorate deprived areas that are now suddenly well connected with major foreign cities. Farm tourism is another possibility of interaction between rural economic activities. As an example in Israel (Fleischer and Tchetchik 2005) shows how rural tourism enterprises combined with active farming differ from those without agricultural activity. Not only tourism benefits from the agrarian activity, but that the converse is also true. Combined forms of agro tourism increase externalities and can help sustainable agriculture in areas lacking a strong market presence.

The analyst should also be aware of aspects such as cultural differences or the trends in rural tourism. Regarding the latter, the 'Tourism Market Trends' publications from the World Tourism Organization are a reliable guide to try to adapt local specificities to trends and to be in touch with new needs and demands.

Coming back to Heneghan's paper, it might be kept in mind that 'rural tourism is a serious instrument of rural development. To enable an area to become a rural tourism destination, not just a stop off point, structured networks and collaboration agreements are essential. Creating inter-community co-operation and collaboration is a complex and difficult process requiring education, training, capacity building and facilitation'.

5.3 Technological Factors of the Agrarian Activity

The agrarian is the predominant economic activity in rural areas. Some authors even define rural by its connection with the agro, and agrarian exploitation affects many other activities, beginning with the landscape values of the area. In this respect, the technological factors of the agrarian activity define how the land is exploited and managed. In the 21st century, the role of new techniques such as biotechnology, improved irrigation, remote sensing or information technologies are the most relevant factors since they are the latest means to increase the productivity of the rural areas, reducing workforce, and therefore improving competitiveness. One aspect to take into account when analyzing these factors is the effect these technologies have in the area they are being applied to. If this is not the case, the analysis should focus on the suitability, costs and trade-off of applying those technologies to the area under scrutiny. It does not always make economic sense to introduce remote sensing technology in an area with fragmented property and very small plots, and it would be unwise to rapidly introduce genetically-modified crop practices in very deprived areas, to name two examples of possible pitfalls.

5.3.1 Agrarian Exploitation and Management

The characters of the agrarian property are relevant as they influence the social and economic relations of the area. They also affect how the land is exploited and managed. It is widely accepted that ownership means better exploitation and management, since more care is usually put in keeping one interest than to the benefit of others. This is not the situation for agribusiness where land is a simple support for economic activity and land ownership, once exploitation is secured, is a marginal issue (corporate farms can either own the land, or rent it). Family farms, on the contrary, usually own the land they work on.

Vertical integration of agricultural activity presents the major advantage of allowing scale economies. Family farms viability depends on many factors, including variable energy cost, input (seeds, fertilizers, water) costs, food market changes, consumers preferences, sanitary regulations or public intervention in the sector, while agribusiness has a greater adaptation capability to this factors, even being able to influence them as in the case of the sanitary regulation of dairy products.

5.3.2 Biotechnology

Biotechnology has become a hot topic in rural analysis for the huge potential implications that its application can have in rural areas. The precedent of the current stage is the 'green revolution' in the 40s., when an unprecedented increase in food production was made possible thanks to technical advances such as the use and development of high yielding varieties, chemical fertilizers, pesticides and improved water management. The green revolution brought a surplus of food to the world as a whole: the fact that three-quarters of the world's population have not got enough food to sustain themselves is no longer due to technical problems or environmental factors, but to the economic decisions of the developed societies to maintain their current quality of life. Also, there is not a direct relationship between economic improvement and social welfare: it widely depends on how wealth is distributed in the society and the redistribution mechanisms in place.

The effects of the green revolution, in terms of social impact are similar to those that biotechnology can have today. The technological and social effects cannot be separated in an analysis, but the difference today is that a greater technical expertise is required to fully understand the implications of biotechnology on rural life. Many of the criticisms against biotechnology are due to a misunderstanding of what it really is, or to cultural or religious apriorisms. It is therefore worthwhile to distinguish between several interlinked concepts.

5.3.2.1 Genetically Modified Organisms

Genetically Modified (GM) organisms are plants and animals with artificially-altered DNA. Thus, GM crops are plants that have genetic material extracted from one living being, either animal or another plant, placed in it. This creates new varieties or even new species, aimed not only to improve food production but also to develop, for instance, new renewable energy crops. Other purposes are to make food stay fresher for longer, to kill pests, to produce more of the crop or to generate new plants to suit a particular demand. GM goats or cows, for instance, can provide milk suitable for those allergic to natural milk, or milk with drugs to treat health disorders.

Following the UK Strategy Unit 2003, several definitions can be established to clarify concepts:

Genetic modification (GM): The technology of altering the genetic material of an organism by the direct introduction or removal of deoxyribonucleic acid (DNA).

Genetically modified organism (GMO): An organism in which the genetic material has been altered by the direct introduction (or removal) of DNA.

GM crops: One type of GMO; crop plants whose genetic material has been altered by the direct introduction or removal of DNA in order to confer particular characteristics on the plant.

GM food: A less well-defined term that applies with different degrees of clarity to a range of foodstuffs, including:

- any food produced from a GM crop, and in which the altered DNA is still present and detectable. E.g. food containing maize flour from GM maize, or whole foods such as GM tomatoes;
- foods containing ingredients derived from GM crops but not themselves containing the altered DNA, e.g. sugar from GM sugar beet or oil from GM oil seed rape.

There are also some foodstuffs that are not commonly classed as 'GM food' although GMOs have some part in their production. These include:

- meat or dairy products from animals fed with GM crops;
- foodstuffs that use products derived from GM microbes either as ingredients (some riboflavin is produced in GM microbes) or as processing aids (a version of chymosin – the active ingredient in rennet – is derived from GM microbes).

GM seed for sowing or **GM grain** for food use: Although they could grow into GM crops they are not strictly speaking GM crops themselves. They are generally described by the more generic term **'GMO'**. GM crop grain which has been milled into flour, and could not grow into a GM crop, would be classed as GM food.

GM feed: GM crops used to provide animal feed.

GM non-food crops: a subset of GM crops that are not used in food or feed. One significant example is GM cotton.

Gene Technology is one type of Biotechnology. In its simplest formulation, it is the use of living things to make or change other living things in a quick way. Nonetheless, mankind has historically modified the genetic makeup of plants and animals by selective breeding. The wheat we all know today is the result of centuries of agriculture and selection of the most productive and more resistant individual grains. The small grains and mean ears found in Neolithic wheat from the Middle East show the degree of improvement achieved. Also, some fruits such as the grapefruit are the result of cross-fertilization and grafts. The difference between this procedure and Biotechnology is that we are now capable to work at molecular level avoiding trials and errors and targeting specific properties.

The first genetically modified crops to be sold in the USA were tomatoes, which were modified to be more resistant, then Soya beans and oilseed rape. They were modified to survive certain herbicides and weed-killers. Then, maize and cotton were modified to carry a poison that kills pests and protects crops against damage. Potatoes were also some of the earliest vegetables to be modified. In the Andes, gene scientists have produced potatoes more resistant to cold weather, avoiding the ruin of the crops in hash years.

The importance of GM crops for rural analysis in the 21st century is paramount. The '*Field Work: weighing up the costs and benefits of GM crops*' report from the UK Strategy Unit identified in 2003 the following key areas of cost and benefits. Some of them are widely applicable to other areas, but a particularized study for the rural area under study may be required:

- *Impacts on the economics of agriculture.* GM crops are designed to offer specific cost or convenience savings to farmers. Savings can be expected at farm-level in areas such as spending on pesticides, along with advantages in terms of convenience of farm management and increased yields. In the rural analysis, those benefits would have to be set against the costs associated with commercial cultivation, including the cost to farmers of buying the technology, and some hidden costs of compliance with regulations and national normative. The balance of costs and benefits would vary between crops, countries, and between types of farm. Some GM crops may only offer advantages to farmers with particularly serious weed problems, for example. Others may offer more general benefits, for example GM sugar beet might allow farmers to make significant cuts in herbicide costs. Farming practice will also influence the balance of costs and benefits. Farmers' decisions will depend critically on the demand for the GM crop. If the price they received for GM crops was much lower than that for non-GM, then this could easily outweigh any cost savings. Ultimately, economic logic would apply, and farmers would have to make individual decisions on whether to adopt the technology.

- *Practicalities of the supply chain.* Many European consumers continue to demand a choice between GM and non-GM food. Reflecting this, some costs are already incurred in order to maintain separate non-GM lines of some imported products. These are likely to rise as the range of available GM foods increases, whether or not GM crops are grown in the UK. Additional costs may arise if GM crops were cultivated in the UK, particularly in keeping crops separated at the farm level. The nature of any measures which may need to be put in place – and hence their likely cost – is not yet clear. The cost may also vary significantly from crop to crop, depending, for example, on the characteristics of the crops and on the extent to which organic varieties are grown. But the analysis suggests that many of the measures that could be required are consistent with existing farm practice, or could be achieved through co-operation between farmers. This means that these costs could be relatively small as a proportion of the total costs of crop growing. However, the detailed design of any measures will have a significant impact on costs for GM, conventional and organic agriculture.
- *Wider developments in agriculture are likely to be more important in the short term.* Any cost savings or increased yields offered by GM crops could contribute to improved competitiveness in UK agriculture. However, any economic benefits from the commercial cultivation of current GM crops are likely to be outweighed by other developments, at least in the short term. UK farmers do not operate in a fully competitive market, and hence their future profitability is more likely to be determined by national and EU policy decisions – for example, on the Common Agricultural Policy (CAP) – rather than smaller-scale cost savings arising from the use of current GM crops.
- *Wider impacts on the rural economy and communities.* There have been concerns that the introduction of GM crops to the UK might have impacts on rural economies more generally, such as effects on the reputation of rural areas. However, there is little evidence on this issue.
- *Agriculture, the environment and human health.* This study has not looked in detail at the scientific evidence on the environmental and human health impacts from GM crops. Some people have concerns about these impacts and the broader uncertainties associated with GM crops. These issues have been dealt with by the Science Review, and are already a key element of the risk assessment procedures applied to all applications for the release of GM crops. However, the study does consider how the changes to farm management associated with GM crops could have an impact. The assessment is inevitably preliminary in nature, particularly in advance of the results of the FSEs, which are looking at one aspect of this issue. There is scope for both positive and negative effects, although the exact scale and nature of these are hard to judge, as the relationship between agriculture and the environment is complex and relatively poorly understood. Impacts would also depend on farmer behaviour, which may vary from individual to individual. In the short term, the biggest changes could come from the use of herbicide tolerant crops, and resulting changes in patterns of herbicide use. These changes could

have environmentally beneficial effects, for example, by replacing persistent or high toxicity herbicides with more benign ones. However there is also scope for costs. Low-cost herbicides that can be used with little or no damage to the main crop could lead farmers to increase herbicide use, to the detriment of biodiversity. Future GM crops could also directly address environmental or human health objectives, such as the production of renewable materials, or of foods with reduced allergenicity. However, non-GM techniques might achieve some of the same goals – for example, a broccoli with enhanced levels of a cancer-preventing substance has been developed using non-GM techniques. And future developments may also increase risks and uncertainties.

- *Impacts on the science base in the UK.* What happens with GM crops over the next few years will have an important influence over the long-term options available to the UK – for example in affecting the UK's expertise in GM and non-GM technologies, the UK's ability to capitalise on future developments and the signals sent to wider biotechnology and science based industries. Commercial GM crop research in the UK has declined, and is a small component of overall scientific activity, with one multinational crop research facility being based in the UK, and very few small businesses working in the area. The direct economic impacts of GM crop policy that reduced this sector further would therefore be fairly limited in the short term. However, UK academic crop research is still strong, and the commercial sector has the potential to expand. Significant economic benefits might accrue to companies providing GM crop technologies – and it may be in the UK's interest for these companies to be sited in the UK. There would be a significant opportunity cost if the UK lost its ability to research, develop and bring GM crops to market.
- *International impacts.* UK policy towards GM crops will have international repercussions, and the stance taken by the EU as a whole is likely to have a still greater impact. Differences in approach between the EU and some other countries on GM policy are already causing trade tensions. There may also be specific impacts on the UK's policy of ensuring that developing countries should be free to decide for themselves whether to adopt GM crop technology. UK and EU policy towards GM crops may impact on the context in which developing countries make these choices about GM crops.

Key issues when considering the effect of GM crops or animals in a rural area are the sometimes uninformed resistance to GMO, and public opinion position to GMO, especially regarding environmental lobbies. While the European Union has indicated that member countries will in the future have to base decisions on whether or not to permit GM agriculture on science rather than public opinion, subjective appraisals still dominate GM production and commercialization.

The risks of GMO have been scientifically analysed in several comprehensive projects. The findings of the BRIGHT project on '*Botanical and Rotational Implications of Genetically Modified Herbicide Tolerance Objectives*', aimed to determine the agricultural implications of herbicide tolerant crops and produce management information for their sustainable production. It found that GM crops are no more harmful to the environment than conventional plant varieties. This project was initiated in the UK in autumn 1998 to study the agricultural implications arising from the use of herbicide tolerant crops, including genetically modified varieties in arable rotations. The four-year program carried out on relatively large plots (0.25–0.5 hectares) studying GM winter rape and beet grown in rotation patterns compared with conventional plots, showed improved GM crops in terms of tolerance to glyphosate or glufosinate herbicides, or cross-bred resistance to imidazolinone in rape, with similar effects on the environment that traditional varieties have. GM crops showed an increase in weed seeds over 4 years and weed seed diversity in soil, which is needed by many birds and other wildlife, were found similarly in GM and non-GM. Advantages were also found in GM in terms of reduced weed control costs and greater flexibility in spray timing.

The study was severely criticized, and other studies have shown a different picture. The Farm-Scale Evaluations or FSEs, found that two GM crops, a sugar beet and a spring rape, were more damaging to biodiversity than conventional crops (few insect types). A GM maize, on the other hand, appeared to be more insect-friendly than its traditional variety.

In relation to public position to GMO, Hall and Moran (2006) have reported how surveys and polls continue to point to consumer opposition to genetically modified (GM) food. It has been claimed that opposition exists because people perceive the technology to be risky to themselves, their families and the environment, without offering any consumer benefits. To investigate this issue, these scientists studied the nature of GM risk perceptions held by anti-GM campaigners and environmental group members in Scotland. They concluded that there does appear to be a link between the perceived risks relating to GM food, and perceived benefits. When benefits are expected to be greater, risks are expected to be smaller. It could be thus concluded that this finding lends weight to the expectation that the public are likely to be more accepting of GM food in the future when consumer benefits are offered by the products. However, respondents appear to be sceptical about the claims of future benefits of GM technology, suggesting that actually the perception of high risks will remain among the most risk-averse section of society, regardless of further scientific evidence on the contrary.

Nevertheless, the results of this study show that anti-GM campaigners and members of environmental groups can be segmented into groups with different perceptions of the risks and benefits of GM food, and should not therefore be viewed as a homogeneous group. Some campaigners may be less inclined to remain staunchly opposed to GM food than others in function of a number of socio-demographic variables affecting campaigners' perception of the risks and benefits of GM food. They found evidence that, in some cases, women and those living in rural areas consider the risks of GM to be greater and the benefits less, compared to men and people living in urban areas. Perhaps, therefore, men and people living in urban areas, even those opposed to the technology at this stage, may be prepared to purchase GM food at some point, or perhaps there is other explanations such as different information and approaches to technology.

But perhaps more relevant than an unlikely scientific evidence against health risk of GMO are the social and economical implications of its use. Concerns about deploying genetically modified crops in developed and underdeveloped countries include rural economies impacts, ethics, and the lack of appropriate regulations: a GM crop being safe under specific conditions may be damaging for the environment if poorly managed in a deprived framework. All these issues have to be considered when confronted with GMO issues in the analysis.

5.3.3 Irrigation

Water is a limiting factor for life. Water resources and irrigation policies are fundamental to agriculture. Water irrigation can help to increase food output and exports, and thus reduce poverty in deprived areas. For developed areas, a sensible water policy can increase competitiveness over other poorly managed rival areas. Also, technological factors related with how water is used have a strong influence on rural areas since advanced methods usually require both political and social support. The mere existence of irrigation policies where water is available indicates a certain degree of social organization and shared interests.

Knowledge on how to irrigate is crucial: irrigation type needs to match soil type and crop, and surface irrigation, or micro-irrigation can be more suited to one or another kind of topography. Timing is as key as the quantity of water supplied. Efficiency is one of the key aspects of irrigation in a scenario were fresh water is almost a commodity. Only a minute fraction of all the water on Earth is directly available for humans and animals.

Examples such as the success of drip irrigation in Israel, illustrate the role of irrigation in rural, regional and even national development. This particular case also illustrates how rural areas also play a role in advanced knowledge transfer from research organizations to companies. One of the first spin-off companies involved in setting up a drip irrigation system sold over 250,000 units in the first years. This shows that the countryside can also be a source of innovation and economic development.

If irrigation is identified as an relevant element of the rural area, either because it is deemed that it would help the economic development of the area or because other issues such as over irrigation practices or social discontent, it is useful to perform an in-depth assessment of this element. Many issues can then be analyzed, such as how sensitive are policy-makers and stakeholders to irrigation problems and needs; how irrigation is managed and by whom; the local knowledge on state-of-the-art technologies; the need of construction works to improve supply; the conflicts or synergisms with neighbouring areas; the accountability of the stakeholders; the legal framework of water use and irrigation; the ownership of the infrastructures; the cross-relationships with other human and environmental elements such as energy availability for pumping; the financing and economics of irrigation; the links between irrigation and the rest of the production chain; or the environmental impacts of this economic activity.

Advanced irrigation is linked with information technologies. Computerized dripping systems save labour and allow a better monitoring of the crops water needs if sensors are used. These systems can be complemented to supply fertilizers to the plants in the required amounts and timing. The combined use of dripping methods and greenhouses dramatically improve the productions at the time they rationalize the use of water.

Geographical Information Systems (GIS) and remote sensing can help to monitor and exploit water resources (Fig. 5.7), and to support irrigation programming. They can also be used to monitor mutual agreements on water use between rural areas, or to assess the real needs of irrigation. A centralized irrigation office providing a joint service to many rural areas can help to both coordinate policies and to provide a more rational use of this scarce natural resource.

In a wider scale, the management of the water resources is crucial not only for agriculture but also for rural societies as a whole. The balance between agrarian use and other uses such as recreation or tourism should be assessed in this part of the analysis. In coastal areas, the possibility of using desalination plants should be considered if the real needs of the area demands so. A potential problem however, is the overuse of this technology to support uncontrolled tourism development instead of agrarian activities. The costs, including the social cost of these policies, require careful consideration by regional and national authorities to avoid creating diseconomies and an eventually costly unsustainable development. In some cases, it can be sensible to provide a generous volume of water to second residence homes or to leisure activities, but in others, the social value of the water may require a more careful approach, even if the technology to provide the required amounts is available.

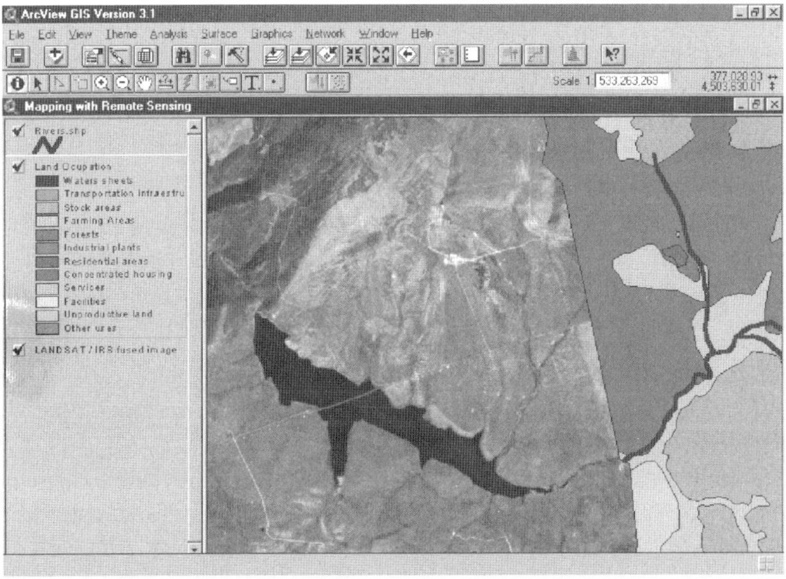

Fig. 5.7 Integration of GIS and remote sensing for water management (see Color Plate on following p. 333)

5.3.4 Fertilizers and Chemical Pollution

Plants need several chemical elements to grow and to store energy. Animals can only get these elements from plants. Some elements are of little use for plants, but are required for animals such as cobalt, iodine and fluorine. These micronutrients are required in variable quantities and if they are not naturally available they need to be provided through fertilization. While it is possible to grow crops in some deprived soils, the resulting plants can lack nutritional value. In other cases, microelements act as limiting factors for growth and impede any crop if lacked. Nitrogen, phosphorus and potassium are required in large amounts and are called macronutrients.

Given the importance of these nutrients, the availability of fertilizers, how are they applied, and the knowledge of how to use them, are technological factors of agrarian activity. A lack of this knowledge impedes scale economies and even the existence of modest economies in weak soils. In developed countries, this is seldom an issue. Farmers usually know better than the analysts the needs of their fields, and their nutritional requirements.

On the other hand, in areas in which fertilizers are applied, the soils become historical products of years of mankind use. They store not only elements from several millions of years of slow weathering but also salts and heavy metals accumulated from year after year of cropping. The problem of over-fertilization is primarily associated with the use of artificial fertilizers, not because of the intrinsic nature of these fertilizers, but because of the massive quantities applied. The same applies to the organic fertilizers. Nonetheless, as these are less efficient, their release rates and solubility are lower and thus their effects on the soil are less critical.

5.3.5 Non-tillage Approaches

No till practices consist of avoiding disturbing the soil from year to year. This is deemed to control soil erosion and the physical and chemical properties of the soil. The idea is not to till and leave the remains of the crop in the fields waiting for natural decomposition. This helps to avoid soil losses by runoff and keeps the soil moist in summer. A good conservation plan however, must also include a sensible use of fertilizers and pesticides, although in some cases they are not used.

Specialized seeding machinery is required to properly place the seed into the soil and rotation may be required in some cases. The results in terms of production are comparable to traditional farming, but the cost of machinery and fuel are lower, thus increasing productivity. Pastor and Guerrero (1990) reported a study comparing conventional tillage and non-tillage of bare soil in olive groves over 15 years. Olive yield increased with non-tillage in 81 of the trials, whereas only in 4 fields the tilled olive trees produced more. Average yield increase in non-tilled olive groves with respect to the tilled ones was almost 16%. Average oil content decreased with non-tillage in 57% of the fields. However, the non-tilled olive groves produced higher amounts of oil.

A mixed alternative to no till is to use cover crops to protect the soil. Cover crops protect the soil from wind and rain erosion; they decrease the compactness of the soil; increase the nitrogen content of the soil and the cation exchange capacity; improve the water holding capacity; and increase the organic material content.

Non-tillage approaches have attracted attention due to public awareness of climate change. Tilling releases carbon to the atmosphere, thus contributing to the greenhouse effect. Therefore, no-till farming is then seen as a way of recycling urban emissions, thus providing a new economic place to the countryside in the global economy.

Fig. 5.8 IKONOS composite image, showing the capability of this satellite to show small features in the crops. Trees are also clearly visible in the image (see Color Plate on following p. 333)

5.3.6 Remote Sensing Management: Precision Agriculture

One of the most useful new technologies in agriculture has been the integration of Geographical Information Systems and Remote Sensing technologies to agricultural management. Precision agriculture or precision farming is the use of information technologies (GIS, satellite imagery, GPS) to analyze the variability of the crops at high spatial resolution (in less than a 30m grid), estimating the needs of the crop in terms of water, fertilizers, pesticides, etc.

Remote sensing technology can also be used to predict crop yields and to then successfully trade the products in the future markets. Earth Observation satellites have evolved from the first meteorological satellites in the Sixties to high-resolution satellites with images rivalling with aero transported sensors. Ikonos (Fig. 5.8) and QuickBird satellites provide data up to one meter resolution that can be used to monitor the growth of the crops and then derive soil properties. This allows a precise and more efficient fertilization. While nitrogen grid sampling using GPS can be very expensive, satellite images can map this variable at a fraction of the cost, and the

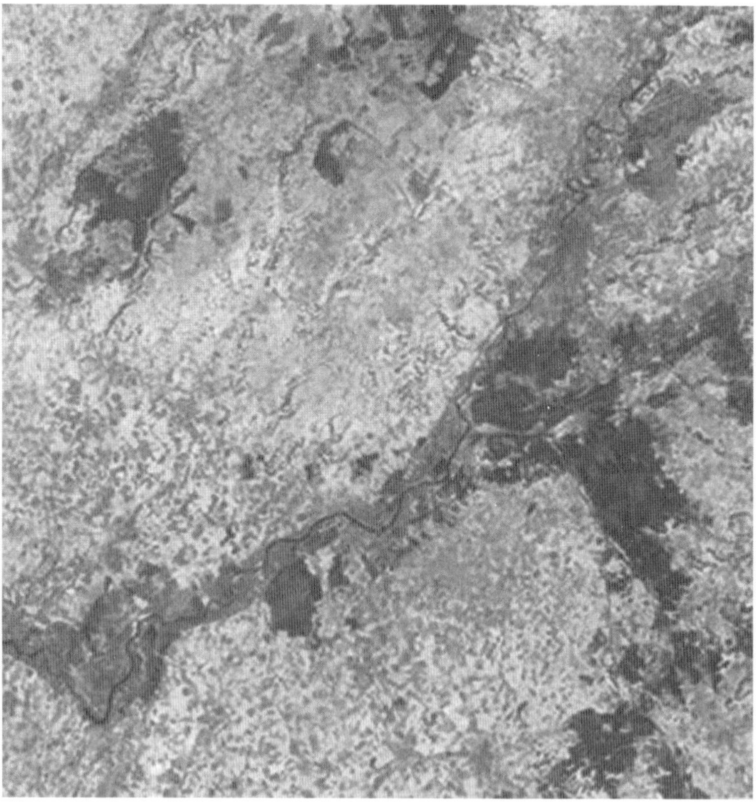

Fig. 5.9 A false-color LANDSAT image. Irrigated land appears in reds, woods in dark green and rivers in dark (see Color Plate on following p. 333)

same applies to other variables. Using several images taken at different vegetative stages, the farmer can improve the crop yield dedicating more effort to the area requiring it most, and also reducing environmental pollution.

Precision agriculture is well suited for large exploitations. In the case of fragmented or small properties, resources can be pooled to acquire the processed images or derived products through specialized remote sensing companies or Universities. Scale economies can be developed even in the smaller farmers associations since satellite coverage permits to survey large areas at the same time. The availability of this technology has to be considered in a rural assessment.

Images of lower resolution such as LANDSAT or SPOT (30 and 20m. spatial resolution) can also be used for assessing the state of the vegetation in larger areas with lower precision (Fig. 5.9). This can be useful for rural planning in the cases in which an overall view is needed.

References and Bibliography Used

J. Abbott and P. Yoong, 'The stages of telecentre development: the case of the Kapiti Telecentre', Technovation 25 (2005), 421–431.

M. K. Ahmed, J. van Ginneken and A. Razzaque, 'Factors associated with adolescent abortion in a rural area of Bangladesh', Tropical Medicine & International Health 10 (2005), 198–205.

S. M. Ahmed, G. Tomson, M. Petzold and Z. N. Kabir, 'Socioeconomic status overrides age and gender in determining health-seeking behaviour in rural Bangladesh', Bulletin of the World Health Organization 83 (2005), 109–117.

R. A. Bell, T. A. Arcury, B. M. Snively, S. L. Smith, J. M. Stafford, R. Dohanish and S. A. Quandt, 'Diabetes foot self-care practices in a rural, triethnic population', Diabetes Educator 31 (2005), 75–83.

M. Boland, A. Staines, P. Fitzpatrick and E. Scallan, 'Urban-rural variation in mortality and hospital admission rates for unintentional injury in Ireland', Injury Prevention 11 (2005), 38–42.

G. H. Brody, V. M. Murry, L. McNair, Y. F. Chen, F. X. Gibbons, M. Gerrard and T. A. Wills, 'Linking changes in parenting to parent-child relationship quality and youth self-control: The strong African American Families Program', Journal of Research on Adolescence 15 (2005), 47–69.

L. A. Bryan and H. J. McLaughlin, 'Teaching and learning in rural Mexico: a portrait of student responsibility in everyday school life', Teaching and Teacher Education 21 (2005), 33–48.

E. Burnes, D. Wichelns and J. W. Hagen, 'Economic and policy implications of public support for ethanol production in California's San Joaquin Valley', Energy Policy 33 (2005), 1155–1167.

E. Burton, 'Measuring urban compactness in UK towns and cities', Environment and Planning B-Planning & Design 29 (2002), 219–250.

B. Carme, 'Substantial increase of malaria in inland areas of eastern French Guiana', Tropical Medicine & International Health 10 (2005), 154–159.

F. Cartmel, 'Young people in rural areas of Europe', Journal of Social Policy 33 (2004), 689–690.

A. K. Chakraborty, 'Epidemiology of tuberculosis: Current status in India', Indian Journal of Medical Research 120 (2004), 248–276.

W. H. Chua, R. Husain, L. Tong, A. Fong, J. F. Cheng, A. How, L. Lee, G. Gazzard, D. Tan and S. M. Saw, 'Prevalence of cataract in rural Indonesia', Investigative Ophthalmology & Visual Science 45 (2004), 1064.

R. Colapietra, 'Mezzogiorno and emigration. The Inchiesta Faina concerning rural living conditions in southern Italian provinces and Sicily, 1906–1911', Rassegna Storica Del Risorgimento 91 (2004), 294–298.

S. Dabbert, W. Grosskopf, F. Heidhues and J. Zeddies, 'Prospects in land use - regions, landscapes, farms, decision- makers and tools - Report on the 43rd annual meeting of the "Gesellschaft fur Wirtschafts- und Sozialwissenschaften des Landbaus (GEWISOLA) 2003 in Hohenheim', Berichte Uber Landwirtschaft 82 (2004), 362–390.

D. De Ronchi, D. Berardi, M. Menchetti, G. Ferrari, A. Serretti, E. Dalmonte and L. Fratiglioni, 'Occurrence of cognitive impairment and dementia after the age of 60: A population based study from northern Italy', Dementia and Geriatric Cognitive Disorders 19 (2005), 97–105.

C. S. Dolan, '"I sell my labour now": Gender and livelihood diversification in Uganda', Canadian Journal of Development Studies-Revue Canadienne D Etudes Du Developpement 25 (2004), 643–661.

F. Echanove, 'Globalisation and restructuring in rural Mexico: The case of fruit growers', Tijdschrift Voor Economische En Sociale Geografie 96 (2005), 15–30.

S. Franceschi, R. Rajkumar, P. J. F. Snijders, A. Arslan, C. Mahe, M. Plummer, R. Sankaranarayanan, J. Cherian, C. Meijer and E. Weiderpass, 'Papillomavirus infection in rural women in southern India', British Journal of Cancer 92 (2005), 601–606.

B. C. Frueh, J. Monnier, J. D. Elhai, A. L. Grubaugh and R. G. Knapp, 'Telepsychiatry treatment outcome research methodology: Efficacy versus effectiveness', Telemedicine Journal and E-Health 10 (2004), 455–458.

K. M. Gebert, D. E. Calkin and E. G. Schuster, 'The Secure Rural Schools Act, federal land payments, and property tax equivalency', Western Journal of Applied Forestry 20 (2005), 50–57.

P. Gerber, P. Chilonda, G. Franceschini and H. Menzi, 'Geographical determinants and environmental implications of livestock production intensification in Asia', Bioresource Technology 96 (2005), 263–276.

I. C. Guerrero, S. M. Ocana and I. G. Requena, 'Thermal-physical aspects of materials used for the construction of rural buildings in Soria (Spain)', Construction and Building Materials 19 (2005), 197–211.

S. L. Handy and D. A. Niemeier, 'Measuring accessibility: An exploration of issues and alternatives', Environment and Planning A 29 (1997), 1175–1194.

S. Hartley, P. Ojwang, A. Baguwemu, M. Ddamulira and A. Chavuta, 'How do carers of disabled children cope? The Ugandan perspective', Child Care Health and Development 31 (2005), 167–180.

N. C. Hazarika, K. Narain, D. Biswas, H. C. Kalita and J. Mahanta, 'Hypertension in the native rural population of Assam', National Medical Journal of India 17 (2004), 300–304.

D. Hegney, S. Pearce, C. Rogers-Clark, K. Martin-McDonald and E. Buikstra, 'Close, but still too far. The experience of Australian people with cancer commuting from a regional to a capital city for radiotherapy treatment', European Journal of Cancer Care 14 (2005), 75–82.

M. Heneghan, 'Structures and Processes in Rural Tourism'. In Signposts to rural change, Proceedings Rural Development Conference, 2002

J. V. Henderson and H. Y. G. Wang, 'Aspects of the rural-urban transformation of countries', Journal of Economic Geography 5 (2005), 23–42.

D. M. Hilty, R. L. Ingraham, S. P. Yang and T. F. Anders, 'Multispecialty telephone and e-mail consultation for patients with developmental disabilities in rural California', Telemedicine Journal and E-Health 10 (2004), 413–421.

F. Hochtl, S. Lehringer and W. Konold, '"Wilderness": what it means when it becomes a reality -a case study from the southwestern Alps', Landscape and Urban Planning 70 (2005), 85–95.

K. Hoggart and C. Hiscock, 'Occupational structures in service-class households: comparisons of rural, suburban, and inner-city residential environments', Environment and Planning A 37 (2005), 63–80.

S. H. Hounton, H. Carabin and N. J. Henderson, 'Towards an understanding of barriers to condom use in rural Benin using the Health Belief Model: A cross sectional survey', Bmc Public Health 5 (2005), art. no.-8.

N. Iiyama, M. Kamada and N. Nakagoshi, 'Ecological and social evaluation of landscape in a rural area with terraced paddies in southwestern Japan', Landscape and Urban Planning 70 (2005), 301–313.

D. Jain, S. Sinha, K. N. Prasad and C. M. Pandey, 'Campylobacter species and drug resistance in a north Indian rural community', Transactions of the Royal Society of Tropical Medicine and Hygiene 99 (2005), 207–214.

M. K. Johnson, G. H. Elder and M. Stern, 'Attachments to family and community and the young adult transition of rural youth', Journal of Research on Adolescence 15 (2005), 99–125.

D. Kaplan, 'Management of the gazelle population in Mediterranean grasslands and rural landscapes in Israel', Landscape and Urban Planning 31 (1995), 303.

E. Klouman, R. Manongi and K. I. Klepp, 'Self-reported and observed female genital cutting in rural Tanzania: associated demographic factors, HIV and sexually transmitted infections', Tropical Medicine & International Health 10 (2005), 105–115.

R. La Rovere, P. Hiernaux, H. Van Keulen, J. B. Schiere and J. A. Szonyi, 'Co-evolutionary scenarios of intensification and privatization of resource use in rural communities of south-Western Niger', Agricultural Systems 83 (2005), 251–276.

A. Martin-Dominguez, M. T. Alarcon-Herrera, I. R. Martin-Dominguez and A. Gonzalez-Herrera, 'Efficiency in the disinfection of water for human consumption in rural communities using solar radiation', Solar Energy 78 (2005), 31–40.

B. McCormick and J. Wahba, 'Why do the young and educated in LDCs concentrate in large cities? Evidence from migration data', Economica 72 (2005), 39–67.

L. McKie, 'Children and domestic violence in rural areas: A child-focused assessment of service provision', Journal of Social Policy 34 (2005), 165–167.

H. J. Miller, 'Measuring space-time accessibility benefits within transportation networks: Basic theory and computational procedures', Geographical Analysis 31 (1999), 187–212.

E. S. Mohamed and G. Papadakis, 'Design, simulation and economic analysis of a stand-alone reverse osmosis desalination unit powered by wind turbines and photovoltaics', Desalination 164 (2004), 87–97.

C. Moshiro, I. Heuch, A. N. Astrom, P. Setel, Y. Hemed and G. Kvale, 'Injury morbidity in an urban and a rural area in Tanzania: an epidemiological survey', Bmc Public Health 5 (2005), art. no.-11.

C. Moshiro, I. Heuch, A. N. Astrom, P. Setel and G. Kvale, 'Effect of recall on estimation of non-fatal injury rates: a community based study in Tanzania', Injury Prevention 11 (2005), 48–52.

B. K. Nallamothu, E. R. Bates, J. Herrin, Y. F. Wang, E. H. Bradley and H. M. Krumholz, 'Times to treatment in transfer patients undergoing primary percutaneous coronary intervention in the United States - National Registry of Myocardial Infarction (NRMI)-3/4 analysis', Circulation 111 (2005), 761–767.

E. M. Nakazono, E. M. Bruna and R. C. G. Mesquita, 'Experimental harvesting of the non-timber forest product Ischnosiphon polyphyllus in central Amazonia', Forest Ecology and Management 190 (2004), 219–225.

M. D. Nellis, 'Geospatial information technology, rural resource development, and future geographies', Annals of the Association of American Geographers 95 (2005), 1–10.

P. K. Nirmalan and K. D. Frick, 'Effectiveness of population intervention models of eye care service delivery for children in rural south India', Investigative Ophthalmology & Visual Science 45 (2004), 1063.

B. Olowokure, S. A. Wardle, M. Beaumont, H. V. Duggal and G. Colling, 'An enquiry into the respiratory health effects on a rural community of a soil mound erected close to residential property', Public Health 119 (2005), 217–222.

N. Ozay, 'A comparative study of climatically responsive house design at various periods of Northern Cyprus architecture', Building and Environment 40 (2005), 841–852.

M. Ozdirenc, A. Ozcan, F. Akin and N. Gelecek, 'Physical fitness in rural children compared with urban children in Turkey', Pediatrics International 47 (2005), 26–31.

M. Pastor and A. Guerrero, A. 'Influence of non-tillage on olive grove production', Acta Hort. 286 (1990), 283–286

A. L. Paugh, 'Multilingual play: Children's code-switching, role play, and agency in Dominica, West Indies', Language in Society 34 (2005), 63–86.

S. D. Pohekar, D. Kumar and M. Ramachandran, 'Dissemination of cooking energy alternatives in India - a review', Renewable & Sustainable Energy Reviews 9 (2005), 379–393.

J. Polaha, S. L. Cooper, T. Meadows and C. J. Kratochvil, 'The assessment of attention-deficit/hyperactivity disorder in rural primary care: The portability of the American Academy of Pediatrics guidelines to the "real world"', Pediatrics 115 (2005), E120–E126.

S. Pramod and R. Narayanan, 'Psychological advantage of traditional faith healers for tobacco de-addiction in tribal/rural parts of India', Psycho-Oncology 13 (2004), 232.

T. V. Ramachandra and B. V. Shruthi, 'Wind energy potential mapping in Karnataka, India, using GIS', Energy Conversion and Management 46 (2005), 1561–1578.

G. Raso, J. Utzinger, K. D. Silue, M. Ouattara, A. Yapi, A. Toty, B. Matthys, P. Vounatsou, M. Tanner and E. K. N'Goran, 'Disparities in parasitic infections, perceived ill health and access to health care among poorer and less poor schoolchildren of rural Coke d'Ivoire', Tropical Medicine & International Health 10 (2005), 42–57.

L. Reese, 'Rural models for chronic care', Hospitals & Health Networks 79 (2005), 14–14.

L. Rudebeck, 'Lineages of state fragility: Rural civil society in Guinea- Bissau', African Affairs 104 (2005), 148–150.

S. Sahay, S. Mehendale, S. Sane, R. Brahme, A. Brown, K. Charron, C. Beyrer, R. Bollinger and R. Paranjape, 'Correlates of HIV vaccine trial participation: an Indian perspective', Vaccine 23 (2005), 1351–1358.

S. Sharma and S. P. Singh, 'Energy Use Pattern And Sustainable Development - A Case-Study In Rural Landscape Of The Central Himalaya', Landscape and Urban Planning 29 (1994), 19–24.

D. Sietz, B. Untied, O. Walkenhorst, M. K. B. Ludeke, G. Mertins, G. Petschel-Held and H. J. Schellnhuber, 'Smallholder agriculture in Northeast Brazil: assessing heterogeneous human-environmental dynamics', Regional Environmental Change 6 (2006), 132–146.

Z. Simsek and M. A. Kurcer, 'Malaria: knowledge and behaviour in an endemic rural area of Turkey', Public Health 119 (2005), 202–208.

A. H. Skelly, T. A. Arcury, B. M. Snively, R. A. Bell, S. L. Smith, L. K. Wetmore and S. A. Quandt, 'Self-monitoring of blood glucose in a multiethnic population of rural older adults with diabetes', Diabetes Educator 31 (2005), 84–90.

R. C. Stedman, J. R. Parkins and T. M. Beckley, 'Forest dependence and community well-being in rural Canada: variation by forest sector and region', Canadian Journal of Forest Research-Revue Canadienne De Recherche Forestiere 35 (2005), 215–220.

M. Tapia and L. Zambrano, 'From aquaculture goals to real social and ecological impacts: Carp introduction in rural Central Mexico', Ambio 32 (2003), 252–257.

J. Tucker, V. Hundley, A. Kiger, H. Bryers, J. Caldow, J. Farmer, F. Harris, J. Ireland and E. van Teijlingen, 'Sustainable maternity services in remote and rural Scotland? A qualitative survey of staff views on required skills, competencies and training', Quality & Safety in Health Care 14 (2005), 34–40.

P. A. Vesk and R. Mac Nally, 'The clock is ticking- Revegetation and habitat for birds and arboreal mammals in rural landscapes of southern Australia', Agriculture, Ecosystems & Environment 112 (2006), 356.

Wahyudi and D. F. Treagust, 'An investigation of science teaching practices in Indonesian rural secondary schools', Research in Science Education 34 (2004), 455–474.

J. Wenzel, S. W. Utz, R. Steeves, I. Hinton and R. A. Jones, 'Descriptions by African Americans living in rural areas with type 2 diabetes', Diabetes Educator 31 (2005), 98–107.

F. Xu, X. M. Yin, M. Zhang, E. Leslie, R. Ware and N. Owen, 'Family average income and body mass index above the healthy weight range among urban and rural residents in regional Mainland China', Public Health Nutrition 8 (2005), 47–51.

Q. F. Zhang, Q. G. Ma and X. Xu, 'Development of land rental markets in rural Zhejiang: Growth of off-farm jobs and institution building', China Quarterly (2004), 1050–1072.

C. Zwerling, C. Peek-Asa, P. S. Whitten, S. W. Choi, N. L. Sprince and M. P. Jones, 'Fatal motor vehicle crashes in rural and urban areas: decomposing rates into contributing factors', Injury Prevention 11 (2005), 24–28.

Chapter 6
Dynamics and Processes of the Anthropic Elements

The dynamics of the anthropic elements are more complicated to analyze than those of the environmental elements. This is due to the uncertainties in the measuring process, the cost of performing such measurements and the privacy policies that precludes obtaining useful but personal data. Even considering that a great deal of information can be derived in this phase of the analysis, as we discussed in the previous chapter, gaps in our knowledge of the anthropic elements dynamic persists.

In any case, we can define a cross-relationship matrix as we did for the environmental elements. The result (Fig. 6.1) contains $12\times12-12=132$ cells to explore. The same kind of discussion and algebra operations can be applied here.

If enough data is available, it is sometimes possible to use mathematical models to analyze the dynamics of the social elements. This is the case of household income and demography. The plots of Fig. 6.2 illustrate one of such analysis using differential equations modelling. Considerable expertise in mathematics is required to establish such models, but their usefulness in a likely planning phase recommends their development by an expert, providing the required data is available.

If a complex model is built, sensitivity analysis is a powerful tool to unveil hidden processes and relationships. If one aspect of the anthropic elements has been modelled using for example a neural network, inversion of the net can help to understand the underlying dynamics. This is simply done by calculating the Jacobian of the trained neural net.

The key factor here is that modelling permits finding very complex relationships that cannot be found using a purely discursive method. These relationships can and should be explained in words once the analysis has identified them, but it is doubtful that they would arise without numerical techniques: mathematics is not a jargon, but a way of processing a large amount of information through a large string of logical steps without committing incoherencies.

6.1 The Rural History

The importance of analyzing rural history for a proper management of rural areas is debatable. There are areas in which history still plays a determinant role, such as those that have suffered recent wars. In other cases, however, historical events play a minor role.

ANTHROPIC ELEMENTS	HABITAT	AGRARIAN LAND	SOCIAL STRUCTURE	LEISURE	UTILITIES AND SERVICES	TRANSPORT AND ACCESSIBILITY	ENERGY	INDUSTRIAL ACTIVITIES	GENDER RELATIONSHIPS	PUBLIC HEALTH ISSUES	POLITICS	RURAL TRADITIONS AND FOLK	RURAL TOURISM
HABITAT		A	A	B	A	A	A	B	B	A	A	A	A
AGRARIAN LAND	A		A	B	C	-	-	C	A	C	A	A	B
SOCIAL STRUCTURE	A	A		B	C	A	-	C	A	C	A	B	A
LEISURE	B	A	A		-	-	-	C	B	B	B	B	A
UTILITIES AND SERVICES	A	B	A	A		-	A	A	-	B	B	C	A
TRANSPORT AND ACCESSIBILITY	A	B	A	A	B		A	A	B	A	A	B	A
ENERGY	B	B	B	C	A	B		A	-	A	B	-	B
INDUSTRIAL ACTIVITIES	B	B	B	B	B	B	B		C	B	B	C	B
GENDER RELATIONSHIPS	B	-	-	-	-	-	-	C		-	A	B	B
PUBLIC HEALTH ISSUES	B	-	A	A	A	B	C	A	C		A	B	A
POLITICS	B	A	A	B	A	A	A	B	B	A		B	A
RURAL TRADITIONS AND FOLK	B	-	-	B	-	-	-	C	B	B	B		A
RURAL TOURISM	B	A	A	A	A	A	A	C	B	B	A	C	

A	DIRECT EFFECT
B	VARIABLE EFFECT
C	INDIRECT OR WEAK EFFECT
-	UNCOMMON OR LONG TERM EFFECT

Fig. 6.1 Cross correlation B matrix for the anthropic elements (cf. Fig. 4.1)

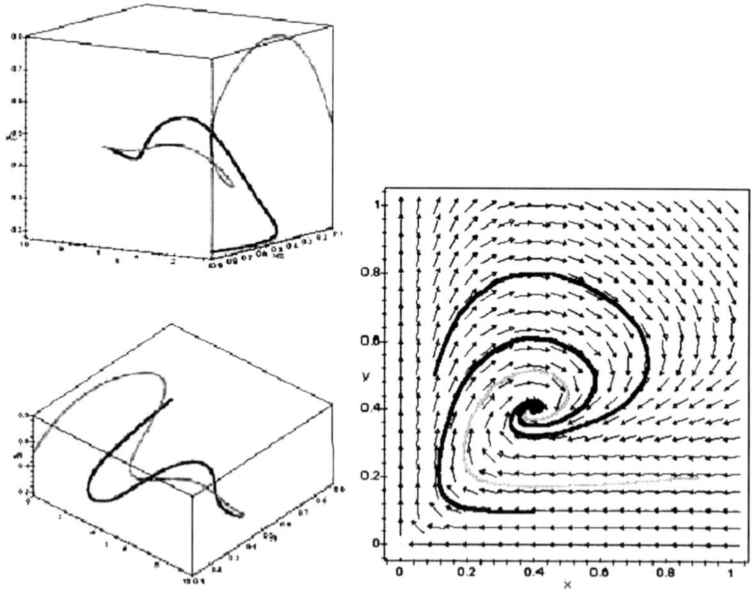

Fig. 6.2 Dynamical modelling of population growth/income with several initial conditions

History can also be understood in a different sense in this context as the time-evolution of the anthropic elements. Tracking the evolution of the variables used to characterize the rural area can be prohibitively expensive. In this case, the evolution can be roughly characterized. Those areas with little or no change in many decades can be described as inertial rural areas. If changes have occurred, an evolution graph can be constructed using the available data gathered in the analysis. A before/after analysis will help to identify the relevant patterns: which element or sum of elements initiated the changes, how the other elements react to this change, and how sensitive is the current situation to variations in those elements.

6.2 Rural Economies

From the point of view of the dynamics, economy plays a similar role to physical relationships in the environmental elements. As those elements are linked by physical and biological feedbacks, the fabric of the rural social structure is made up of capital fluxes. Exogenous inputs from regional or national governments modify the anthropic elements, as endogenous values do.

The rural economies have been thoughtfully studied and excellent handbooks and manuals exist on the economy of farms or small villages. The capital availability, cash flows, the actual economic value of land and properties and the economic orientation of the village have a central effect on both spatial and non-spatial elements. This is the reason of conceptually considering rural economies within the dynamics of the anthropic elements, and not as a separate component. Economies are the result of how the anthropic elements perform, and of the mutual relationship between them. It could be considered as another non-spatial element strongly related with each of the others, but it is more practical from the analysis point of view, to study economy as a dynamical aspect of the rural areas. Using a well-known simile, economy is the blood of the landscapes. In fact, many of the variables studied (land use, leisure, tourism) can be valued, that is, quantified in economical terms.

Rural areas have some potential competitive advantages, such as their proximity to natural resources, the perceived high quality of life, the environmental and socio-cultural resources and lower labour and land costs. However, they also have also weaknesses such as the sparse distribution of the population which increases the cost of providing services, limits the availability of infrastructures and hinders the development of social capital due to the lack of the critical mass needed.

It is obvious that rural economies differ from area to area and thus a specific analysis should be done. Bear in mind that if statistics are aggregated, rural economies may appear similar to their urban counterparts as similar analytical methods to the urban cases can be used.

An economic study of the area is more useful in the planning and management phases, where cost-benefit analyses are mandatory in order to evaluate the interest and feasibility of the proposal. Multi-criteria analyses and more complex methods such as cellular automata may help to encapsulate the economic dynamics in a model, illuminating how the elements are mutually related. Benefits from biotechnological inversions may span to better gender relationships through an improvement in education, or by making the case for public inversion in telecommunication infrastructure in the area, to name just one possible process that can be economically quantified.

References and Bibliography Used

J. P. Brans, C. Macharis, P. L. Kunsch, A. Chevalier and M. Schwaninger, 'Combining multicriteria decision aid and system dynamics for the control of socio-economic processes. An iterative real-time procedure', European Journal of Operational Research 109 (1998), 428–441.

M. O. Campbell, 'Actor networking, technological planning and conceptions of space: The dynamics of irrigation farming in the coastal savanna of Ghana', Applied Geography 25 (2005), 367–381.

J. Q. Cheng and I. Masser, 'Understanding spatial and temporal processes of urban growth: cellular automata modelling', Environment and Planning B-Planning & Design 31 (2004), 167–194.

S. A. O. Cousins and O. Eriksson, 'Plant species occurrences in a rural hemiboreal landscape: effects of remnant habitats, site history, topography and soil', Ecography 24 (2001), 461–469.

S. A. O. Cousins and O. Eriksson, 'The influence of management history and habitat on plant species richness in a rural hemiboreal landscape, Sweden', Landscape Ecology 17 (2002), 517–529.

I. Dyner, R. A. Smith and G. E. Pena, 'System Dynamics Modeling for Residential Energy Efficiency Analysis and Management', Journal of the Operational Research Society 46 (1995), 1163–1173.

V. Higgins and S. Kitto, 'Mapping the dynamics of new forms of technological governance in agriculture: methodological considerations', Environment and Planning A 36 (2004), 1397–1410.

M. Hoard, J. Homer, W. Manley, P. Furbee, A. Haque and J. Helmkamp, 'Systems modeling in support of evidence-based disaster planning for rural areas', International Journal of Hygiene and Environmental Health 208 (2005), 117–125.

K. E. Maani and V. Maharaj, 'Links between systems thinking and complex decision making', System Dynamics Review 20 (2004), 21–48.

Chapter 7
A Synthetic Approach

The rationale behind the analysis phase is to divide the rural reality into pieces to be able to do an in-deep study. The next step is to gather all this information to analyse the cross-relationships between the elements that are relevant for the area. Then, it is necessary to generate a synthesis of the rural area. This helps to fully characterize the area being studied.

The first step of this synthesis is to examine the visual elements of the area, that is, the landscape. After the completion of the landscape documentation, the next step is to generate a final analysis report including all the relevant aspects of the project. This constitutes the first step of the subsequent planning task.

7.1 The Rural Landscapes

The effects on our senses of the environmental and anthropic elements in a given area define a rural landscape. While each and every landscape is unique in that it always differs from similar landscapes, patterns and similarities can be found in how those elements combine in a rural area and how we perceive this combination. Classifying the rural area in terms of one dominant of several combined landscapes can help the analysis of the potentialities and opportunities of the area, as well as to identify threats and weaknesses. Also, the taxonomy of rural landscapes offers an opportunity to compared analyses. It is seldom found that very similar rural landscapes have markedly different problems, dynamics, and/or defining elements. It is the case of several mountain landscapes in the Alps and in the Pyrenees; areas with vast socioeconomic and cultural differences and thus subject to different management approaches, but with reciprocal practices of interest to the mass-tourism challenges.

Analyzing the landscapes of the area under scrutiny is vital for recreation and tourism, but also for land use zoning aiming to preserve diversity, or to the whole process of rural and regional planning. To do that in a sensible and objective manner, it is required to follow a defined and explicit procedure that could be easily reproduced by other researchers or practitioners. It is therefore advisable to use some sort of landscape classification scheme.

There are as many landscape classifications as there are geographers. It is not surprising then that even the definition of what a landscape is has been a recurrent discussion since the early Sochava's definition of *geosystems* in 1953. The original pictorial term refers to the representation of natural scenery by itself that modernly started in the early Renaissance. The dictionary definition is a single aspect of the land characteristic of a particular region. The term 'landscape architecture' (in the sense of structure) is also used to name the planning of gardens. The FAO definition explicitly includes the human action: 'The landscape is a product of the interaction of human beings with the natural environment often over a period of many centuries. The landscape therefore reflects the impacts of social, economic and political changes on the natural environment'. This definition, however, is not as general as it might seem, but it can be used as a working definition.

The classification of the landscapes or landscape taxonomy has many different coexisting systems depending on the aim of the research. A system devised for ecological preservation may not be applicable in the context of regional planning, or landscape architecture. Here, we will refer only to rural landscapes as different from natural and urban ones.

In order to facilitate the analysis, we have devised a simple classification specifically aimed to the analysis and management of rural areas. It uses the environmental and human elements previously discussed to classify each landscape using a very simple protocol (cf. Table 7.1). This classification is intended to be as general as possible, but there are other alternatives. What should be avoided, however, is to assign local names to the landscapes. This practice is useless from a rural science perspective, since the whole point of classification is to extract the common elements and not just label the otherwise unique characteristics of each landscape.

The protocol is as follows:

1. First, the analyst has to select the strongest element that could be used as a single element to define the landscape. The choice can be done individually or on an expert panel system. This classifies the landscape in one of the 24 types that we have defined. The 'base geosystem' label can help to decide if the landscape has a greater natural or human character.
2. The second part is to select an element to clarify or to be more accurate in the definition, using again one of the 24 elements. If the landscape clearly belongs to a category, the same factor can be used for the secondary factor choice, generating a 'pure landscape'. This procedure generates 576 possible rural landscapes, which are deemed to encompass virtually all the practical situations found in practice. Landscapes can then by named as 'type 3–10'.
3. A further repetition of the process (if the landscape is perceived in more than one way) can help to more precisely define the area. Then, the landscape can be defined as 'type 3/10 (alt. 10/3, 22/10)'. This can be stated in a matrix such as the one in Table 7.1 using a code: the first choice can be marked with a 1 and successive appraisal with increasing figures, trying to minimize the number of choices.

Depending on the scale and instrumentation (naked eye, camera, and binoculars), the same landscape could be classified in several ways depending on different

Table 7.1 Typology matrix for classifying rural landscapes

PRIMARY FACTOR	RURAL LANDSCAPE MATRIX	SECONDARY FACTOR 1 Geology	2 Mineral Resources	3 Soils	4 Geomorphology	5 Meteorology	6 Climate	7 Hydrology	8 Land cover	9 Ecology	10 Natural resources	11 Scenic values	12 Natural and environmental risks	13 Agrarian land	14 Habitat	15 Social structure	16 Leisure	17 Utilities and services	18 Transport and accessibility	19 Energy	20 Industry	21 Gender relationships	22 Public health issues	23 Political issues	24 Rural culture, traditions and folklore	25 Rural tourism	Base Geosystem
1	Geology																										Abiotic
2	Mineral resources																										Abiotic
3	Soils																										Abiotic
4	Geomorphology											1															Abiotic
5	Meteorology																										Abiotic
6	Climate																										Abiotic
7	Hydrology																										Abiotic
8	Land cover variety																										Abiotic
9	Ecology																										Biotic
10	Natural resources																										Biotic
11	Scenic values				2																						Mixed
12	Natural and environmental risks																										Anthropic
13	Habitat																										Anthropic
14	Agrarian land																										Anthropic
15	Social stucture																										Anthropic
16	Leisure																										Anthropic
17	Utilities and services																										Anthropic
18	Transport and accessibility																										Anthropic
19	Energy																										Anthropic
20	Industry																										Anthropic
21	Gender relationships																										Anthropic
22	Public health issues																										Anthropic
23	Political issues																										Anthropic
24	Rural culture, traditions and folklore																										Anthropic
25	Rural tourism				4							3															Anthropic

expert's appraisals. Just as a medical diagnosis is a categorical but established and objective method, deciding what category a landscape belongs to is a difficult procedure requiring a good knowledge of the elements involved and a lot of field work. Moreover, an area can coexist with very different landscapes and a landscape map may be required. This map (which can also be a GIS layer) consists of a base map and a legend depicting the different landscapes found with a graphical support associated.

Considering that there are not simple recipes to analyze an area even when the rural landscape has been defined, there are several common issues that link rural areas with similar landscapes. What follows is a description of some types provided to help both classify and identify the major issues arising from the classification in terms of analysis. It should not be forgotten that the ultimate aim of any rural analysis is for it to be used from an anthropic point of view, so if there is a clear alternative between two classes it is advisable to choose the one with the strongest human component.

The scale is another issue to be considered. Here, we assume that the rural area has been fully delimited and we characterize the landscape within those limits. An oasis on a desert is a desert landscape if we took a large area, but becomes a hydrology/agrarian (6/12) landscape case within its limits. Firstly, an attempt should be done to characterize the area as a whole. Only if the area is too varied, several classifications can be assigned to different spatial areas.

Type 1. Geological Landscapes

Geological landscapes have a twofold potential: as mining sites, they provide an economic resource for rural areas, albeit environmental impacts must be tackled. Geosites and geological national parks have an undoubted tourist potential, but their potential is also to be used to locate scientific or higher education centres. A few examples of this kind of landscapes are:

- Type 1/1: Examples of pure geological landscape comprises of quarries and extensive mining areas. Tourism opportunities exist for these sites if abandoned (industrial archaeology tours), apart from their economic importance if currently at work.
- Type 1/6: Karsts, such as the hot calcium-laden water spring in Pamukkale (Turkey) with terraced travertines of calcium that form pools is a clear example of Type 1/6.
- Type 1/10: Geosites, geological parks such as the Songshan Mountain Stratum Structure National Geological Park in China, where millions of years of the geological record are visible in an area of more than 450 sq. km, attracting tourists from overseas. These areas can also be used to locate scientific research institutes, technical schools or higher education centres.
- Type 1/22: The geological features of Iceland clearly define not only natural but also rural areas where routes and logging are used as a complementary income resource.

Type 2. Soil Landscapes

There are a few examples of landscapes clearly defined by the soil. Most steppe and tundra are natural, not rural landscapes. Nonetheless, areas of the Mongolian steppe are used by nomadic husbandry and are thus rural.

- Type 2/2: This may include steppe landscapes, defined by the soil but with endemic flora and birds.
- Type 2/3: Deltas and some alluvial planes, created by sediment deposition, define areas such as some of the Danube River in Serbia and Montenegro.
- Type 2/5: Tundra landscapes (with the permafrost) are a good example of this category.
- Type 2/22: Canary Islands black beaches, from volcanic material, are an attractive resort and can categorize within this theme.

Type 3. Geomorphologic Landscapes

Geomorphologic landscapes in rural areas have great tourism potential, but other elements such as scenic values or energy generation are also important.

- Type 3/3: (Pure type) Dune landscapes are a pure geomorphologic landscape. Rural areas such as Douz in Tunisia are both rural and dominated by geomorphologic processes.
- Type 3/17: This category may include areas of hydroelectric power generation or wind turbines over anfractuous relief.
- Type 3/14: The leisure and tourism category concentrates most of the geomorphologic rural landscapes. The 'Ciudad Encantada' in Cuenca, Spain, is a fine example of this type.

Type 4 and 5. Meteorological and Climate landscapes

Meteorology and climatology seldom define a rural landscape. They are more often associated with the strongest component. Nonetheless, some landscapes can be associated with certain meteorological situations. Climate also defines some other landscapes (desert fringes, polar regions, etc.)

- Type 4/4: (Pure type): Some rural landscapes are renowned with the occurrence of meteorological phenomena such as auroras, diamond dusts or sunshowers. Also, transient landscapes such as the devastation left by tornadoes, forest fires and floods may be included within this category.
- Type 4/19: Foehn characterizes some mountains and has a strong effect on health. Recurrent winds such as the Sirocco that characterizes some areas may be a serious health issue (it is known that this wind affects paroxysmal atrial fibrillation in the elderly population).

- Type 4/22: Extreme-weather can be a tourist attraction by itself, as is the case of the rainiest place on Earth (Cherrapunji, India), where this fact is combined with other attractions such as vegetation and hydrological elements (waterfalls, etc.).

Type 6. Hydrological Landscapes

The hydrological landscapes are those in which water plays a major role. Water is indeed crucial for life and some rivers sustain a varied economic activity.

- Type 6/6: (Pure type) Inhabited river estuaries and deltas qualify as pure hydrological landscapes.
- Type 6/8: Mangroves are the classical example of areas with a strong hydrological and ecological component.
- Type 6/14: Includes dams, sportive ports in lakes, and some spas.
- Type 6/16: Rivers such as the Rhine or the Danube have been used for fluvial transport for centuries, and large areas in the shore belong to this category if there is not any other stronger element.
- Type 6/19: Ponds and marshes may represent a serious health risk not only in developing countries but also in developed ones. Some malaria rural areas can be characterized as a hydrological/health issue landscape, which is a categorization preferable to a hydrological/ecological characterization due to the strongest human component of the former class.

Type 7. Land Cover Landscape

A varied land cover may characterize an area. It is the case of highly diversified areas where economic activities and natural elements coexist.

- Type 7/7: Includes transition zones between other marked landscapes. The shores of the Nile are a good case of this kind of landscape.
- Type 7/10: The rural fields of Normandy are an example of diverse landcover with a scenic vocation.
- Type 7/20: The whole Principality of Liechtenstein is a fine example of a rural area defined by a varied (within the Central Europe limits) landscape blended with a political singularity that characterizes the space.

Type 8. Ecological Landscapes

- Type 8/8: (Pure type) Natural areas with a minimum human intervention. As the cities, they are of the interest of rural analysis in that they are not rural, and in the relationship with the rural areas near by.
- Type 8/14: Primarily-ecological landscapes with a leisure use include some national parks, such as Doñana in Southern Spain.

- Type 8/21: Ecological landscapes related with traditions, has an example in the area inhabited by the Saharia people in of Rajasthan, where the links between ecology, rural life and traditions are apparent.

Type 9. Natural Resources Landscapes

- Type 9/9: Tree plantations for timber are an example of this kind of rural landscape.
- Type 9/12: Areas devoted to charcoal production or lignite extraction (Ballymoney in Northern Ireland) may be included into this category.

Type 10. Scenic Values Landscapes

- Type 10/10: (Pure type) The quintessential scenic value landscape is that landscape than can immediately be defined as a 'postcard' area. It may refer either to a natural beauty or to a highly humanized area.
- Type 10/3: Such as the scenery from rural coastal areas.
- Type 10/7: English grazed moorlands are an example of this sort of rural landscape with a land cover variety interest.

Type 11. Natural and Environmental Risk Landscapes

- Type 11/11: (Pure type) This category includes potentially dangerous landscapes near rural areas such as avalanche sites. The economical significance of these areas cannot be diminished: they offer opportunities such as extreme sports areas. Sports such as ice climbing are usually done on areas catalogued as risky areas.
- Type 11/6: Such as some sectors of the flooding plains used for water storage.
- Type 11/13: In some areas, the habitat is located on potentially risky areas due to historical reasons.

Type 12. Habitat Landscapes

This category is for rural settlements.

- Type 12/12: (Pure type) Rural places with architectural interest.

Type 13. Agrarian Landscapes

This category embedded more rural landscapes in the sense of the agrarian land being a characteristic that defines what a rural area is.

- Type 13/13: (Pure type) The extensive farming landscape in the corn belt of the USA is a good example of a pure agrarian landscape.
- Type 13/8: Non-tillage farming areas, where a balance between farming practices and natural life exists are agrarian/ecological areas.

Type 14. Social Landscapes

This category can be used for rural settlements where social strcuture or population themselves are the dominant factors. Depending on the scale, population may be the dominant feature of the area.

- Type 14/14: (Pure type) This very particular type of anthropic landscape is defined by the population component. This includes refugee camps, semi-permanent nomad settlements, or periurban areas (some Brazilian favelas, etc.)
- Type 14/15: This category might include many periurban settlements used as secondary residence.

Type 15. Leisure Landscapes

Contrary to the last category, leisure landscapes are so common that care must be taken in discarding other elements before tackling the leisure indicators. This is particularly relevant if one of the aims of the project is rural tourism, since this orientation may distort the analysis in favour of categorizing as leisure landscape an area with a different main orientation. Also, it should be remembered that different landscapes may coexist in the area.

- Type 15/15 (Pure type) A pure type is represented by the diverse forms of thematic parks if they are associated to the rural area.
- Type 15/10 An area devoted to leisure in which the scenic values are central. This type of landscape can be a small, secluded part of the rural area ('window landscape'), and should not be mistaken with a landscape that is visible from this landscape.

Type 16. Utilities and Services Landscapes

These landscapes may present any scale, for tens of square kilometres to a few hectares.

- Type 16/16: (Pure type) Those rural areas such as rail intersections (e.g. Crew in England) or other utilities.
- Type 16/18: Such as sites of extreme density of (La Mudarra, Spain) or some wind farms.

Type 17. Transport and Accessibility Landscapes

This is a category with a limited number of cases in most rural areas. Nonetheless, a transport landscape has the potential of representing a competitive advantage for the area.

- Type 17/17 (Pure type) Regional airports, rural rail stations or large parking areas for other activities.

Type 18. Energy Landscapes

- Type 18/18: (Pure type) Most rural wind farms, electric generation sites (nuclear plants, etc.).

Type 19. Gender Relationships Landscapes

Rural landscapes dominated by gender issues are those in which a clear segregation of male/female tasks, occupations or even location exists.

- Type 19/19 (Pure type) Some rural areas in which either women or men live apart from each other.

Type 20. Public Health Landscapes

- Type 20/20: (Pure type) Areas unsuitable for life contaminated soils or waste disposal sites.
- Type 19/22: Including spas, health resorts or health farms.

Type 21. Political Landscapes

- Type 21/21: (Pure type) Rural Palestine is an example of landscape dominated by political issues. The category may also include land divided by artificial, political barriers, walls (Germany during the Cold War), or hot borders (India/Pakistan, etc).
- Type 21/22: Such as some battlefields or memorial sites.

Type 22. Cultural Landscapes

- Type 22/22: Such as most of UNESCO's cultural heritage sites. The pure class corresponds with those areas in which the cultural aspects are stronger than any other secondary aspect.

- Type 22/23: Most of the cultural landscapes attract tourism; so many cultural landscapes have this secondary value.
- Type 22/8: Such as some biosphere reserves.

Type 23. Rural Tourism Landscapes

- Type 23/23: (Pure type) Some areas of the Mediterranean coast have become so linked with tourism that they have become rural tourism landscapes. Also, areas that have been revamped as rural tourism areas may have a primary tourism landscape character rather than a secondary one.
- Type 23/4: To name a non-obvious case, a tourism-meteorology landscape may be represented by Mont St-Michel in France, a rural area where tides play such a central role that technical projects exist to maintain the isolation of the rock.

Type 24. Industrial Landscapes

- Type 24/24: (Pure type) Landscapes in which productive activities-other than agrarian/forestry and tourism- are central. Since a rural area is characterized by the absence of these elements, pure industrial landscapes are less frequently found in rural analysis works. Nonetheless, they are sometimes a secondary factor (e.g. Industrial archaeological sites).

Fig. 7.1 A potential 14/15 landscape (Mainly a social landscape with a leisure component). Periurban settlements and part-time agriculture plots near a large city (see Color Plate on following p. 333)

Table 7.2 Expertise and sciences involves in the analysis of the cross-relationships between environmental and anthropic elements

	Environmental Elements	Anthropic Elements
Environmental Elements	Environmental and Natural Sciences	Geography
Anthropic Elements	Geography	Social Sciences

One of the more rewarding exercises that can be done in a rural analysis academic course is creating a worksheet with Table 7.1 and fill-in all the cells with examples, including pictures such as Fig. 7.1. The greater the difficulty is in finding a suitable example (e.g. a type 4/2: meteorological landscape with soil factor), the greater the learning reward.

7.2 Landscape Documentation

In order to document the landscape analysis conclusion, it is advisable to compile a landscape report. This may include the landscape typology matrix (Table 7.2), geographical data, and cartographic and photographic support. If the area comprises several different landscapes, a landscape map could be required. The information gathered in this part of the analysis should be as precise as the available technical means permit.

Figure 7.2 shows an example of the minimum contents of the graphical part: a picture for each view and a location map with the co-ordinates and season of the take will help to the member of the multidisciplinary panel, who does not necessarily have to carry out a detailed field trip. Photographs do not have to be professional, pleasing pictures, but rather being illustrative of the elements in the landscape. Recording the field of vision of the picture, the azimuth and height, the geographical position of the viewer, and the direction of sight will help to organize the information. If several landscapes can be defined, it is sometimes helpful to trace several views in the same map, generate 360° views or digital videos. Georeferenced oblique aerial pictures can also be used if the area to be analyzed is large enough (Fig. 7.3).

7.3 The Geographical Synthesis

Once the previous phases have been carried out the next step is to gather all the information and start the synthesis. This task can be even more time consuming than the analysis phase, since it requires a thoughtful insight into all the cross-relationships operating in the rural area. We already discussed some related issues in Chaps. 4 and 6. What we did there was a preliminary approach to this task. The difference

Fig. 7.2 An example of the graphical documentation required for the landscape report. Type 3.1 (or 10.3) landscape. Map by Endoxon AG (see Color Plate on following p. 333)

here is that we now aim for an overarching, more comprehensive approach to fully characterize the rural area.

The relationships between environmental and anthropic elements by themselves are of the interest to environmental sciences and social sciences respectively, and expertise in those fields should be sought after in order to acquire insight in to the well-defined aspects of those two major areas. When humans are a critial part of those fields, physical and human geographers may also help. A classical example is the difference in which climate is analysed in Physics (a natural science) and in Geography, which is principally interested in how climate affects humans and human activities.

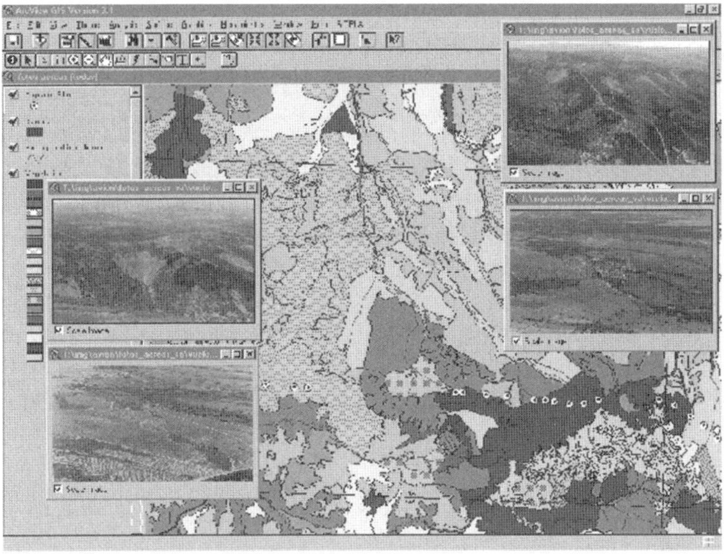

Fig. 7.3 GIS support for oblique aerial photographs of the landscapes. Photographs are linked to specific coordinates in a land-use map to help interpretation (see Color Plate on following p. 333)

The analysis of the effects of anthropic elements on environmental elements, and vice-versa, corresponds almost exclusively to geography (Table 7.2). Within this science the synthesis methods have been developed and it is geography the science whose primary interest is establishing the links between the environment and the humans.

The synthesis is a very difficult task, but a rural project missing this part cannot be properly used for either planning or management. Following with the example of the Airbus 350 in the first chapter, a description of the elements without a synthesis is like describing the components of the airplane without providing an insight on how they work when put together; how an airplane flies, and how this airplane compares with, say, a Boeing 747.

Moreover, it is in this phase when the analyst realizes what is worth to be known of the environmental and anthropic elements in terms of rural analysis.

7.3.1 First Order Cross-relationships

The final goal of rural analysis should be providing a useful model on the dynamics of the rural area. A good criterion for complexity is that the model has to be close enough to reality to provide better results than can be obtained by common sense. If so, the model is useful. In the case of rural models, the new knowledge can be used to assess the impact of regional policies, new infrastructures, economic initiatives; to improve the quality of life of the rural population with targeted, optimized, cost-effective actions; or to help maintain and manage natural areas. The trophic levels of the riparian invertebrates are relevant in an ecology study, but they are seldom useful for rural analysis unless a singular case deserving protection occurs in the area. If the analysis is always done with an eye on that final goal, much time can be saved devoting resources to what is really useful.

This selection process is fundamental in rural analysis. Otherwise, the analysis becomes an endless task. It may take an entire life to fully analyze just the ecology or the rural culture of an area. The goal of the first part of the book was to help the analysis identify the key elements, which are highly variable in space. What is relevant for Bolivia (the importance of indigenous culture and traditions in defining viable economic solutions) may be of marginal importance in rural England (where indigenous culture is constrained to folk manifestations and leisure). Thus, it is crucial for the analyst in a real project to devote time to think rather than devoting time to apply a set of general recipes. There is no handbook capable of containing the immense variety of situations. All what can be done is to create the structure of the study (which is given by the scheme of the analysis) and the analysis patterns (which are given by the way the elements are related with the human interests in some known cases).

The same applies for the synthesis part. As we will soon see, the number of possible interactions is very high, but not all of them are always relevant. The key to success in the task is following a rational and explicit methodology and using quantitative data to make the process fully comparable. Information technologies can assist in the synthesis task. We will see in Chap. 21 that a dynamic database is a useful tool for planning development, or to manage a rural area. Since one of the possible outcomes of the analysis is a subsequent management phase, it is always helpful to perform this cross-analysis phase using information system tools if available.

The first task, as in Chaps. 4 and 6, is building the cross-relationships matrix. Table 7.3 reflects the final structure. The 25 elements make $25\times25-25=600$ binary relations to explore. It is desirable to avoid just assigning a weight to each cell. A full justification in terms of objective evidence should be provided. A conservative estimate of the time required to fully analyse them is 30 minutes by relation, making 300 hours of work just for this task. About half of the work can be done by environmentalist and social experts in their reports.

The amount of information that can be extracted from a well-constructed matrix is not limited to binary relationships. Even more interesting (since they are not obvious from the matrix) are the tertiary and *n-degree* relationships.

7.3.2 n-degree Cross-relationships

The *n-degree* relationships are the closed chains of n elements that are sequentially affected. For instance, in the rural area represented by Table 7.3 the fact that Land Cover affects Soils has been considered as relevant for the area for a particular reason (that should be justified in the report). At the same time, Soils affect

Table 7.3 A general cross-relationships table to synthesize the analysis

CROSS RELATIONSHIPS		ON THIS ELEMENT																								
		1	2	3	4	5	6	7	8	9	10	11	12	13	14	15	16	17	18	19	20	21	22	23	24	25
THE EFFECT OF THIS ELEMENT		Geology	Mineral Resources	Soils	Geomorphology	Meteorology	Climate	Hydrology	Land cover	Ecology	Natural resources	Scenic values	Natural and environmental risks	Habitat	Agrarian land	Social structure	Leisure	Utilities and services	Transport and accessibility	Energy	Industrial activities	Gender relationships	Public health issues	Political issues	Rural culture, traditions and folklore	Rural tourism
1	Geology		A	A	A	B	A	A	B	B	B	B	A	A	C	C	C	C	A	-	B	-	-	-	-	-
2	Mineral resources	-		-	-	-	-	-	-	-	A	A	B	B	-	-	-	-	-	B	A	-	-	B	-	B
3	Soils	-	-		C	-	-	B	A	A	A	A	A	A	A	-	-	-	-	-	C	-	-	-	-	-
4	Geomorphology	-	B	A		C	C	A	A	A	A	A	A	A	-	A	A	A	A	-	C	-	-	-	-	A
5	Meteorology	-	-	A	A		A	A	A	A	A	-	A	C	A	-	A	-	B	B	C	-	-	-	-	A
6	Climate	-	-	A	A	-		A	A	A	A	B	B	C	C	C	C	C	B	B	C	-	-	-	-	B
7	Hydrology	-	-	A	A	A	A		A	A	A	A	B	B	A	A	A	-	-	-	B	-	-	A	-	B
8	Land cover variety	-	-	A	B	C	C	B		A	B	A	B	-	-	-	-	-	-	-	-	-	-	-	-	B
9	Ecology	-	-	A	B	B	-	B	A		B	A	B	B	B	-	A	-	-	-	C	-	-	-	-	A
10	Natural resources	-	-	-	-	-	-	-	-	-		-	-	B	-	A	A	A	A	A	B	-	-	-	-	B
11	Scenic values	-	-	-	-	-	-	-	-	B	B		-	B	-	-	A	A	A	-	B	-	-	A	A	A
12	Natural and environmental risks	-	-	-	-	-	-	-	-	B	B	B		B	-	A	A	-	-	-	B	-	A	A	-	A
13	Habitat	-	-	-	-	-	-	B	B	A	-	A	C		A	A	B	A	A	A	B	B	A	A	A	A
14	Agrarian land	-	-	-	-	-	-	-	A	A	-	A	-	A		A	B	C	-	-	C	A	C	A	A	B
15	Social structure	-	-	-	-	-	-	-	A	A	B	A	-	A	A		B	C	A	-	C	A	C	A	B	A
16	Leisure	-	-	-	-	-	-	-	A	-	-	B	-	B	A	A		-	-	-	C	B	B	B	B	A
17	Utilities and services	-	-	-	-	-	-	-	A	A	-	A	-	A	B	A	A		-	A	A	-	B	B	C	A
18	Transport and accessibility	-	-	-	-	-	-	-	A	A	-	A	-	A	B	A	A	B		A	A	B	A	A	B	A
19	Energy	-	-	-	-	-	-	-	C	-	-	-	-	B	B	B	C	A	B		A	-	A	B	-	B
20	Industrial activities	-	-	B	-	-	-	B	C	B	-	B	-	B	B	B	B	B	B	B		C	B	B	C	B
21	Gender relationships	-	-	-	-	-	-	-	C	-	-	-	-	B	-	-	-	-	-	-	C		-	A	B	B
22	Public health issues	-	-	-	-	-	-	-	C	-	-	-	-	B	-	A	A	A	B	C	A	C		A	B	A
23	Political issues	-	-	-	-	-	-	-	C	-	-	-	-	B	A	A	B	A	A	A	B	B	A		B	A
24	Rural culture, traditions and folklore	-	-	-	-	-	-	-	C	-	-	-	-	B	-	-	B	-	-	-	C	B	B	B		A
25	Rural tourism	-	-	-	-	-	-	-	C	-	-	-	-	B	A	A	A	A	A	A	C	B	B	A	C	

A DIRECT EFFECT B VARIABLE EFFECT INDIRECT OR WEAK EFFECT - UNCOMMON OR LONG TERM EFFECT

Ecology, and Ecology affects Land Cover. Also, there is a binary relation in each step. Schematically:

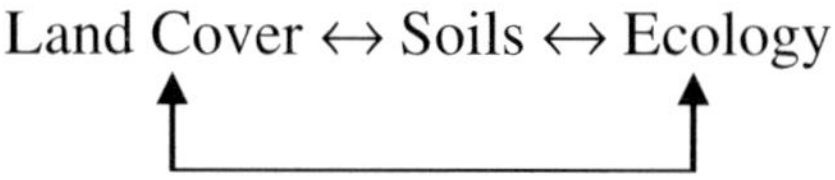

This is a 3-degree relationship as it involves 3 elements. Notice that there is no need for binary relations to build up a 3-degree relationship. The following also is a 3-degree relationship:

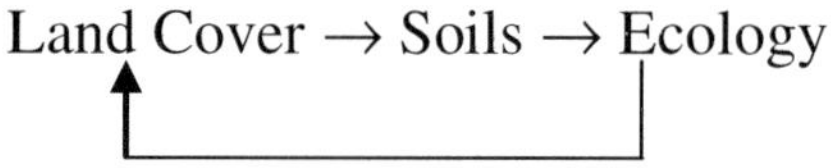

Since the effects, indicated by the arrow, are sequential and the path is closed. In other words, it is possible to travel from any element and back in three-steps. A 4-degree relationship involves 4 elements, and so forth.

These relationships are the fabric of the rural areas. They represent how the elements we have been describing relate with each other. They embedded most of the work done in identifying the relevant environmental and anthropic elements. If the team suddenly realizes a process, and that process has not been stated in the matrix, then it is likely that something was left out in the first analysis. Also, the relationships are the objective expression of processes that are seldom fully explicated in rural analysis.

It is clear that a numerical approach is needed to explore the *n-degree* relationships. This is the rationale behind ranking the cells and building up a mathematical matrix.

The corresponding matrix to Table 7.3 is a 25×25 matrix. As a matter of convenience, the main diagonal of the matrix can be assigned a '1' or a '0' value. The justification for a '1' would be that each element affects itself, creating inner-dynamics. Using standard notation, the resulting matrix is:

$$G \equiv \begin{pmatrix} a_{1,1} & \dots & a_{1,25} \\ \dots & a_{i,i} & \dots \\ a_{1,25} & \dots & a_{25,25} \end{pmatrix}$$

A nice feature of this matrix is that it represents what is known as an adjacency matrix in graph theory. In fact, an alternative way to visualize these matrices is using a graph such as the one depicted in Fig. 7.4.

The adjacency matrix represents the arcs between the chosen nodes in the graph. If an arc (also called edge or path) exists, the adjacency matrix reads a 1 in that cell and a 0 otherwise.

We can use the strongest cross-relationships to illustrate our analysis. We assign a 1 to those cells in which a direct effect has been recorded, and a 0 if not. Two things

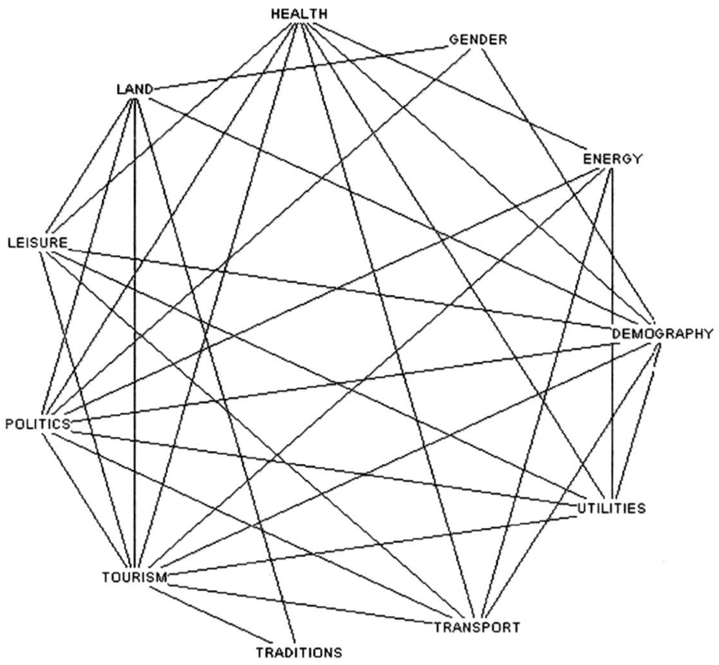

Fig. 7.4 The graph rendering of most of the anthropic elements matrix in Fig. 6.1

are clear: this matrix would generate too messy a graph to be practical and a by-hand analysis of *n-degree* relationships using that graph would be highly difficult.

The huge number of relationships to analyse justifies some of the comments on the "infiniteness" of the synthesis task and the need of an apprehension, more than an analysis, of the rural areas. However, this approach is misguided. No matter how cumbersome the task could be, it is not infinite. The binary relations are limited to (at most) 600, and many of them can soon be discarded for a given area. The number of *n-degree* relationships is also high, as we will see, but the analysis can be done. In any case, this analysis is inescapable and avoid it is indicative of a lazy approach to the work.

In what follows, we will use a sub-matrix of G (a matrix of G in which adjacent rows and columns have been removed). It is clear that both the environmental and the anthropic element matrices A and B are submatrices of G. Let us use B for simplicity to illustrate some features of this matrix, which are applicable to G with all generality.

A graph in which the edges are orientated is called a 'digraph' (for directed graph). The matrices we are dealing with can be thought as adjacency matrices for digraphs.

An advantage of modelling the rural processes in graph language theory is that we can now apply a battery of theorems and formulas to our problem. For example, if we want to know the number of n-degree relationships between elements we find that that is readily done by multiplying n-times the adjacency matrix.

Matrix operations can be done with ad hoc programs using mathematical libraries. This requires basic mathematical training and programming skills. An alternative is to use mathematical packages such as Maple of Matlab to do the operations in a more visual way. What follows is the Maple code to find out the number of n-degrees relationships in the anthropic elements sub-matrix. We know from graph theory that if A is the adjacency matrix for a digraph, the cell $a^n_{i,j}$ of A^n is the number of cycles of length n from j to i:

#Restart

```
restart:
with(networks):
with(linalg):
```

#Define the graph B for a rural area without industrial activities and external inputs in form of tourism

```
new(B):
```

#Build up the graph defining vertices and arcs. This matrix depends on the rural area being considered

```
addvertex([LAND, DEMOGRAPHY, LEISURE, UTILITIES, TRANSPORT,
ENERGY, GENDER, HEALTH, POLITICS, TRADITIONS, TOURISM],B):

connect(LAND, DEMOGRAPHY,'weights'=1,‘directed’,B):
connect(LAND, GENDER,‘weights’=1,‘directed’,B):
connect(LAND, POLITICS,‘weights’=1,‘directed’,B):
connect(LAND, TRADITIONS,‘weights’=1,‘directed’,B):

connect(DEMOGRAPHY, LAND,‘weights’=1,‘directed’,B):
connect(DEMOGRAPHY, TRANSPORT,‘weights‘=1,‘directed’,B):
connect(DEMOGRAPHY, GENDER,‘weights’=1,‘directed’,B):
connect(DEMOGRAPHY, POLITICS,‘weights’=1,‘directed’,B):
connect(DEMOGRAPHY, TOURISM,‘weights’=1,‘directed’,B):

connect(LEISURE, LAND,‘weights’=1,‘directed’,B):
connect(LEISURE, DEMOGRAPHY,‘weights’=1,‘directed’,B):
connect(LEISURE, TOURISM,‘weights’=1,‘directed’,B):
```

```
connect(UTILITIES, DEMOGRAPHY,'weights'=1,'directed',B):
connect(UTILITIES, LEISURE,'weights'=1,'directed',B):
connect(UTILITIES, ENERGY,'weights'=1,'directed',B):
connect(UTILITIES, TOURISM,'weights'=1,'directed',B):

connect(TRANSPORT, DEMOGRAPHY,'weights'=1,'directed',B):
connect(TRANSPORT, LEISURE,'weights'=1,'directed',B):
connect(TRANSPORT, ENERGY,'weights'=1,'directed',B):
connect(TRANSPORT, HEALTH,'weights'=1,'directed',B):
connect(TRANSPORT, POLITICS,'weights'=1,'directed',B):
connect(TRANSPORT, TOURISM,'weights'=1,'directed',B):

connect(ENERGY, UTILITIES,'weights'=1,'directed',B):
connect(ENERGY, HEALTH,'weights'=1,'directed',B):

connect(GENDER, POLITICS,'weights'=1,'directed',B):

connect(HEALTH, DEMOGRAPHY,'weights'=1,'directed',B):
connect(HEALTH, LEISURE,'weights'=1,'directed',B):
connect(HEALTH, UTILITIES,'weights'=1,'directed',B):
connect(HEALTH, POLITICS,'weights'=1,'directed',B):
connect(HEALTH, TOURISM,'weights'=1,'directed',B):

connect(POLITICS, LAND,'weights'=1,'directed',B):
connect(POLITICS, DEMOGRAPHY,'weights'=1,'directed',B):
connect(POLITICS, UTILITIES,'weights'=1,'directed',B):
connect(POLITICS, TRANSPORT,'weights'=1,'directed',B):
connect(POLITICS, ENERGY,'weights'=1,'directed',B):
connect(POLITICS, HEALTH,'weights'=1,'directed',B):
connect(POLITICS, TOURISM,'weights'=1,'directed',B):

connect(TRADITIONS, TOURISM,'weights'=1,'directed',B):

connect(TOURISM, LAND,'weights'=1,'directed',B):
connect(TOURISM, DEMOGRAPHY,'weights'=1,'directed',B):
connect(TOURISM, LEISURE,'weights'=1,'directed',B):
connect(TOURISM, UTILITIES,'weights'=1,'directed',B):
connect(TOURISM, TRANSPORT,'weights'=1,'directed',B):
connect(TOURISM, ENERGY,'weights'=1,'directed',B):
connect(TOURISM, POLITICS,'weights'=1,'directed',B):
```

#This generates Fig.~7.3

```
draw(B);
```

#This calculates the adjacency matrix (which gives the 2-degree relationships)

```
B_adj:=adjacency(B);
```

$$B\_adj = \begin{bmatrix} 0 & 1 & 0 & 0 & 0 & 0 & 1 & 0 & 1 & 1 & 0 \\ 1 & 0 & 0 & 0 & 1 & 0 & 1 & 0 & 1 & 0 & 1 \\ 1 & 1 & 0 & 0 & 0 & 0 & 0 & 0 & 0 & 0 & 1 \\ 0 & 1 & 1 & 0 & 0 & 1 & 0 & 0 & 0 & 0 & 1 \\ 0 & 1 & 1 & 0 & 0 & 1 & 0 & 1 & 1 & 0 & 1 \\ 0 & 0 & 0 & 1 & 0 & 0 & 0 & 1 & 0 & 0 & 0 \\ 0 & 0 & 0 & 0 & 0 & 0 & 0 & 0 & 1 & 0 & 0 \\ 0 & 1 & 1 & 1 & 0 & 0 & 0 & 0 & 1 & 0 & 1 \\ 1 & 1 & 0 & 1 & 1 & 1 & 0 & 1 & 0 & 0 & 1 \\ 0 & 0 & 0 & 0 & 0 & 0 & 0 & 0 & 0 & 0 & 1 \\ 1 & 1 & 1 & 1 & 1 & 1 & 0 & 0 & 1 & 0 & 0 \end{bmatrix}$$

#To calculate the 3-degree relationships, simply multiply the adjacency matrix

```
deg_3:=multiply(B_adj,B_adj);
```

$$def\_3 = \begin{bmatrix} 2 & 1 & 0 & 1 & 2 & 1 & 1 & 1 & 2 & 0 & 3 \\ 2 & 4 & 2 & 2 & 2 & 3 & 1 & 2 & 4 & 1 & 2 \\ 2 & 2 & 1 & 1 & 2 & 1 & 2 & 0 & 3 & 1 & 1 \\ 3 & 2 & 1 & 2 & 2 & 1 & 1 & 1 & 2 & 0 & 2 \\ 4 & 4 & 2 & 4 & 3 & 2 & 1 & 2 & 3 & 0 & 4 \\ 0 & 2 & 2 & 1 & 0 & 1 & 0 & 0 & 1 & 0 & 2 \\ 1 & 1 & 0 & 1 & 1 & 1 & 0 & 1 & 0 & 0 & 1 \\ 4 & 4 & 2 & 2 & 3 & 3 & 1 & 1 & 2 & 0 & 4 \\ 2 & 5 & 4 & 3 & 2 & 3 & 2 & 2 & 5 & 1 & 4 \\ 1 & 1 & 1 & 1 & 1 & 1 & 0 & 0 & 1 & 0 & 0 \\ 3 & 5 & 2 & 2 & 2 & 3 & 2 & 3 & 3 & 1 & 5 \end{bmatrix}$$

#And the same for next levels, 3-degree relationships

```
deg_4:=multiply(B_adj,deg_3);
```

$$def\_3 = \begin{bmatrix} 6 & 11 & 7 & 7 & 6 & 8 & 3 & 5 & 10 & 2 & 7 \\ 12 & 16 & 8 & 11 & 10 & 10 & 6 & 9 & 13 & 2 & 17 \\ 7 & 10 & 4 & 5 & 6 & 7 & 4 & 6 & 9 & 2 & 10 \\ 7 & 13 & 7 & 6 & 6 & 8 & 5 & 5 & 11 & 3 & 10 \\ 13 & 22 & 13 & 11 & 11 & 14 & 8 & 8 & 18 & 4 & 18 \\ 7 & 6 & 3 & 4 & 5 & 4 & 2 & 2 & 4 & 0 & 6 \\ 2 & 5 & 4 & 3 & 2 & 3 & 2 & 2 & 5 & 1 & 4 \\ 12 & 18 & 10 & 10 & 10 & 11 & 8 & 8 & 17 & 4 & 14 \\ 18 & 22 & 11 & 14 & 14 & 14 & 7 & 10 & 17 & 2 & 22 \\ 3 & 5 & 2 & 2 & 2 & 3 & 2 & 3 & 3 & 1 & 5 \\ 15 & 20 & 12 & 14 & 13 & 12 & 8 & 8 & 20 & 3 & 18 \end{bmatrix}$$

The result gives us the number of closed paths (cycles or circuits) of length n from node i to node j. Therefore, the main diagonal of the matrix gives us the number of n-degree relationships to analyze. This easy exercise is required to avoid missing any relationship.

The same calculation has to be done to the whole matrix in Table 7.3.

The difference between this comprehensive method and a more qualitative one is clear: this method ensures that all the relevant aspects are to be considered in the analysis. This down-top approach also has the advantage of improving the knowledge on how the rural area works, since it implies dedicating some attention to all the cross-relationships. In this process, new processes are usually found, and the matrix then refined until a stable form is agreed. This may involve several feedback cycles with experts, the realization of new measurements or even a complete rethinking of the rural area in light of new information.

7.3.3 Numerical Modelling

Once the cycles have been found, the next step is to attempt modelling the system as a whole. This approach is known as 'system dynamics' and it is a powerful mean of not only to analyze but also manage a rural area. The idea is to simulate the behaviour and the dynamics of the rural area in purely quantitative terms.

Several software tools have been developed to help in this task. A graphical interface can be of use since a rural model soon becomes very intricate. Nonetheless, the primitive equations of the models are simple enough to be directly implemented by an expert geographer.

An example would clarify the method to follow and how to model. Let us take a hypothetical rural area of 4,000 inhabitants in which the analysis (for the sake of the argument) has identified just one relevant relationship. Namely, we have found that the relationship between tourism and population is the only relevant factor. This is indeed an oversimplified case, but a real case would be too large to exemplify the process here.

Figure 7.5 shows a possible modelling of this single relationship taking into account what can actually be measured or parameterized in the area, and the factors and characteristics that were identified in the analysis phase for those two anthropic elements. This figure is the visual part of the numerical model and serves to illustrate the relationships involved. A real model is comprised of many variables and relationships in form of non-linear difference or differential equations. A critical point, once the model has been defined, is to make sure that all the equations are dimensionally consistent.

What we want to understand with this model is how tourism affects rural life, and how an orientated tourist marketing policy can help to minimize the negative effects on the rural area.

The dynamics of the models are as follows: the number of tourists arriving depends on proactive tourism policies (see below) and on a natural attraction exerted by the village to neighbouring areas. We assume that this attraction follows

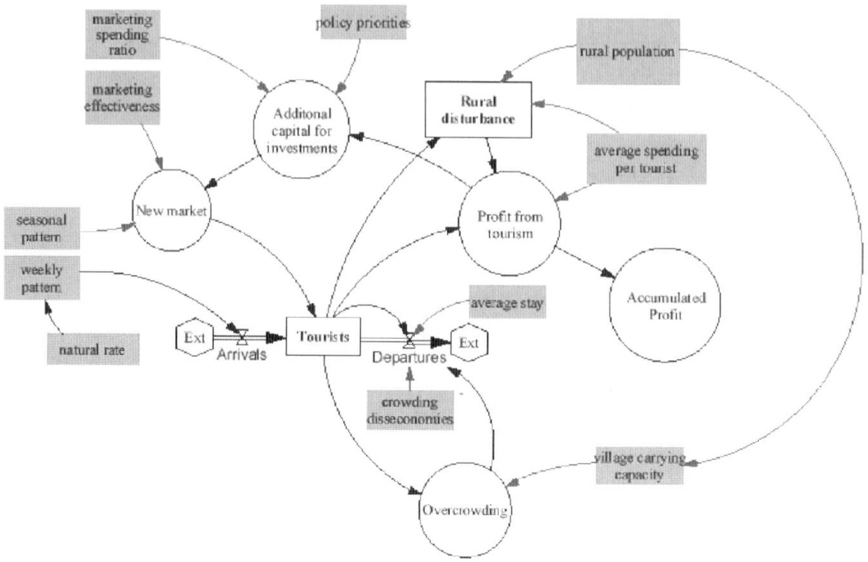

Fig. 7.5 An example of rural tourism-rural population relationship modelling

a weekly pattern: from Friday to Sunday the number of tourist increases, with far less tourists visiting the village in week days. The actual shape of this pattern can be either modelled using a sinusoidal function or be calculated using existing data.

If no actual data is available, this pattern can be parameterized as the combination of a sinusoid with a random Gaussian function around an average value of 100. The result is illustrated in Fig. 7.6.

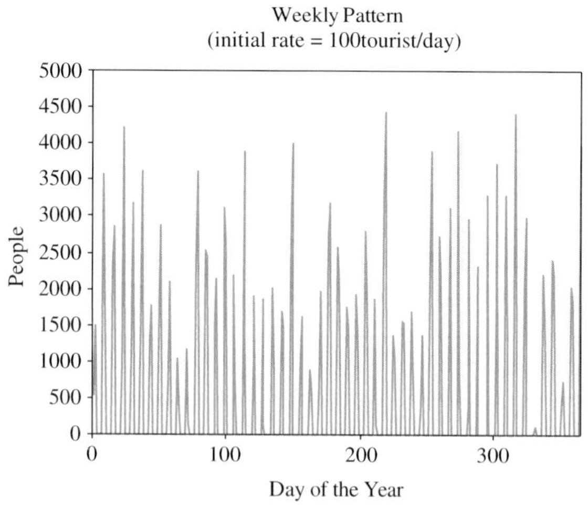

Fig. 7.6 The weekly pattern of visitors without tourism policies in place. Peak indicate the week ends

This initial human capital spends some time in the village. The average stay of the tourist depends on several factors, so it is considered a variable of the system spanning from 0 days (passing tourist) to an arbitrary limit of one month. We have assumed a value of 1.5 days for the simulation. During the time the tourist is in the area he spends money. The pattern of this spending depends on the kind of tourist attracted and on the offer. Here, we have used a seasonal function combined with a random Gaussian to make the values pivot around a realistic 40 euros per person (Fig. 7.7). Once again, if we had this actual data from previous years they could be integrated into the model.

The increase in the number of tourist also increases the overcrowding effect (Fig. 7.8). For some areas, too many tourists make other tourists to shorten their stay or decrease the appeal of the village (Fig. 7.9). The number of tourists a rural area can absorb depends on the population. Thus, we consider that the village has a carrying capacity of a tenth of the rural population. The accumulative overcrowding effect can be parameterized using an exponential function.

The actual profit for the area (Fig. 7.10) comes from the money spent by the tourists (Fig. 7.7) minus the cost of the disturbance these tourists bring to the area.

This new variable, the disturbance, can be quantified as the relationship between the difference between the number of tourists and locals and the expected spending per tourist.

If the tourists spend below an expected value of 70 euros (a value that would compensate the annoyance of hosting the newcomers), and the number of tourists is below the local population, the disturbance is positive: it negatively affects the economy.

The greater the number of tourists, the greater the disturbance and thus the chance of a potential diseconomy. The accumulated profit, that is, the integral of the profit, gives a more accurate idea of the overall picture (Fig. 7.11).

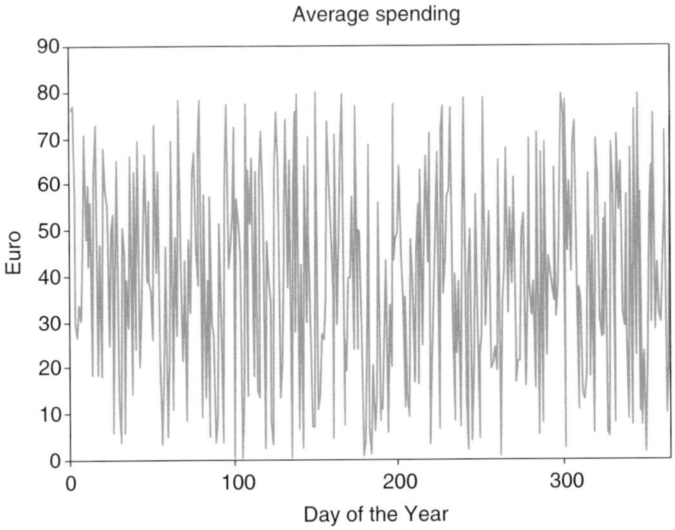

Fig. 7.7 Average spending of the tourists

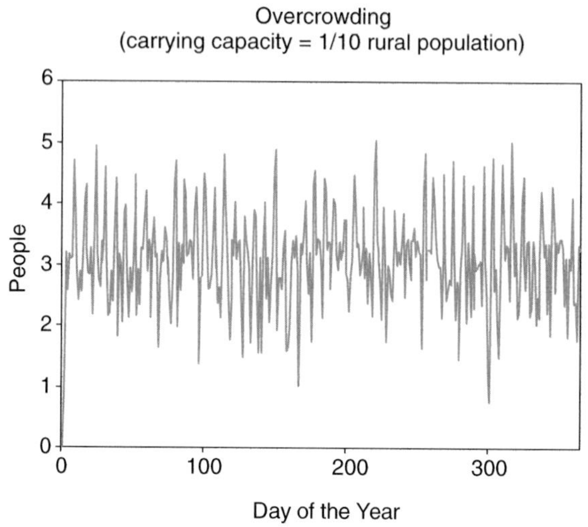

Fig. 7.8 Overcrowding index. Notice the zero value at the beginning of the simulation period

On the other hand, if the tourists spend a reasonable amount of money the benefits for all the population (in form of business revenues and tax returns) compensate for the trouble. Other situations include when the tourists spend at a reasonably rate but they are too many of them compared with the local population. In this case, the disturbance is negative (Fig. 7.12). But if the number of tourists is low, even if they

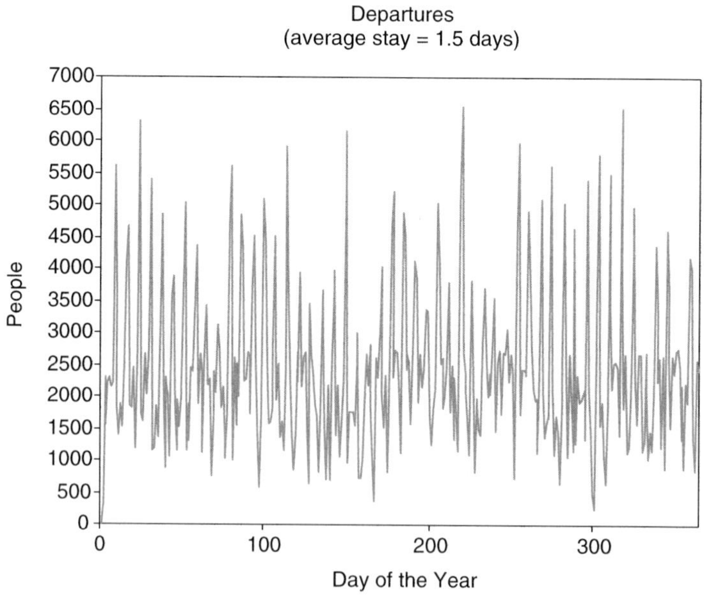

Fig. 7.9 Tourist departures in the area

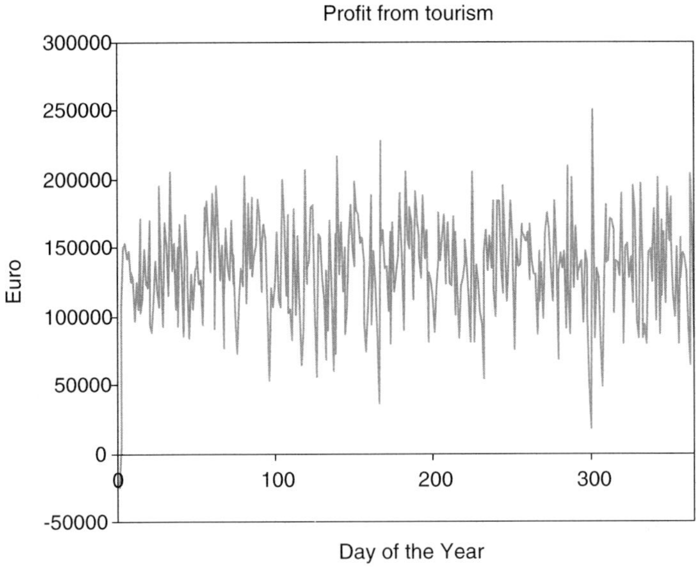

Fig. 7.10 Evolution of profit from tourism in the area

spend little, the disturbance has to be low (that is, a few people cause not much disturbance). This reasoning can be easy included as an equation into the model.

The policy part of the model assumes that part of the money that tourists bring to the area can be used in an attempt to attract more tourists (Fig. 7.13). Providing the

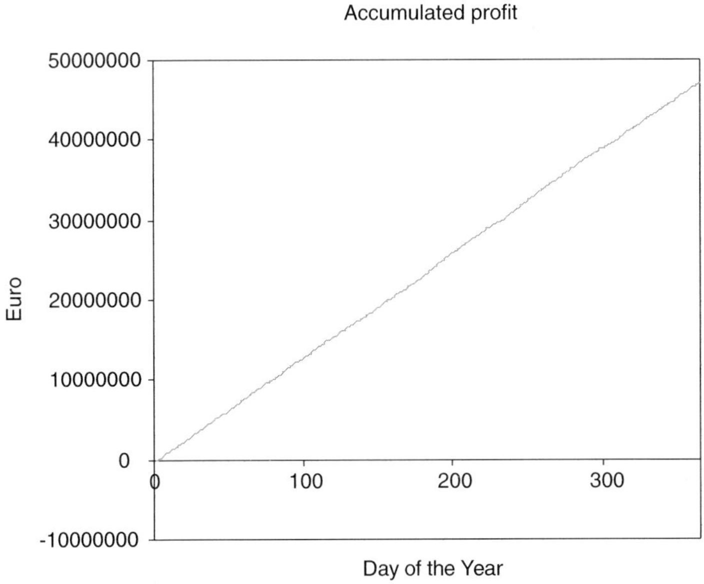

Fig. 7.11 Accumulated (integral) benefit from tourism for the rural area, in a year

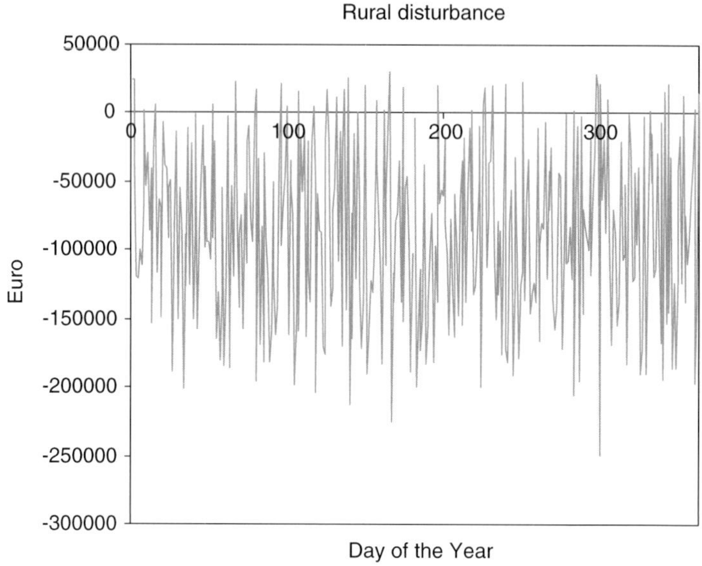

Fig. 7.12 Rural disturbance. Negative values indicate a benefit for the area, while positive values indicate local nuissance

profit is positive, the capital available for marketing depends on the policy priorities and on how much money is to be directly devoted to attract more tourists (marketing spending ratio in the scheme).

A sensible policy would try to compensate the seasonal patterns of the new market (Fig. 7.14), which is a population with different characteristics than the regular tourist market that visit the area, with targeted and timely marketing campaigns (Fig. 7.15).

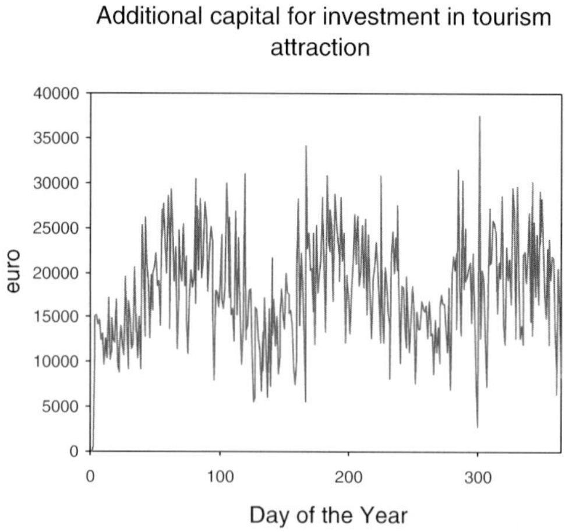

Fig. 7.13 Additional capital coming from existing tourism used for tourism attraction policies

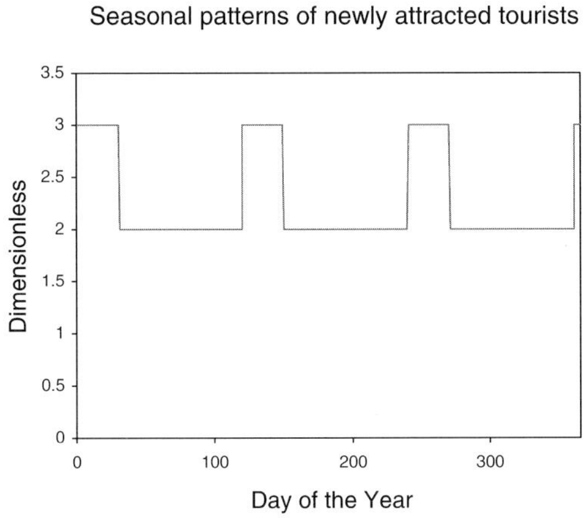

Fig. 7.14 Seasonal travelling behaviour of the target market

A coupled scheme such as this aims to share the tourist influx along the year, but this is not always the case. Policies in mountain areas may prefer to compensate a summer preference of the potential market with a strong winter campaign. The result of the policies depends on the marketing effectiveness, which can also be fine tuned for the area. This creates a new influx of tourists (Fig. 7.16) contributing to the economy of the area.

The possible modifications to this simple model are endless. The average spending per tourist may be modified if a marketing policy to attract more High Spending People (HSP) is put into practice. The disturbance can be weighted with the population pyramid, since not all the ages are equally affected by tourism. The marketing policy might be less dependent on the revenues from tourists, or be accumulated, etc. All depends on the knowledge of the area acquired in the analysis phase and

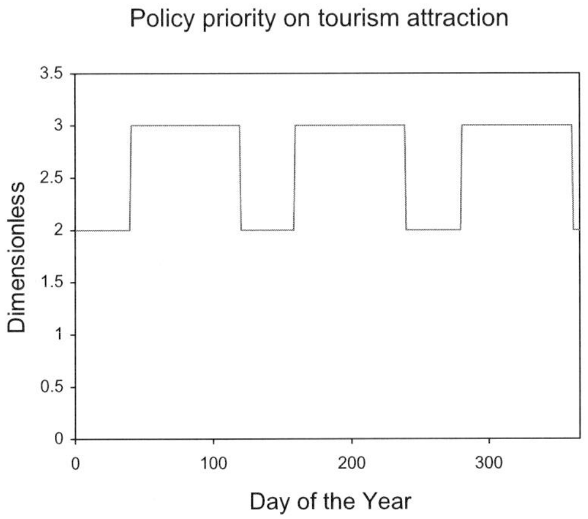

Fig. 7.15 Pattern of the intensity on tourism attraction policies

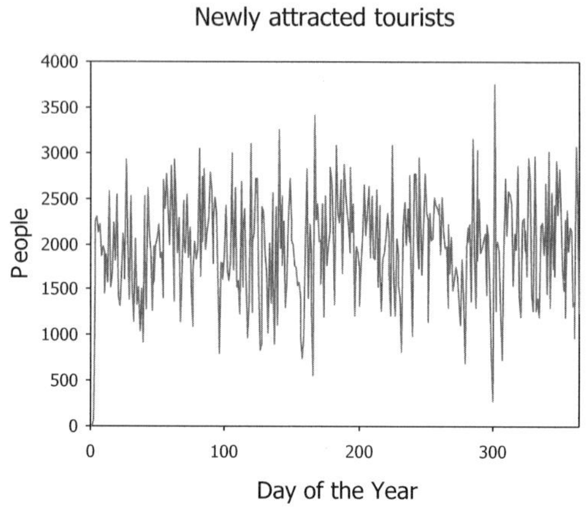

Fig. 7.16 Result of the marketing policies in terms of newly attracted tourist to the area

in the ability of the team to encapsulate the area into a model and on the particular case. At the end, however, the reality is the ultimate test, and the model has to be at least plausible. If the results are not those expected, it may be due to a missing effect, a deficient parameterization or a bug in the equations.

Models as simple as the one in the example may be part of a more comprehensive model that include environmental elements effects of environmental elements or even the complete set of relations. A modular approach, in which independent units are linked together to generate a bigger structure is always advisable. This will help to debug problems and to find out which parts are working properly and which are not.

Additional studies with the model include sensitivity analysis (how a variable changes when another is modified), real time monitoring coupled with expert systems, or planning using several scenarios. In the next section we will explore this last possibility.

7.3.4 Scenario Definition

Until now, all we have done is set up a reasonable model on how the rural area might work. This is relevant by itself, but the main interest of modelling is the ability of simulating scenarios. What would happen if, for example, the marketing spending ratio is increased? How then would the rural disturbance evolve? Would that increase the overall benefit of the village?

Using a discursive argument, we could say that an increase in the marketing spending ratio would certainly increase the number of visitors. But this increase may overcrowd the area, increasing the rate of departures. Also, a large number of visitors affect rural life. The model could be fine tuned to include differences in the spending habits between the traditional tourists and those attracted by the new policies. But assessing the degree of such effect is, without a proper model, a matter of pure speculation.

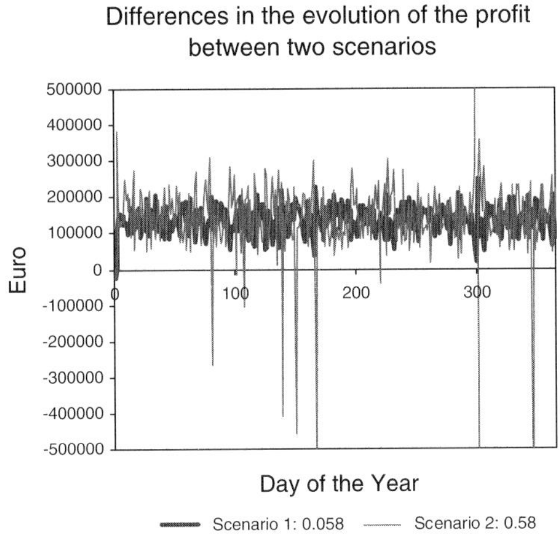

Fig. 7.17 Differences in the evolution of the profit in function of the marketing spending. Values below zero indicates a momentary loss

We can easily simulate this scenario with the model. If the dynamics are correct, and the relevant relationships are embedded in the model, the equations would yield an answer. For this case, Figs. 7.15 and 7.16 show the differences. The new increased spending scenario generates very different dynamics (Fig. 7.15) indicating the non-linearity of the processes involved. The variability is high, and it is difficult to ascertain if the large losses in some days are somehow compensated in the long run. Fig. 7.16 gives us a positive answer to this question. The new scenario clearly increases the benefits (Figs. 7.17, 7.18).

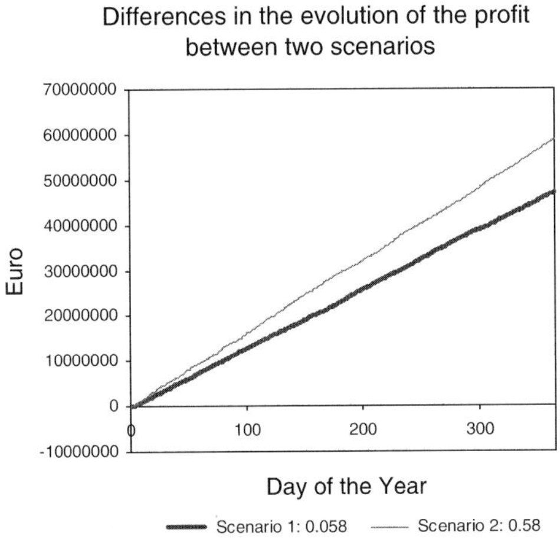

Fig. 7.18 Accumulated profit difference for the two marketing scenarios. The spending ratio of 0.58 outperforms the 0.058 scenario in the long run, even when fluctuations appear on a daily basis. This difference would be hard to be obtained without a model

Other possible scenarios are increasing the average stay of the tourists (because of improved hotels or new guest houses), the effectiveness of the marketing (using improved strategies), or adjusting the policy factors to find an optimum. Optimization packages can assist this task.

7.4 The Synthesis Report

The synthesis report is the material form of a rural analysis project. It is the final outcome of a project (task 27 in Table 1.1) and may be the starting point of a planning

Table 7.4 Minimum contents of the synthesis report

1.	Aim and description of the project	
	a.	How commissioned the analysis
	b.	Aims of the analysis
	c.	Projected budget and final costs
	d.	Timetable and schedule, both projected and final
	e.	Team composition, work breakdown and responsibilities
	f.	Notes from the team coordinator
	g.	Notes from the team members
2.	Location map of the area in the region and country	
	a.	Detailed local maps used in the analyses
	b.	Situation map
	c.	Basic indicators
3.	Measurement methods used	
	a.	Measurement protocols
	b.	Methods
4.	Database description and access directions	
	a.	GIS description
5.	Environmental elements description and maps	
	a.	(one epigraph for element)
6.	Anthropic elements description and maps	
	a.	(one epigraph for element)
7.	Geographical modelling	
	a.	Model diagram
	b.	Geographical model description. Rationale of the elements
	c.	Parameterizations used
	d.	Scenario definition
8.	Synthesis	
	a.	Generalities
	b.	Landscapes analysis
	c.	Landscape report
	d.	Results of the modelling
	e.	Notes for development and management
9.	Final notes	
ANNEXES:		
	a.	Public information documentation
	b.	Additional resources

or management strategy. A complete analysis report might contain at least the items shown in Table 7.4.

Documentation must include all the databases in digital form, the originals of the enquiries and field work notes, the maps used, etc. Eeach and every aspect of these items has to be considered. The aim of the analysis is to provide all the relevant information for the reader.

A printed report is almost always required. If the study has been done in digital form the report can also be delivered as a web page. Even if the information is sensitive, or reserved, there are ways in which the web can be protected. This format helps to disseminate the work, saves money (paper, ink, etc.), permits easy modifications and additions, and allows interaction from people and rural agents.

The printed version of the report can be considered as the original reference, but new editions of the report can routinely be published from time to time including feedbacks from users, additional information or updates. It is advisable that the analysis team is in charge of the site to help to maintain the structural integrity of the information. This can easily be done by including a small maintenance fee in the budget. This is obviously indicated for developed countries where IT communications are generalized.

References and Bibliography Used

D. Alard and I. Poudevigne, 'Factors controlling plant diversity in a rural landscape: a functional approach', Landscape and Urban Planning 46 (1999), 29.

K. Appleton and A. Lovett, 'GIS-based visualisation of rural landscapes: defining "sufficient" realism for environmental decision-making', Landscape and Urban Planning 65 (2003), 117–131.

K. Appleton, A. Lovett, G. Sunnenberg and T. Dockerty, 'Rural landscape visualisation from GIS databases: a comparison of approaches, options and problems', Computers, Environment and Urban Systems 26 (2002), 141.

M. Arriaza, J. F. Canas-Ortega, J. A. Canas-Madueno and P. Ruiz-Aviles, 'Assessing the visual quality of rural landscapes', Landscape and Urban Planning 69 (2004), 115–125.

J. P. Brans, C. Macharis, P. L. Kunsch, A. Chevalier and M. Schwaninger, 'Combining multicriteria decision aid and system dynamics for the control of socio-economic processes. An iterative real-time procedure', European Journal of Operational Research 109 (1998), 428–441.

R. Brush, R. E. Chenoweth and T. Barman, 'Group differences in the enjoyability of driving through rural landscapes', Landscape and Urban Planning 47 (2000), 39.

F. Burel and J. Baudry, 'Social, aesthetic and ecological aspects of hedgerows in rural landscapes as a framework for greenways', Landscape and Urban Planning 33 (1995), 327.

P. Claval, 'Reading the rural landscapes', Landscape and Urban Planning 70 (2005), 9–19.

D. D'haeze, J. Deckers, D. Raes, T. A. Phong and H. V. Loi, 'Environmental and socio-economic impacts of institutional reforms on the agricultural sector of Vietnam: Land suitability assessment for Robusta coffee in the Dak Gan region', Agriculture, Ecosystems & Environment 105 (2005), 59–76.

D. L. Erickson, R. L. Ryan and R. De Young, 'Woodlots in the rural landscape: landowner motivations and management attitudes in a Michigan (USA) case study', Landscape and Urban Planning 58 (2002), 101.

J. M. Fedorowick, 'A landscape restoration framework for wildlife and agriculture in the rural landscape', Landscape and Urban Planning 27 (1993), 7.

Y. N. Fu, H. J. Guo, A. G. Chen and J. Y. Cui, 'Fallow agroecosystem dynamics and socioeconomic development in China: Two case studies in Xishuangbanna Prefecture, Yunnan Province', Mountain Research And Development 25 (2005), 365–371.

R.D. Brown, I. Washitani, A. Tsunekawa, M. Yokohari (Eds.), Satoyama. The Traditional Rural Landscape of Japan, Springer, Berlin, Heidelberg, New York, 2003, 229 pp., ISBN 4-431-00007-0 (85 figures)',

Z. W. Guo, X. M. Xiao, Y. L. Gan and Y. J. Zheng, 'Landscape planning for a rural ecosystem: case study of a resettlement area for residents from land submerged by the Three Gorges Reservoir, China', Landscape Ecology 18 (2003), 503–512.

D. P. Hanley, C. L. Mason and J. B. McCarter, 'Landscape management system: Bringing training to rural forest managers', Journal Of Forestry 100 (2002), 5–5.

T. Herben, Z. Munzbergova, M. Milden, J. Ehrlen, S. A. O. Cousins and O. Eriksson, 'Long-term spatial dynamics of Succisa pratensis in a changing rural landscape: linking dynamical modelling with historical maps', Journal Of Ecology 94 (2006), 131–143.

S. K. Hong, N. Nakagoshi and M. Kamada, 'Human Impacts On Pine-Dominated Vegetation In Rural Landscapes In Korea And Western Japan', Vegetatio 116 (1995), 161–172.

T. Hovardas and G. P. Stamou, 'Structural and narrative reconstruction of rural residents' representations of "nature", "wildlife", and "landscape" ', Biodiversity And Conservation 15 (2006), 1745–1770.

N. Iiyama, M. Kamada and N. Nakagoshi, 'Ecological and social evaluation of landscape in a rural area with terraced paddies in southwestern Japan', Landscape and Urban Planning 70 (2005), 301–313.

N. Iiyama, M. Kamada and N. Nakagoshi, 'Ecological and social evaluation of landscape in a rural area with terraced paddies in southwestern Japan (vol 70, pg 301, 2005)', Landscape and Urban Planning 73 (2005), 59–71.

Y. Kato, M. Yokohari and R. D. Brown, 'Integration and visualization of the ecological value of rural landscapes in maintaining the physical environment of Japan', Landscape and Urban Planning 39 (1997), 69–82.

J. J. Kennedy and N. E. Koch, 'Viewing and managing natural resources as human-ecosystem relationships', Forest Policy and Economics 6 (2004), 497–504.

P. Kunsch, A. Chevalier and J. P. Brans, 'Comparing the adaptive control methodology (ACM) to the financial planning practice of a large international group', European Journal of Operational Research 132 (2001), 479–489.

R. Lafortezza and R. D. Brown, 'A framework for landscape ecological design of new patches in the rural landscape', Environmental Management 34 (2004), 461–473.

J. A. LaGro, 'Landscape context of rural residential development in southeastern Wisconsin (USA)', Landscape Ecology 13 (1998), 65–77.

S. Lee, F. Pena-Mora and M. Park, 'Quality and change management model for large scale concurrent design and construction projects', Journal of Construction Engineering and Management-Asce 131 (2005), 890–902.

S. Lee, F. Pena-Mora and M. Park, 'Web-enabled system dynamics model for error and change management on concurrent design and construction projects', Journal of Computing in Civil Engineering 20 (2006), 290–300.

S. H. Lee, F. Pena-Mora and M. Park, 'Dynamic planning and control methodology for strategic and operational construction project management', Automation in Construction 15 (2006), 84–97.

J. A. Lynch and R. H. Gimblett, 'Perceptual values in the cultural landscape: A computer model for assessing and mapping perceived mystery in rural environments', Computers, Environment and Urban Systems 16 (1992), 453.

J. A. Lynch and R. H. Gimblett, 'Perceptual Values In The Cultural Landscape - A Computer-Model For Assessing And Mapping Perceived Mystery In Rural Environments', Computers Environment And Urban Systems 16 (1992), 453–471.

K. E. Maani and V. Maharaj, 'Links between systems thinking and complex decision making', System Dynamics Review 20 (2004), 21–48.

J. M. Makhzoumi, 'The changing role of rural landscapes: olive and carob multi-use tree plantations in the semiarid Mediterranean', Landscape and Urban Planning 37 (1997), 115.

M. R. Moss and L. S. Davis, 'Measurement of spatial change in the forest component of the rural landscape of southern Ontario', Applied Geography 14 (1994), 214.

T. Nagaike and T. Kamitani, 'Factors affecting changes in landscape diversity in rural areas of the Fagus crenata forest region of central Japan', Landscape and Urban Planning 43 (1999), 209–216.

Y. Natori, W. Fukui and M. Hikasa, 'Empowering nature conservation in Japanese rural areas: a planning strategy integrating visual and biological landscape perspectives', Landscape and Urban Planning 70 (2005), 315–324.

S. Nautiyal, R. K. Maikhuri, R. L. Semwal, K. S. Rao and K. G. Saxena, 'Agroforestry systems in the rural landscape - a case study in Garhwal Himalaya, India', Agroforestry Systems 41 (1998), 151–165.

S. Neal, 'Rural landscapes, representations and racism: examining multicultural citizenship and policy-making in the English countryside', Ethnic And Racial Studies 25 (2002), 442–461.

P. Olsson and C. Folke, 'Local ecological knowledge and institutional dynamics for ecosystem management: A study of Lake Racken Watershed, Sweden', Ecosystems 4 (2001), 85–104.

H. Palang, S. Helmfrid, M. Antrop and H. Alumae, 'Rural Landscapes: past processes and future strategies', Landscape and Urban Planning 70 (2005), 3–8.

H. Palang, A. Printsmann, E. K. Gyuro, M. Urbanc, E. Skowronek and W. Woloszyn, 'The forgotten rural landscapes of Central and Eastern Europe', Landscape Ecology 21 (2006), 347–357.

S. Paquette and G. Domon, 'Changing ruralities, changing landscapes: exploring social recomposition using a multi-scale approach', Journal of Rural Studies 19 (2003), 425.

F. Pauwels and H. Gulinck, 'Changing minor rural road networks in relation to landscape sustainability and farming practices in West Europe', Agriculture, Ecosystems & Environment 77 (2000), 95.

T. Pinto-Correia, 'Future development in Portuguese rural areas: how to manage agricultural support for landscape conservation?' Landscape and Urban Planning 50 (2000), 95.

I. Poudevigne and D. Alard, 'Landscape and agricultural patterns in rural areas: a case study in the Brionne basin, Normandy, France', Journal Of Environmental Management 50 (1997), 335–349.

I. Poudevigne, S. vanRooij, P. Morin and D. Alard, 'Dynamics of rural landscapes and their main driving factors: A case study in the Seine Valley, Normandy, France', Landscape and Urban Planning 38 (1997), 93–103.

T. Purtauf, J. Dauber and V. Wolters, 'Carabid communities in the spatio-temporal mosaic of a rural landscape', Landscape and Urban Planning 67 (2004), 185–193.

P. Selman and N. Doar, 'An Investigation Of The Potential For Landscape Ecology To Act As A Basis For Rural Land-Use Plans', Journal of Environmental Management 35 (1992), 281–299.

E. Skowronek, R. Krukowska, A. Swieca and A. Tucki, 'The evolution of rural landscapes in mid-eastern Poland as exemplified by selected villages', Landscape and Urban Planning 70 (2005), 45–56.

L. Tahvanainen, L. Tyrvainen and I. Nousiainen, 'Effect of afforestation on the scenic value of rural landscape', Scandinavian Journal Of Forest Research 11 (1996), 397–405.

B. Tress, G. Tress and G. Fry, 'Integrative studies on rural landscapes: policy expectations and research practice', Landscape and Urban Planning 70 (2005), 177–191.

L. Tyrvainen and L. Tahvanainen, 'Using computer graphics for assessing the aesthetic value of large-scale rural landscapes', Scandinavian Journal Of Forest Research 14 (1999), 282–288.

J. H. P. van der Vaart, 'Towards a new rural landscape: consequences of non-agricultural re-use of redundant farm buildings in Friesland', Landscape and Urban Planning 70 (2005), 143–152.

I. Vogeler, 'State hegemony in transforming the in rural landscapes of eastern Germany: 1945-1994', Annals Of The Association Of American Geographers 86 (1996), 432–458.

K. Volker, 'Adapted Farming Systems For A Rural Landscape - A Social Typology Of Dutch Farmers', Sociologia Ruralis 32 (1992), 146–162.

K. Volker, 'Local commitment for sustainable rural landscape development', Agriculture, Ecosystems & Environment 63 (1997), 107.
Z. R. Yu, J. Baudry, B. P. Zhao, H. Zhang and S. Q. Li, 'Vegetation components of a subtropical rural landscape in China', Critical Reviews In Plant Sciences 18 (1999), 381–392.
Z. Z. Zhou, 'Landscape changes in a rural area in China', Landscape and Urban Planning 47 (2000), 33–38.

Chapter 8
A Cartographic Overview of the Agrarian World

This chapter comprises a set of maps aimed to situate the rural analysis in its appropriate agrarian framework. The maps are organized by geographical areas, and plot several variables of interest. They are built so they have the same grey scale for each variable, thus permitting a direct comparison between areas. Also, they all have 7 levels in a quantile interval. This means that each class contains the same number of countries. The reason to do it so is to avoid introducing logarithmic scales when countries present very different situations: if were to plot the population using an equal-interval scale, countries such as China or India would be distinguishable, but all the countries of the European Union would belong to the same interval. This is unsuitable for comparison purposes, and therefore we have preferred the less used equal-interval scale.

The data source of the maps is the FAOSTAT web that can be found at the http://faostat.fao.org/ URL. This database comprises many valuable variables, which have been estimated using contrasted methodologies being therefore suitable as a reference source to compare dissimilar areas. Here, we have selected the most relevant agrarian variables of the database for rural analysis. Many of the maps reflect percentual per capita changes from a baseline period. Thus, *index p.c. 1999/2001 in 2004* indicates a change in the variable in 2004 compared with the period 1999/2001, normalized by person to take into account the demographic variation, and in a 100 base –values below 100 indicate a decrease with respect to the 1999/2001 situation. These index-base maps are meant to be used to compare recent tendencies and not baseline situations.

The aim of the chapter is twofold. First, it is intended to help the analyst to frame the study area in its regional context, a topic that exceeds the limits of this book. The second aim is to provide an idea of the importance of the agrarian in different countries, something that has to be considered when proposing any development or management program. Some areas may require an in-deep analysis to solve fundamental issues such as fresh water availability, rural health care or simply food availability. In those cases, rural analysis might become highly biased towards an urgent development strategy in which all the elements of the analysis are to be considered. It may sound shocking that the analyst pays attention to urban leisure in the most deprived areas of Africa, but an open view may help to find creative solutions for chronic problems. Knowledge matters and the ability of finding synergisms between disparate elements can be more helpful than providing unrealistic solutions.

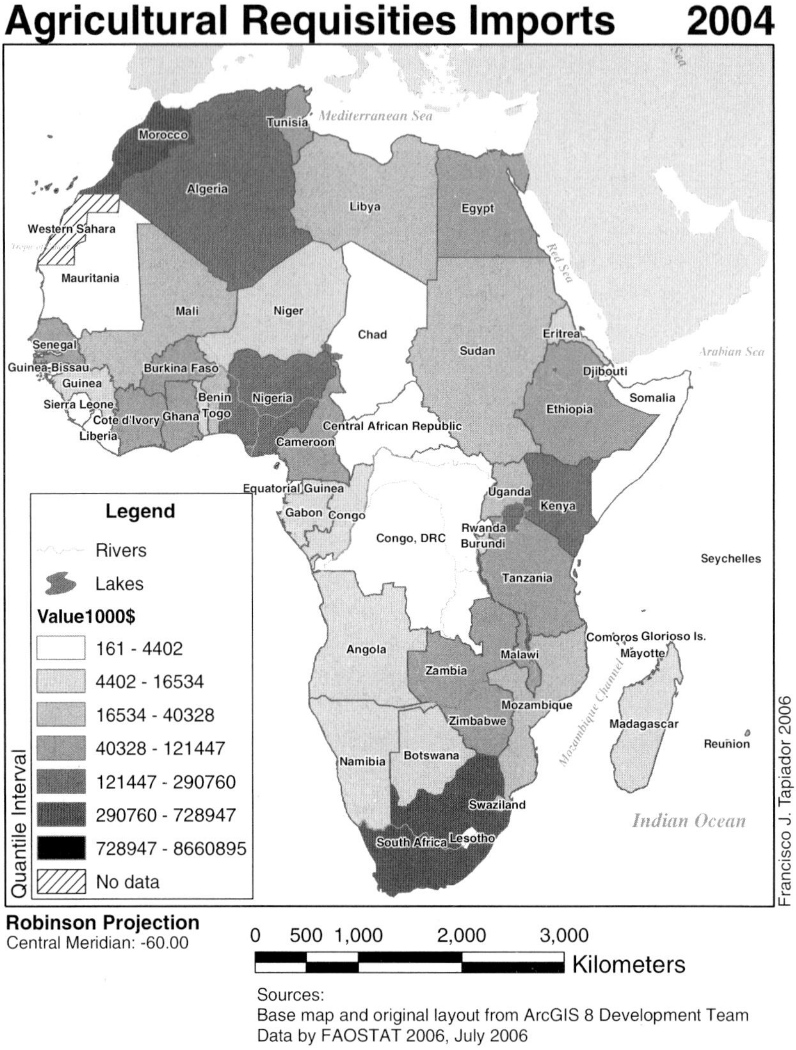

Fig. 8.1 Agricultural requisites imports in Africa in 2004

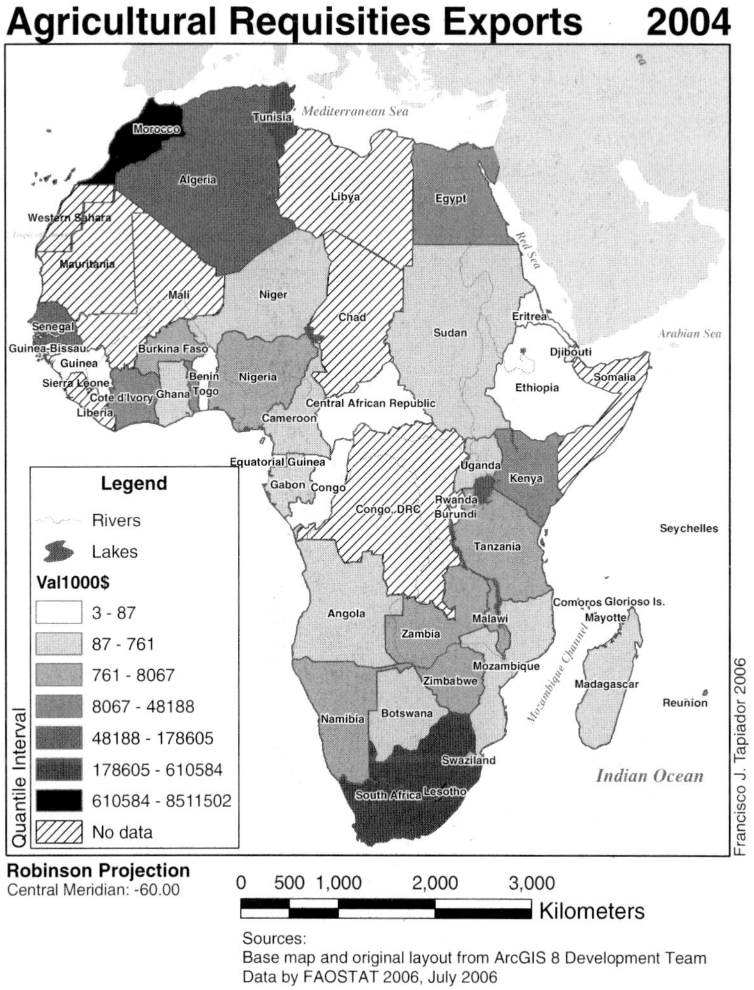

Fig. 8.2 Agricultural requisites exports in Africa in 2004

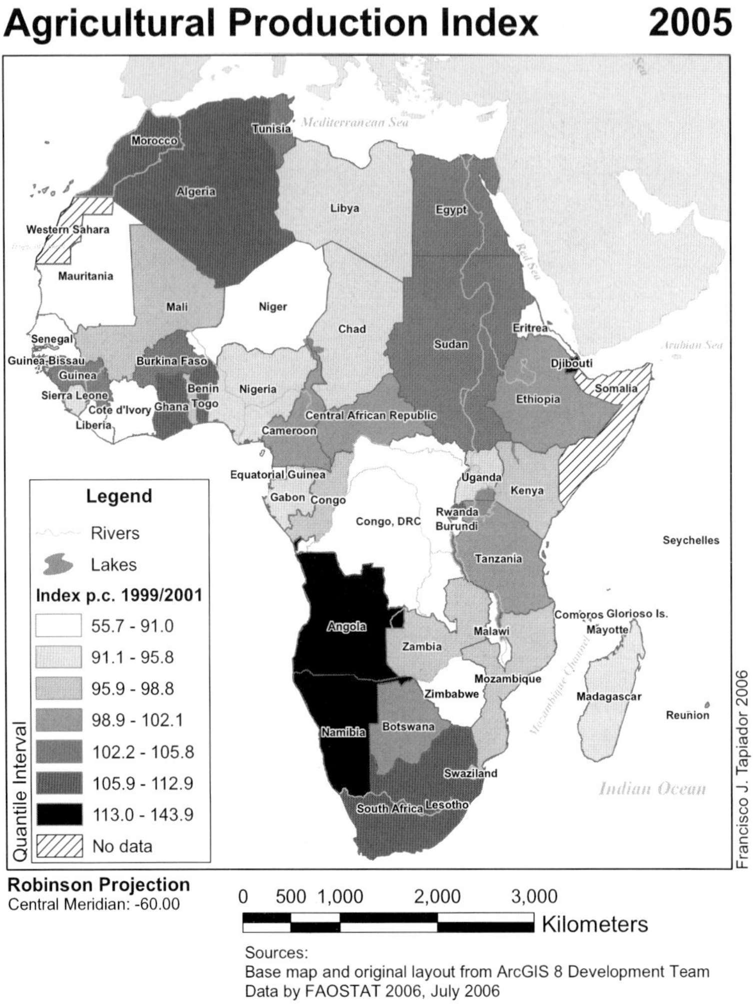

Fig. 8.3 Agricultural production index in Africa in 2005 compared with 1999/2001

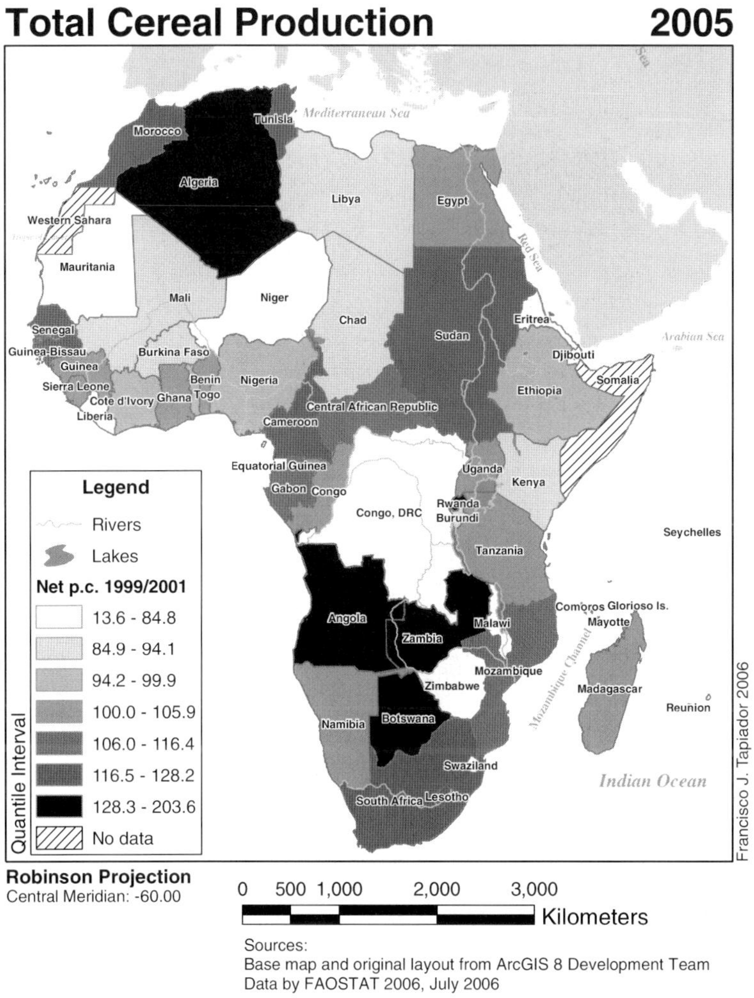

Fig. 8.4 Total cereal production index in Africa in 2005 compared with 1999/2001

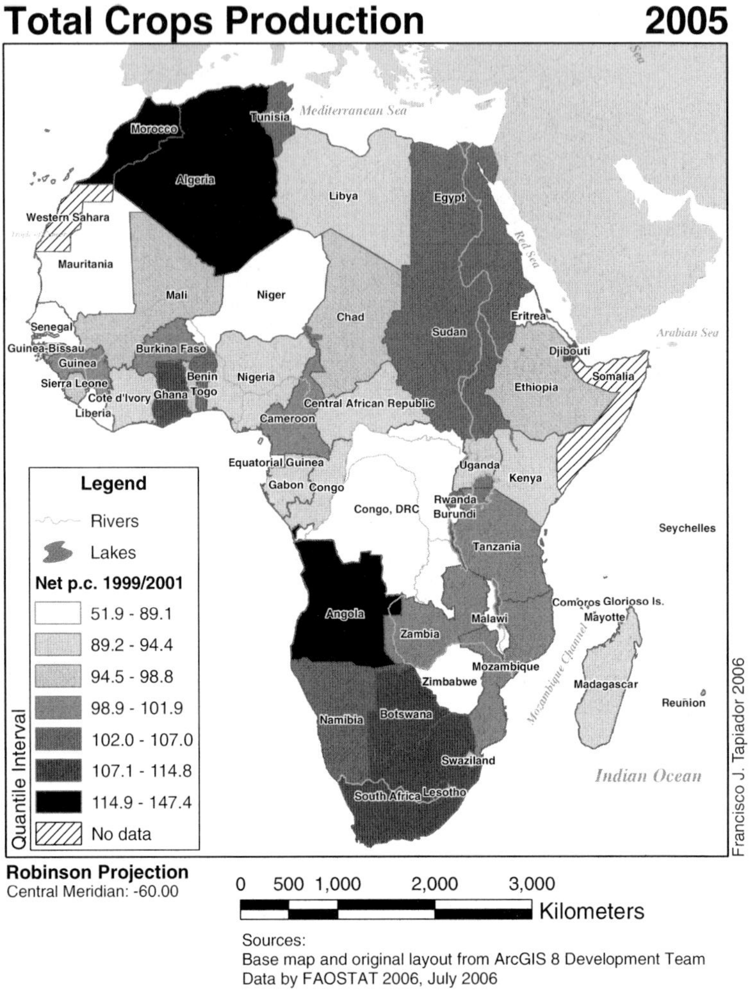

Fig. 8.5 Total crops production index in Africa in 2005 compared with 1999/2001

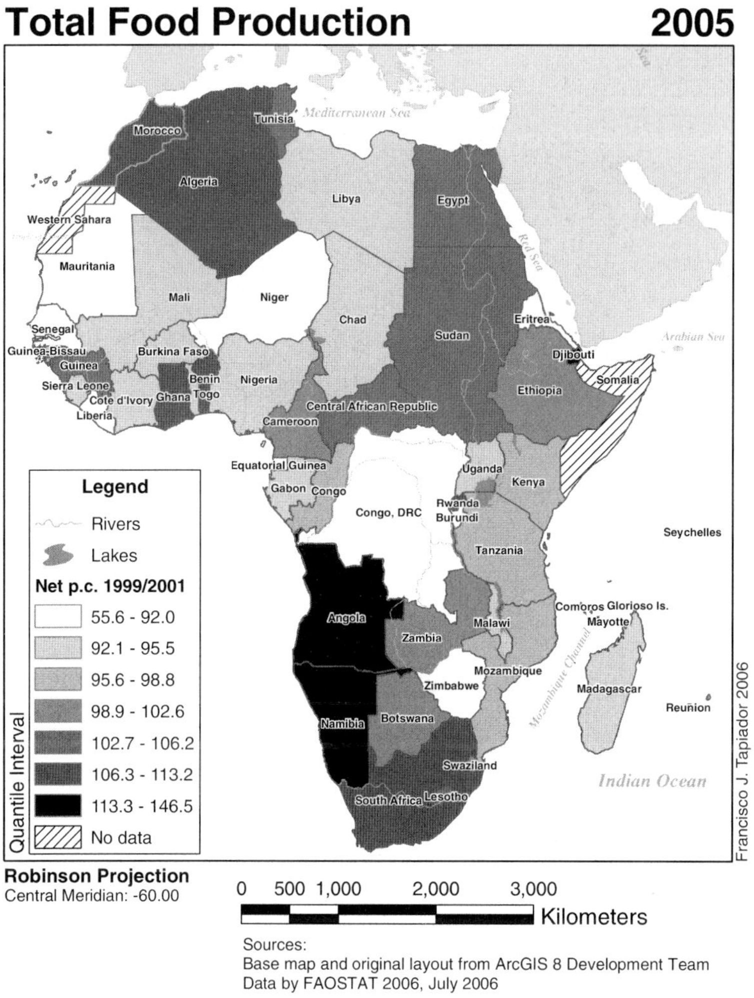

Fig. 8.6 Total food production index in Africa in 2005 compared with 1999/2001

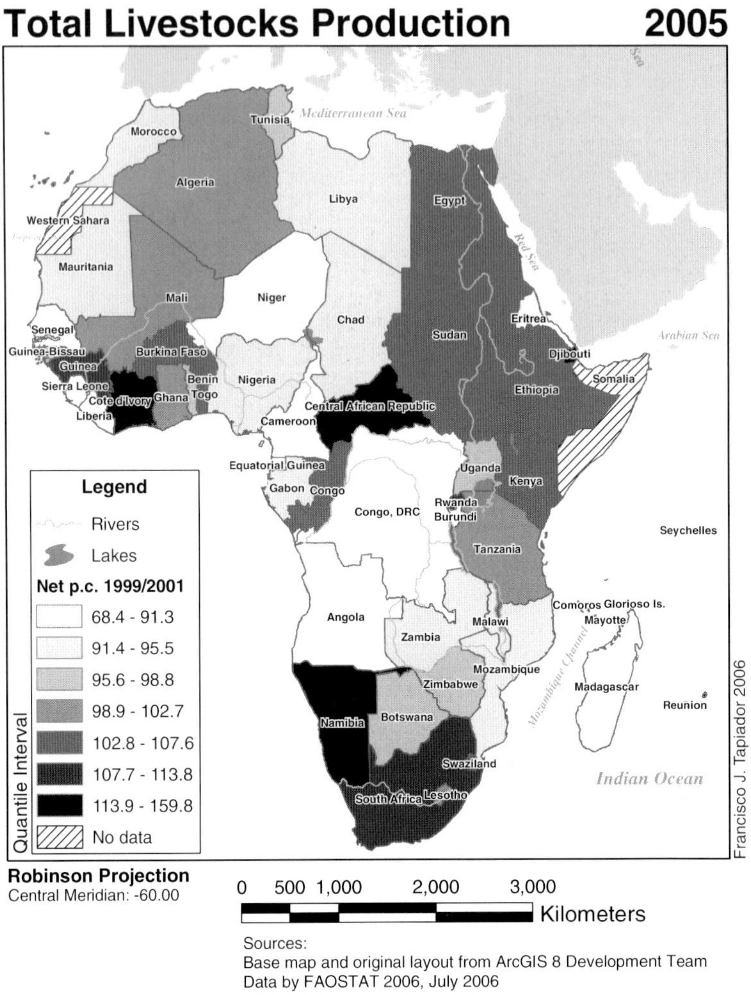

Fig. 8.7 Total livestock production index in Africa in 2005 compared with 1999/2001

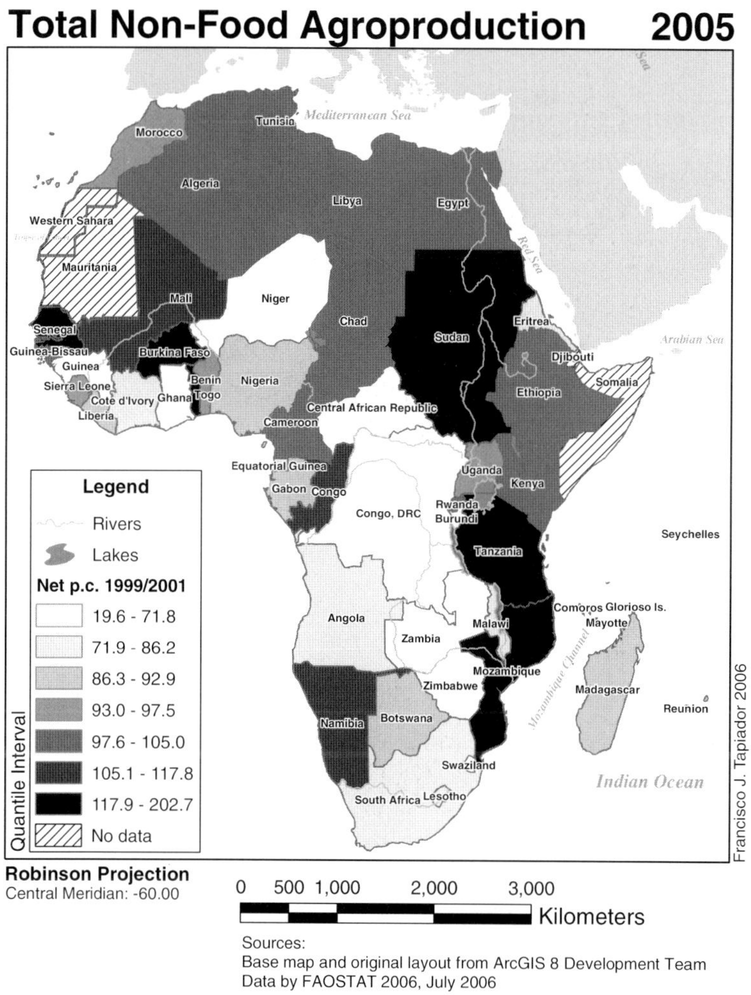

Fig. 8.8 Total non-food agricultural production index in Africa in 2005 compared with 1999/2001

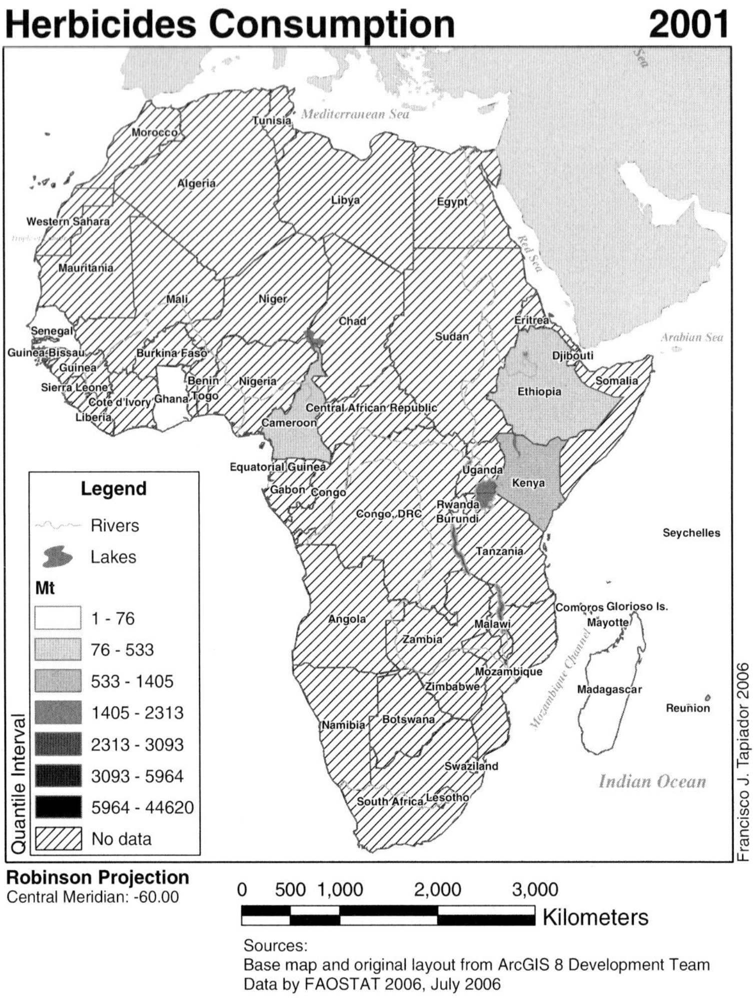

Fig. 8.9 Herbicides consumption in Africa in 2001

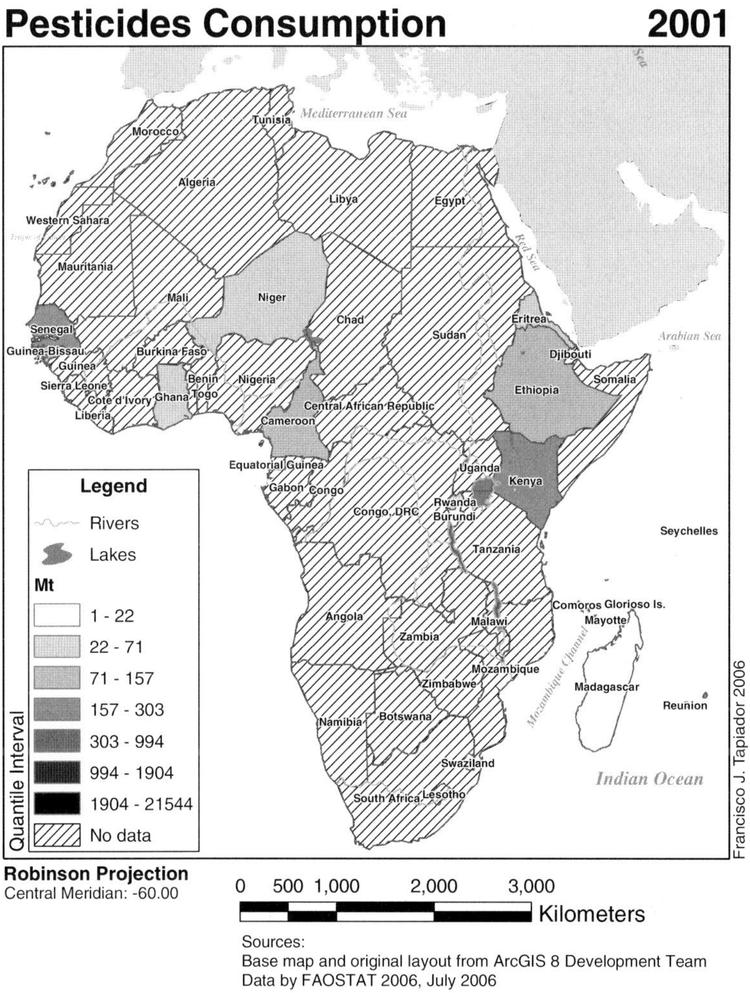

Fig. 8.10 Pesticides consumption in Africa in 2001

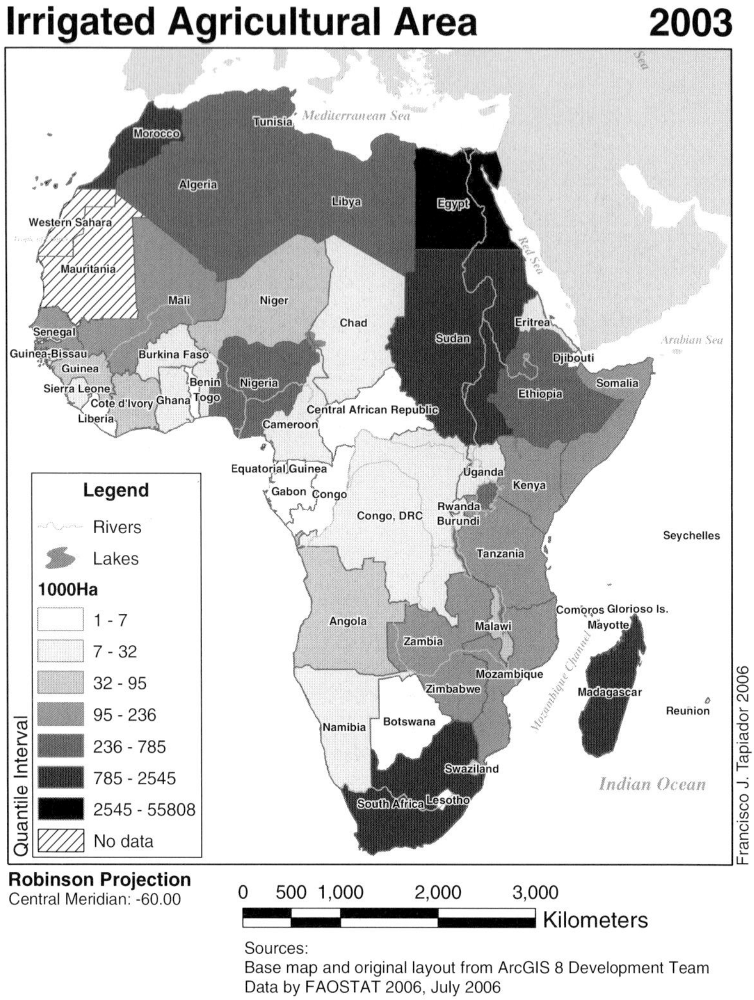

Fig. 8.11 Irrigated agricultural area in Africa in 2003

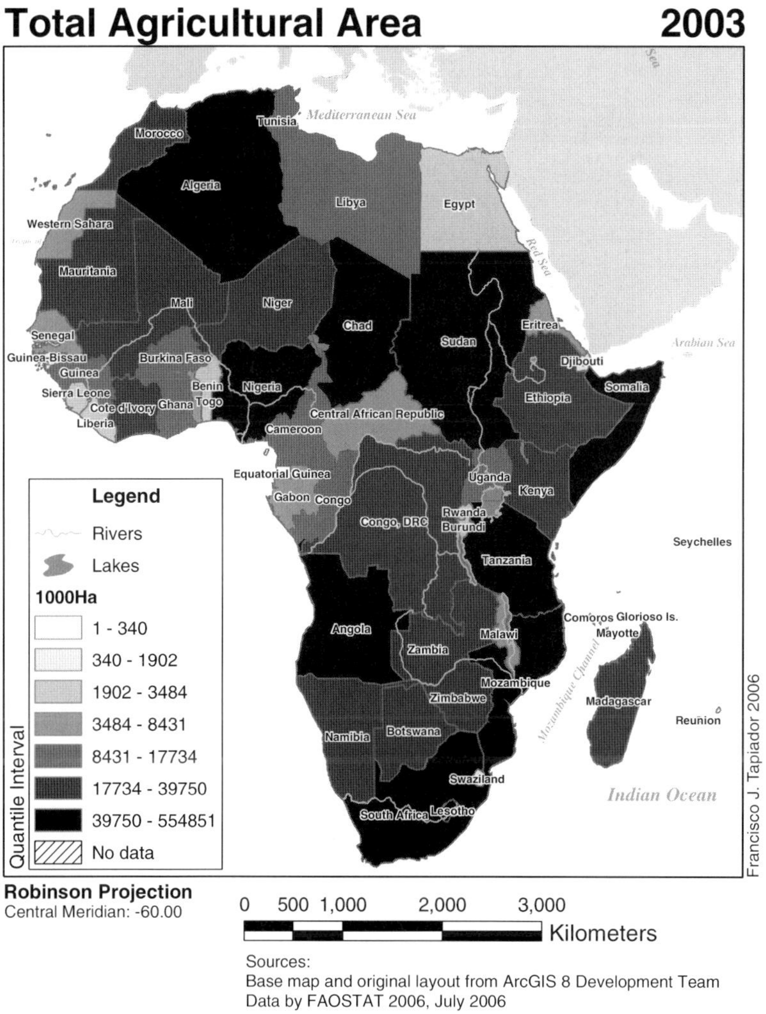

Fig. 8.12 Total agricultural area in Africa in 2003

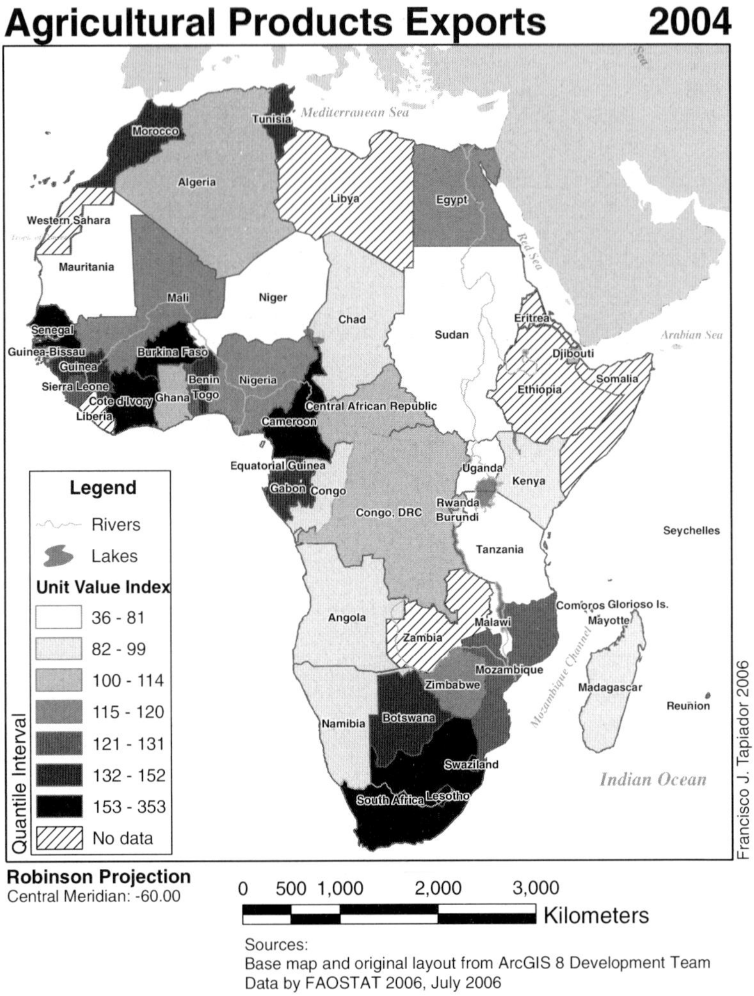

Fig. 8.13 Agricultural products exports (unit value index) in Africa in 2004

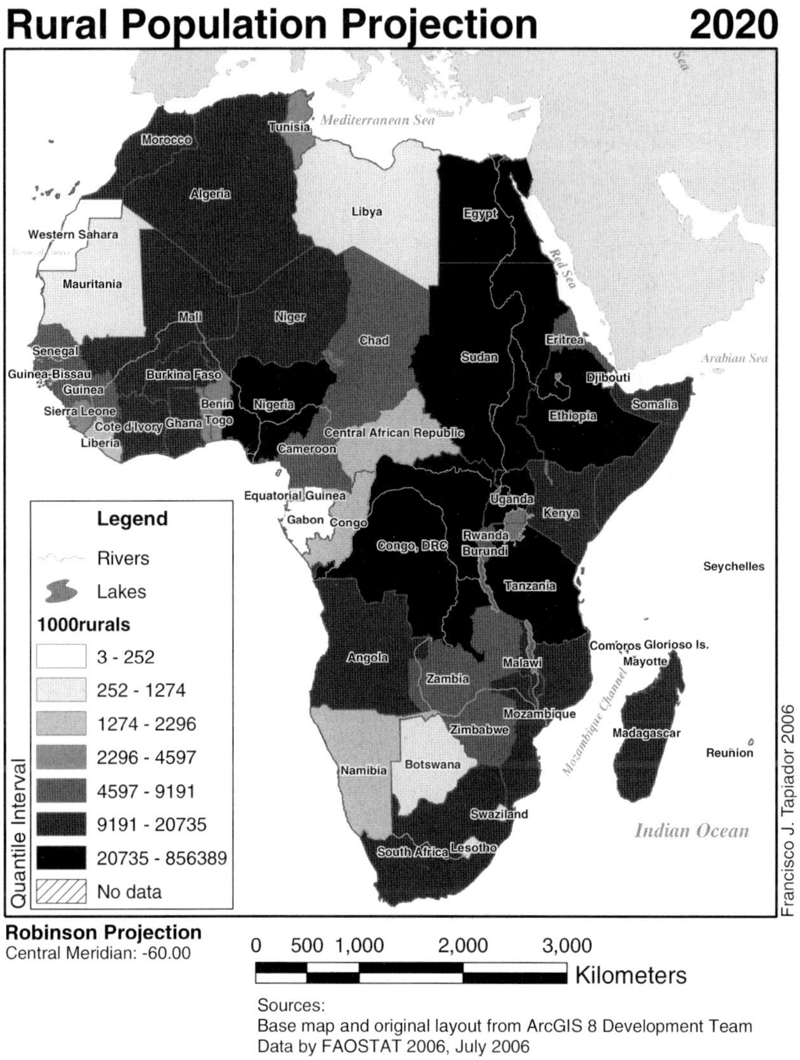

Fig. 8.14 Rural population projection for 2020 in Africa

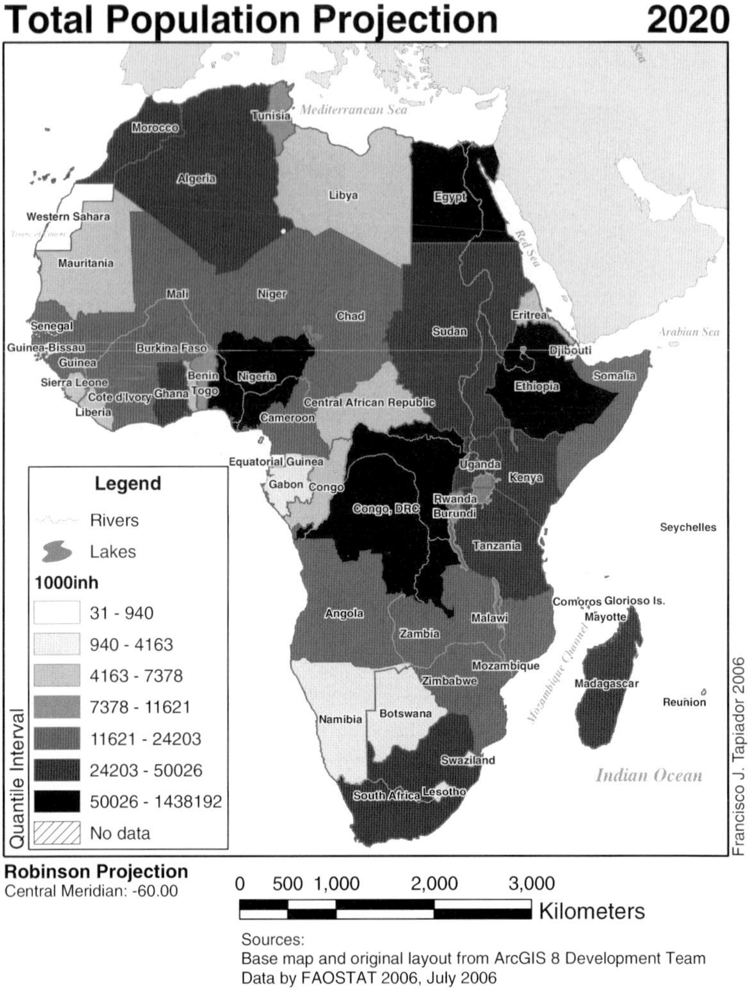

Fig. 8.15 Total population projection for 2020 in Africa

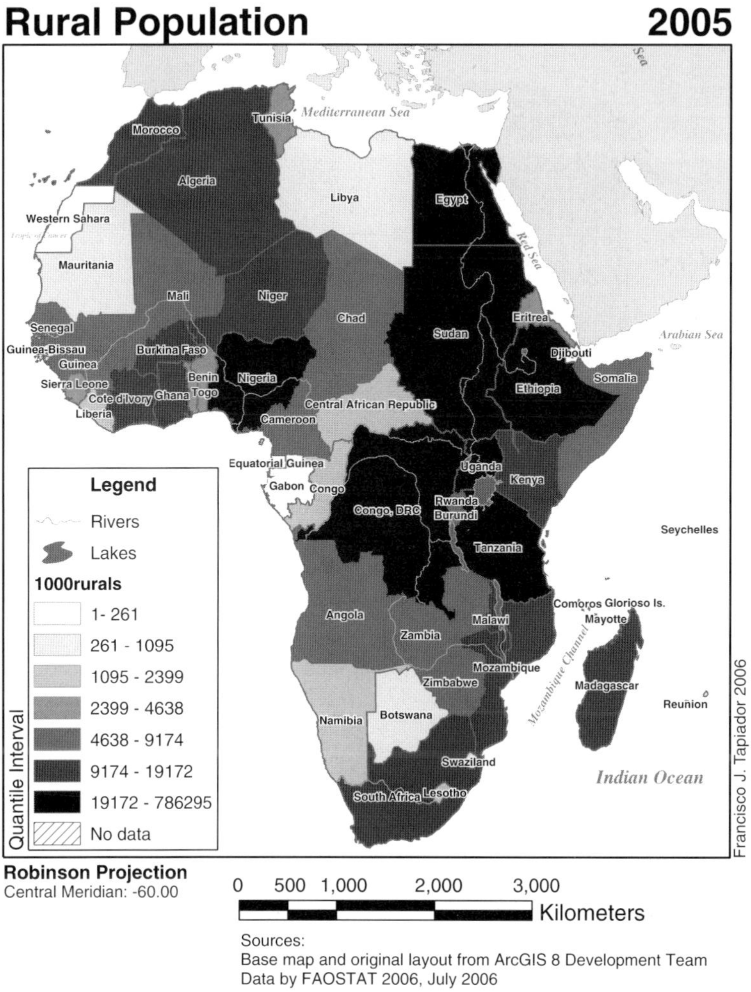

Fig. 8.16 Rural population in Africa in 2005

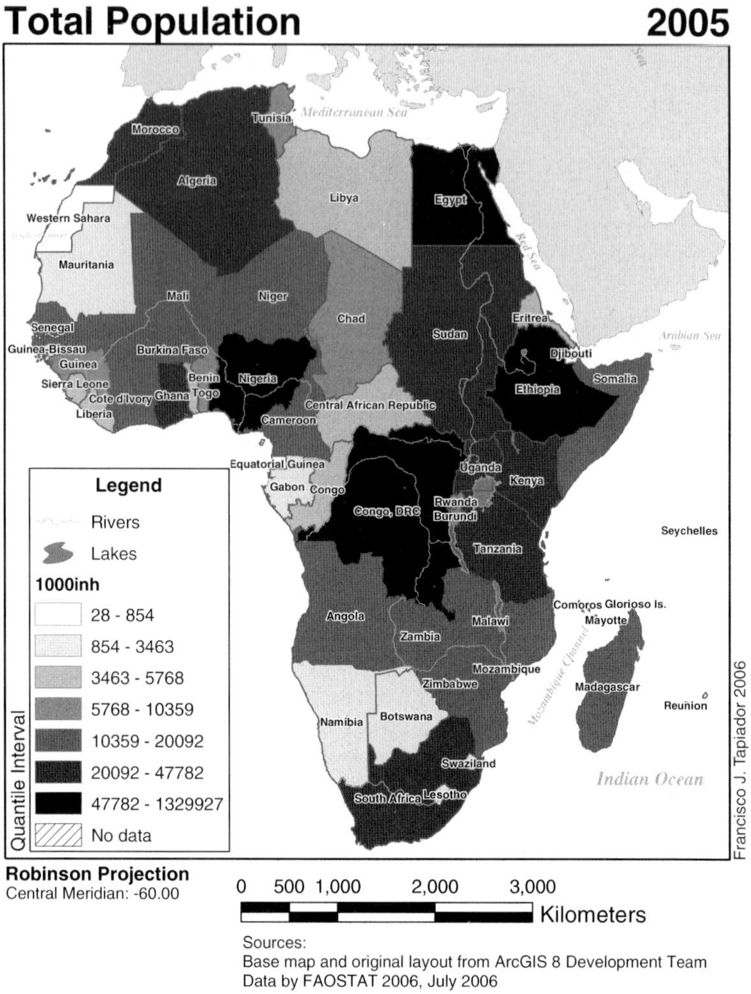

Fig. 8.17 Total population in Africa in 2005

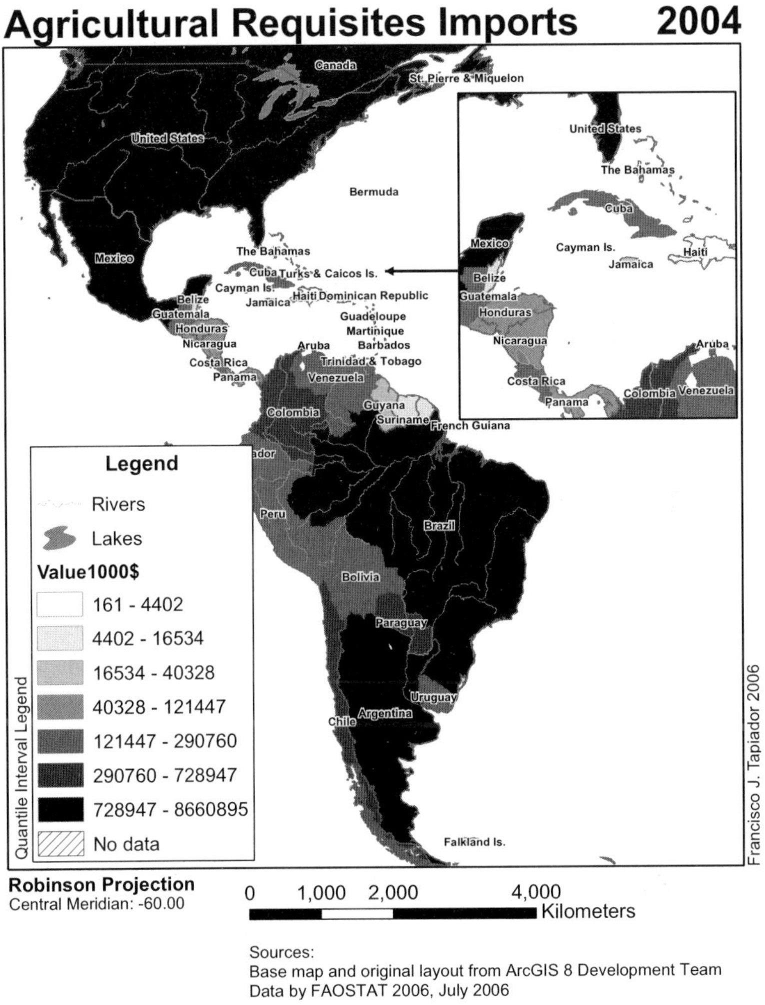

Fig. 8.18 Agricultural resquisites imports in 2004

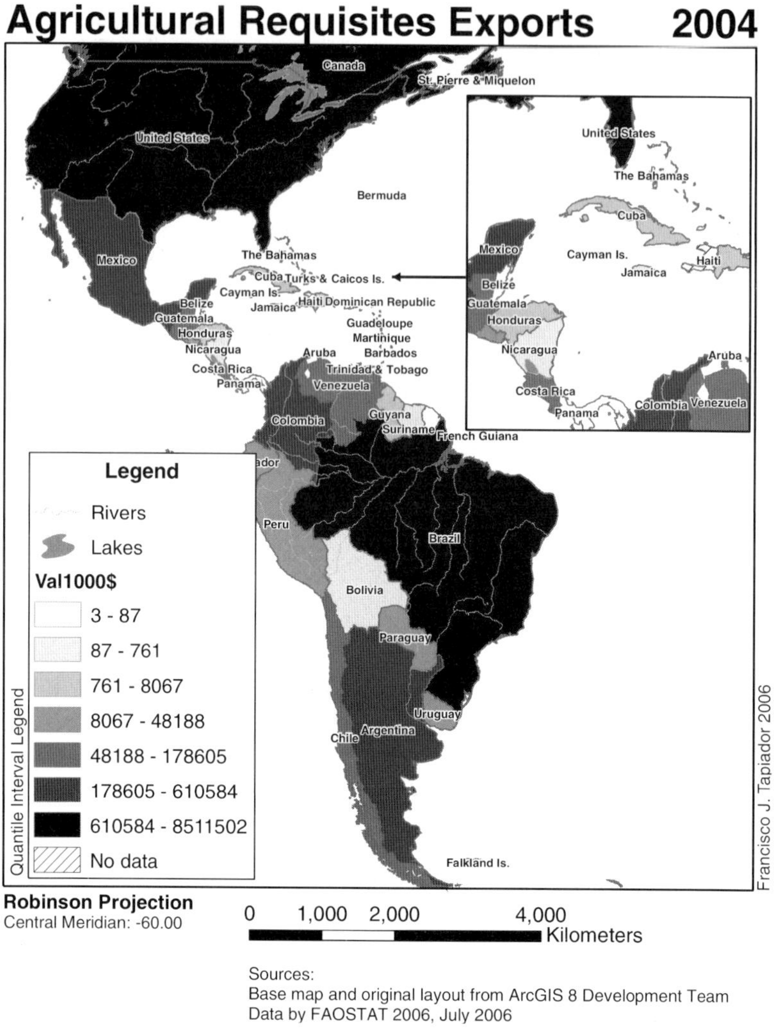

Fig. 8.19 Agricultural requisites exports in 2004

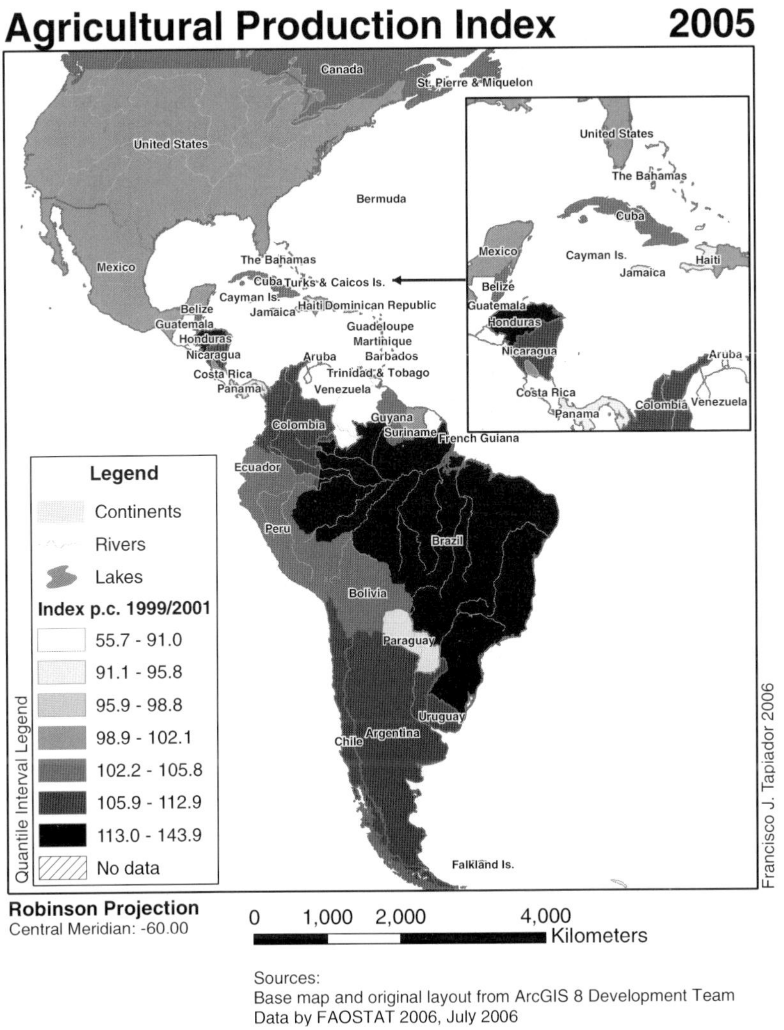

Fig. 8.20 Agricultural production index in 2005 compared with 1999/2001

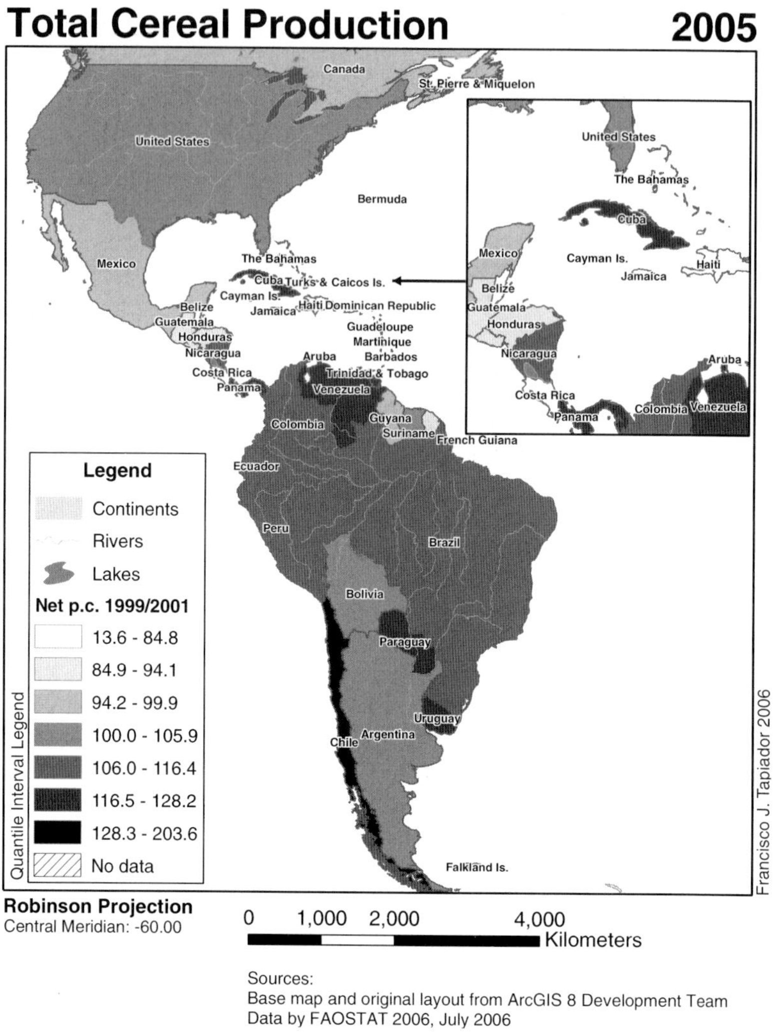

Fig. 8.21 Total cereal production in 2005 compared with 1999/2001

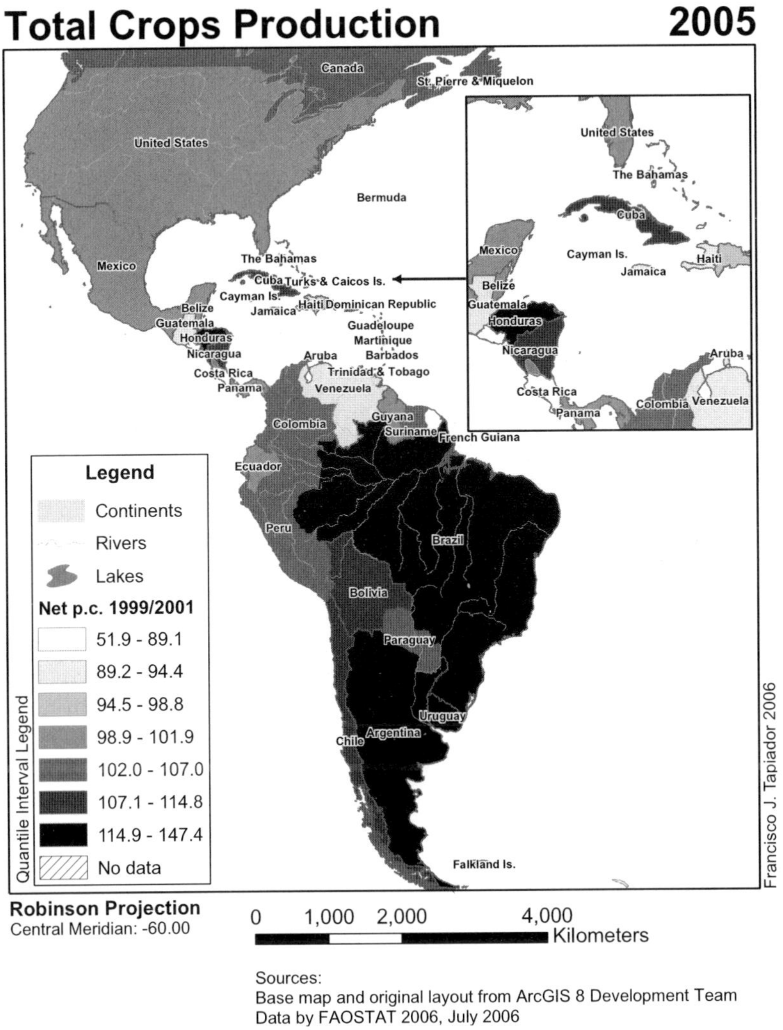

Fig. 8.22 Total crops production in 2005 compared with 1999/2001

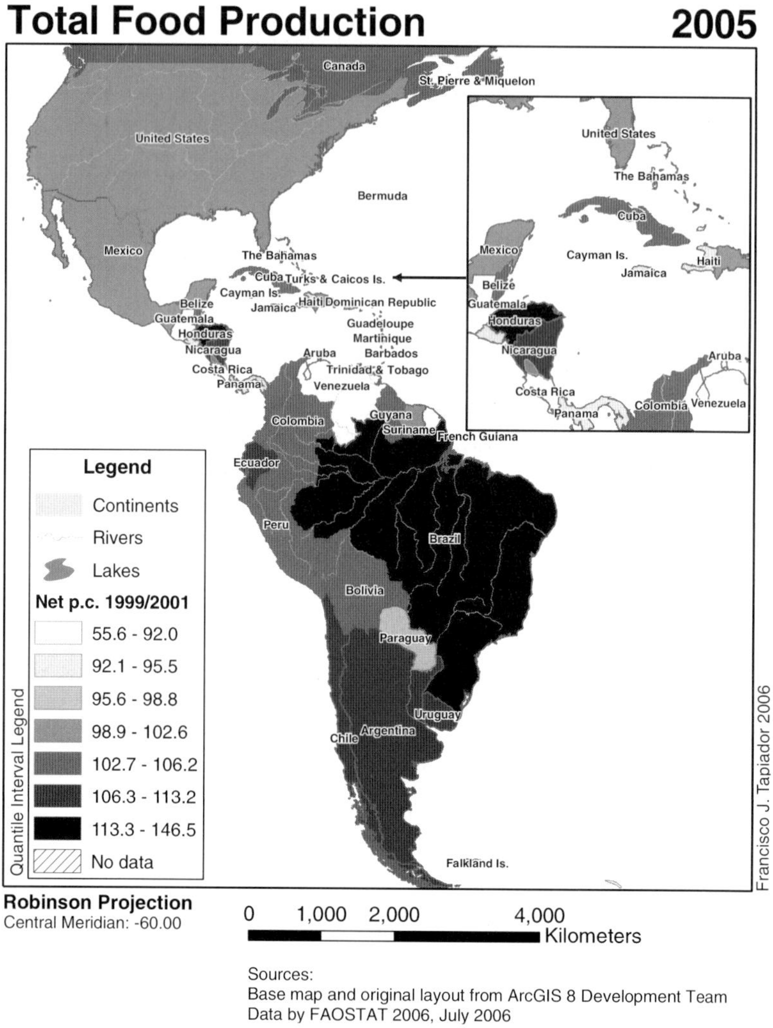

Fig. 8.23 Total food production in 2005 compared with 1999/2001

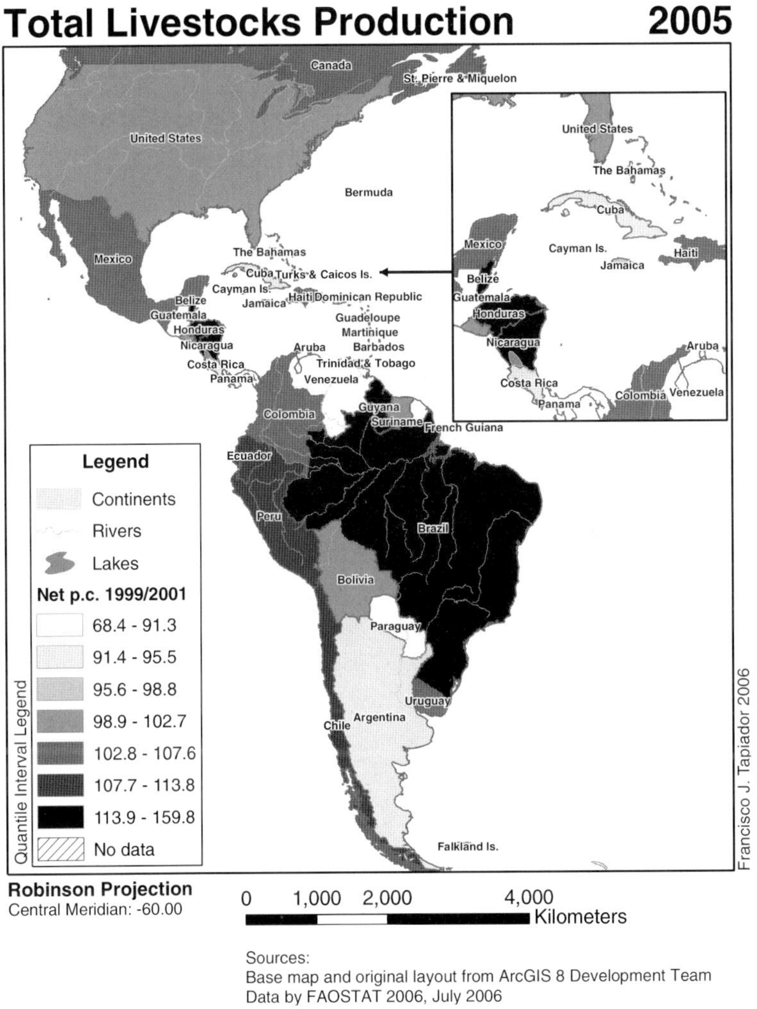

Fig. 8.24 Total livestocks production in 2005 compared with 1999/2001

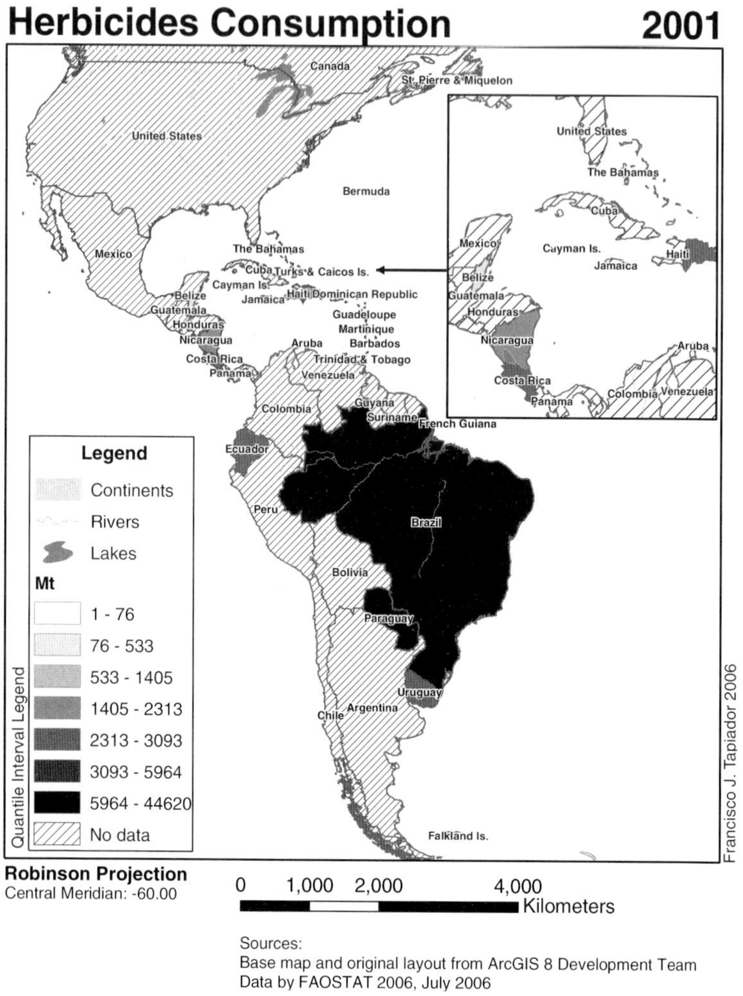

Fig. 8.25 Herbicides consumption in 2001

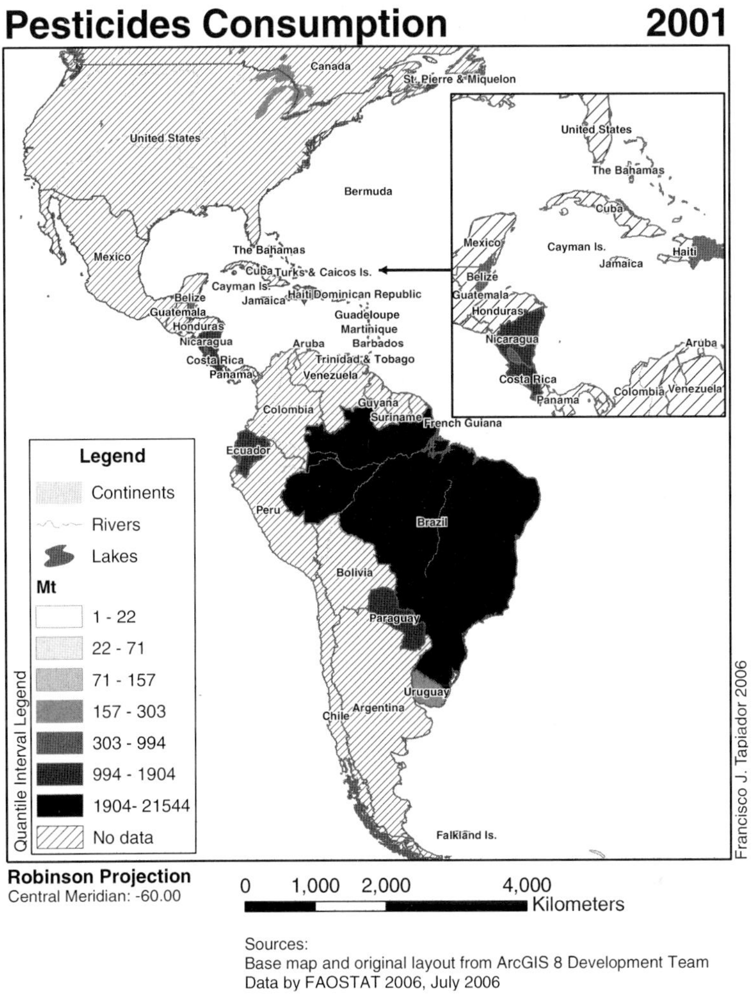

Fig. 8.26 Pesticides consumption in 2001

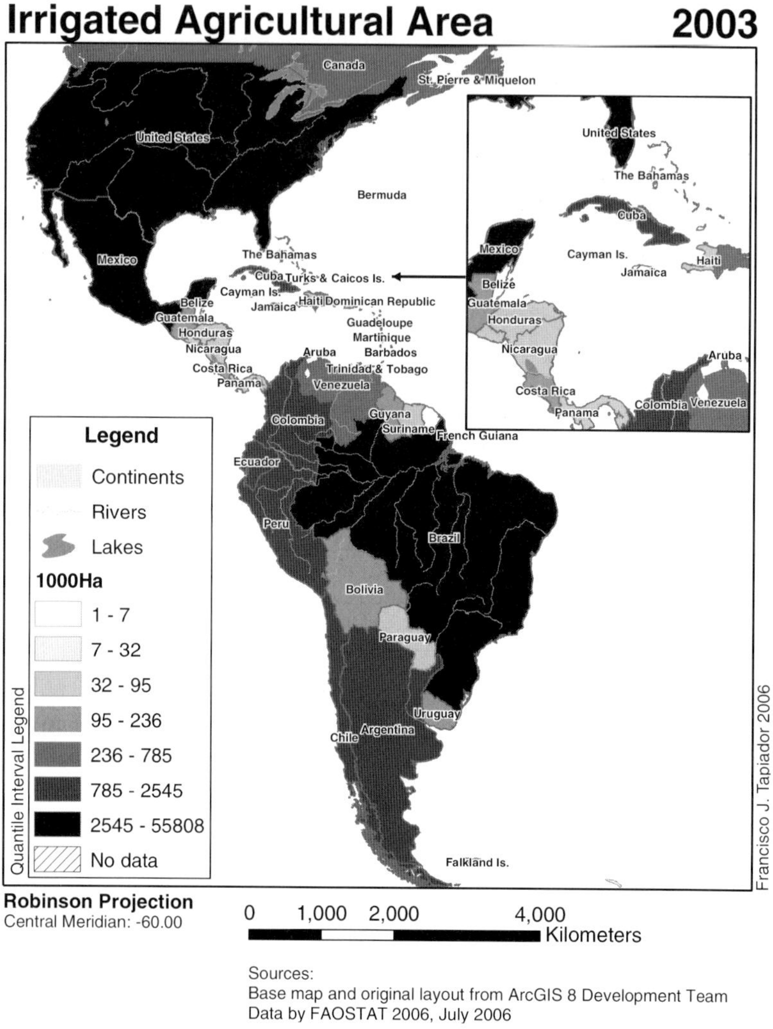

Fig. 8.27 Irrigated agricultural area in 2003

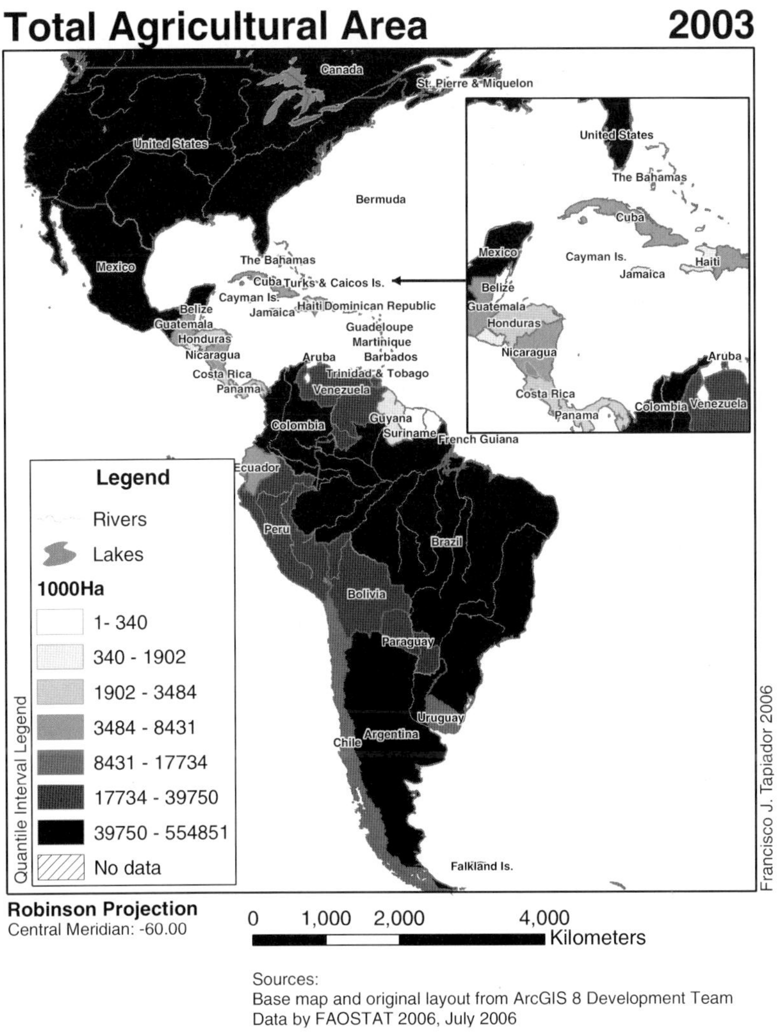

Fig. 8.28 Total agricultural area in 2003

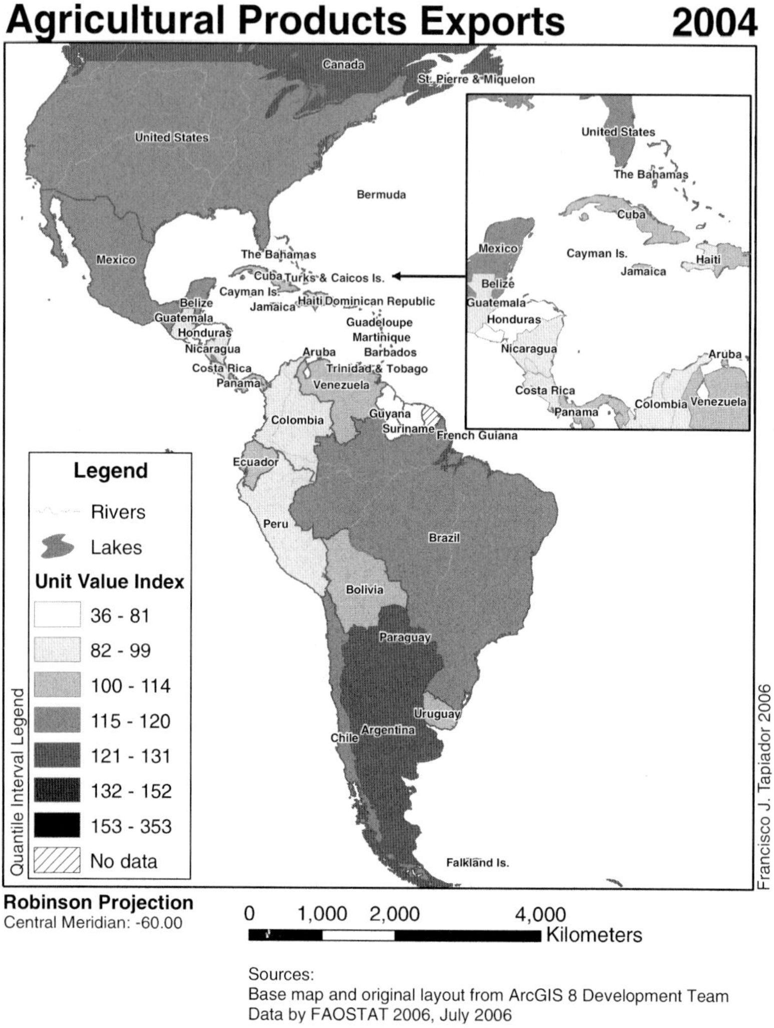

Fig. 8.29 Agricultural products exports in 2004

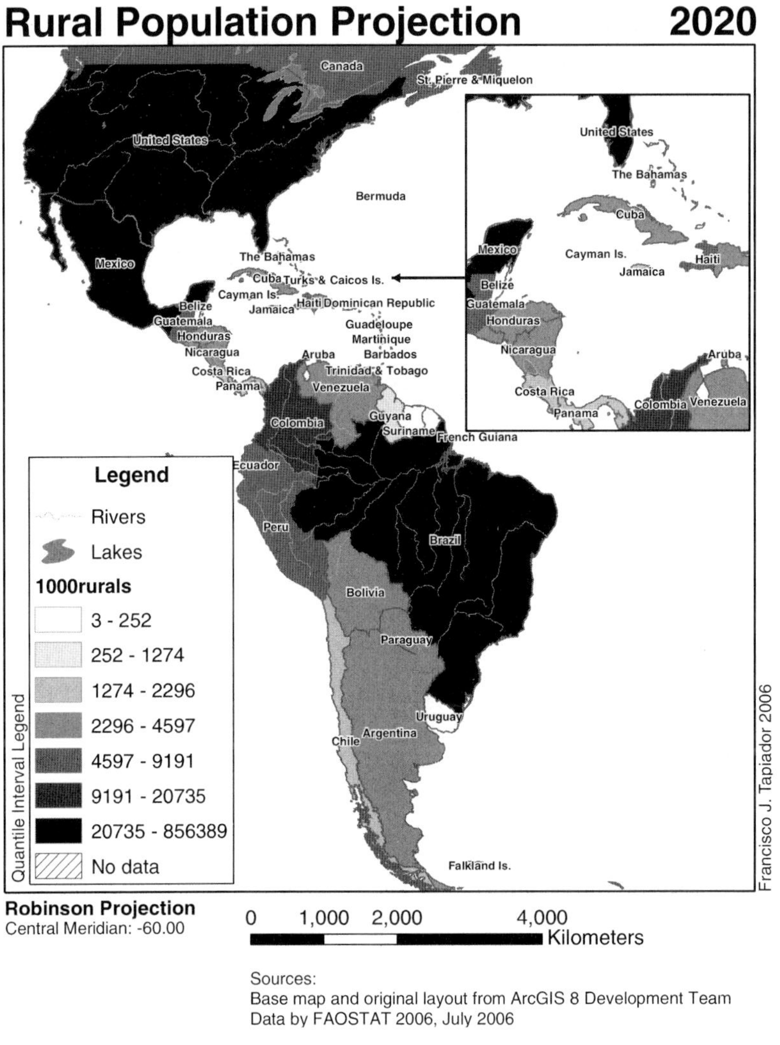

Fig. 8.30 Rural population projection in 2020

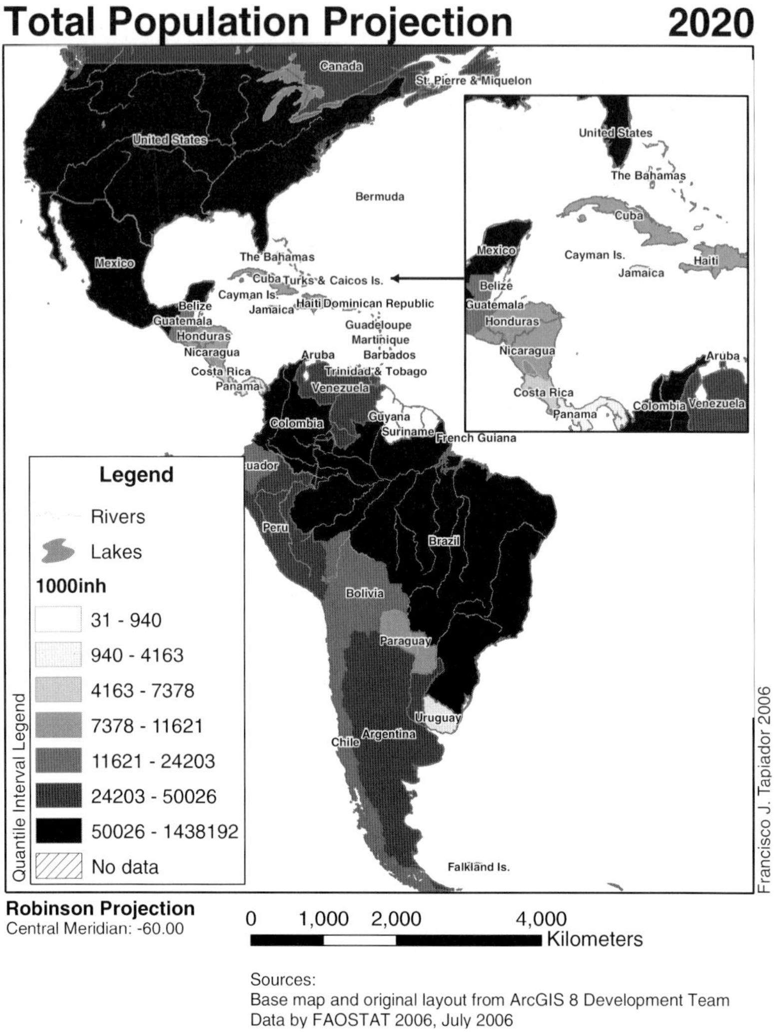

Fig. 8.31 Total population projection in 2020

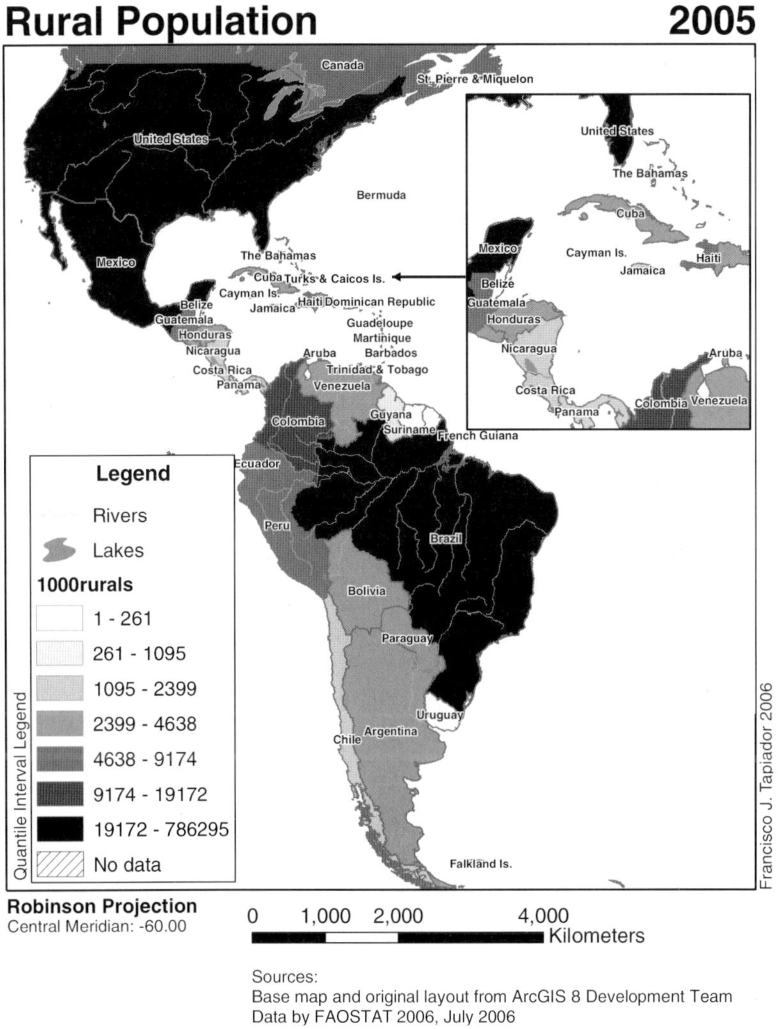

Fig. 8.32 Rural population in 2005

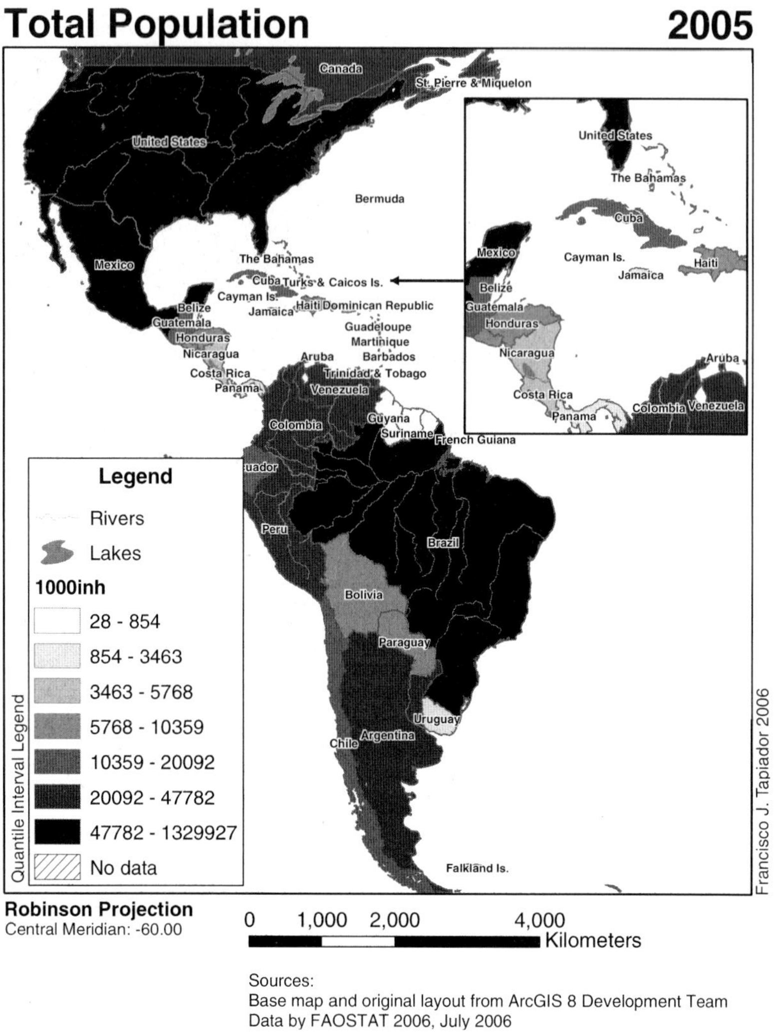

Fig. 8.33 Total population in 2005

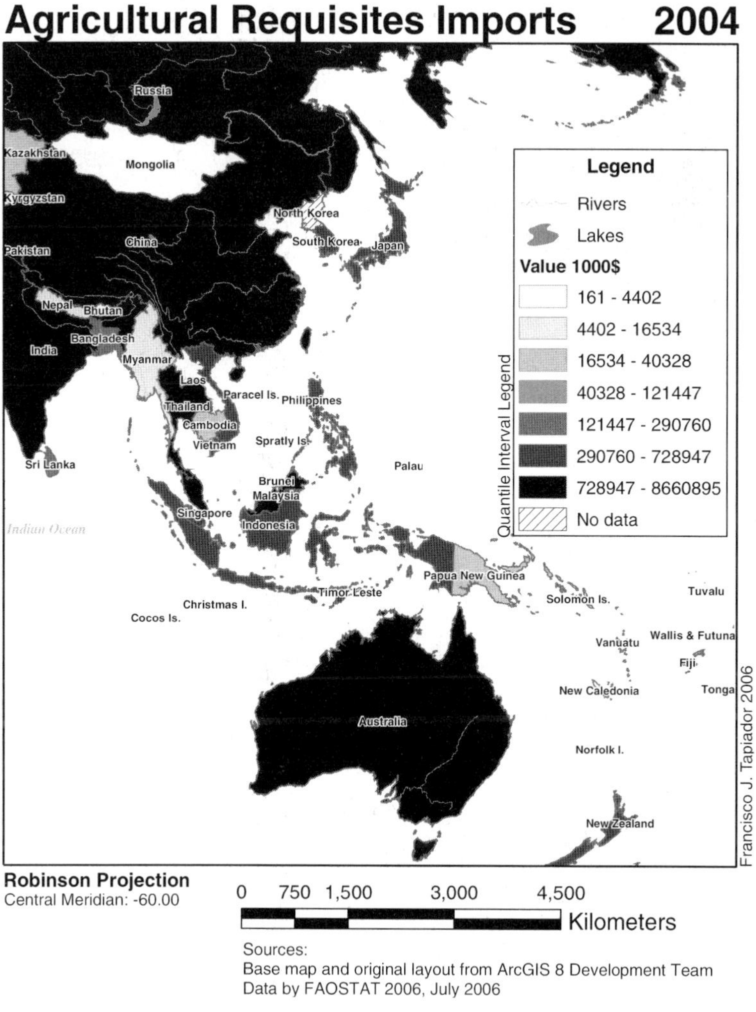

Fig. 8.34 Agricultural requisites imports in 2004

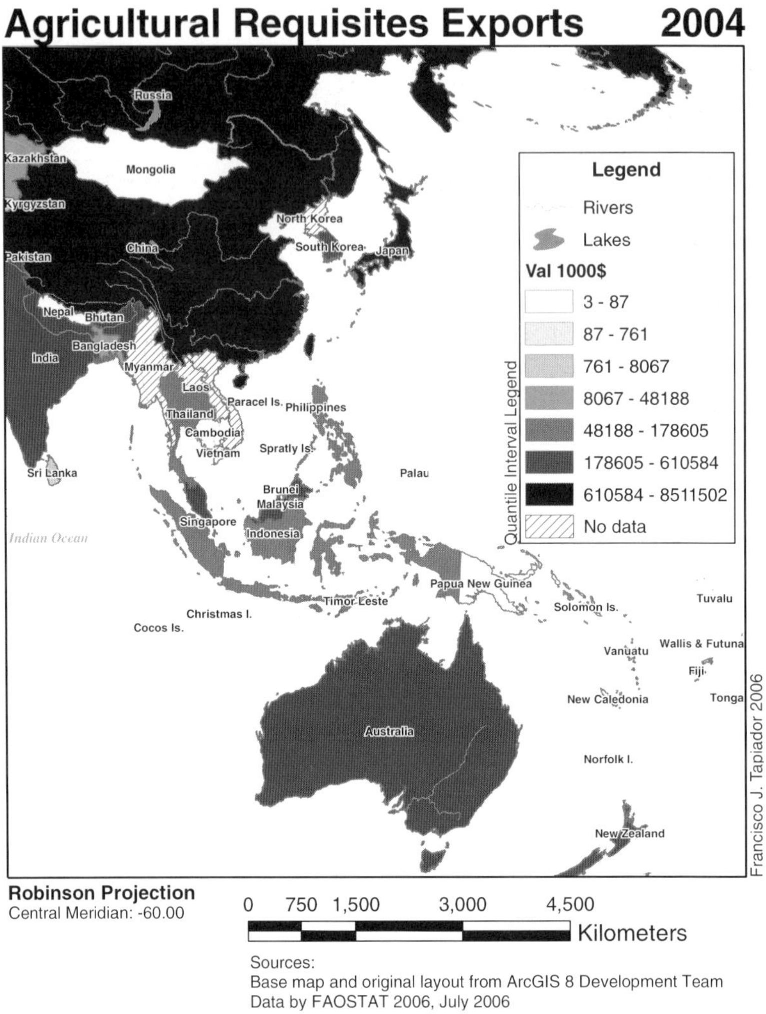

Fig. 8.35 Agricultural requisites exports in 2004

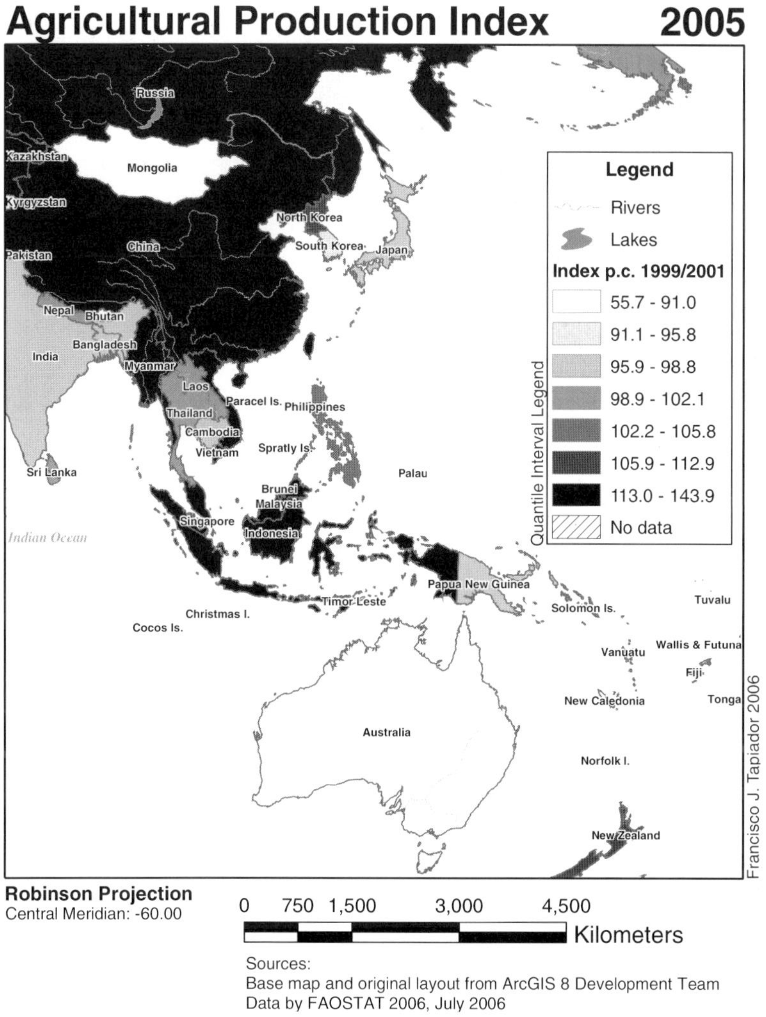

Fig. 8.36 Agricultural production index in 2005 compared with 1999/2001

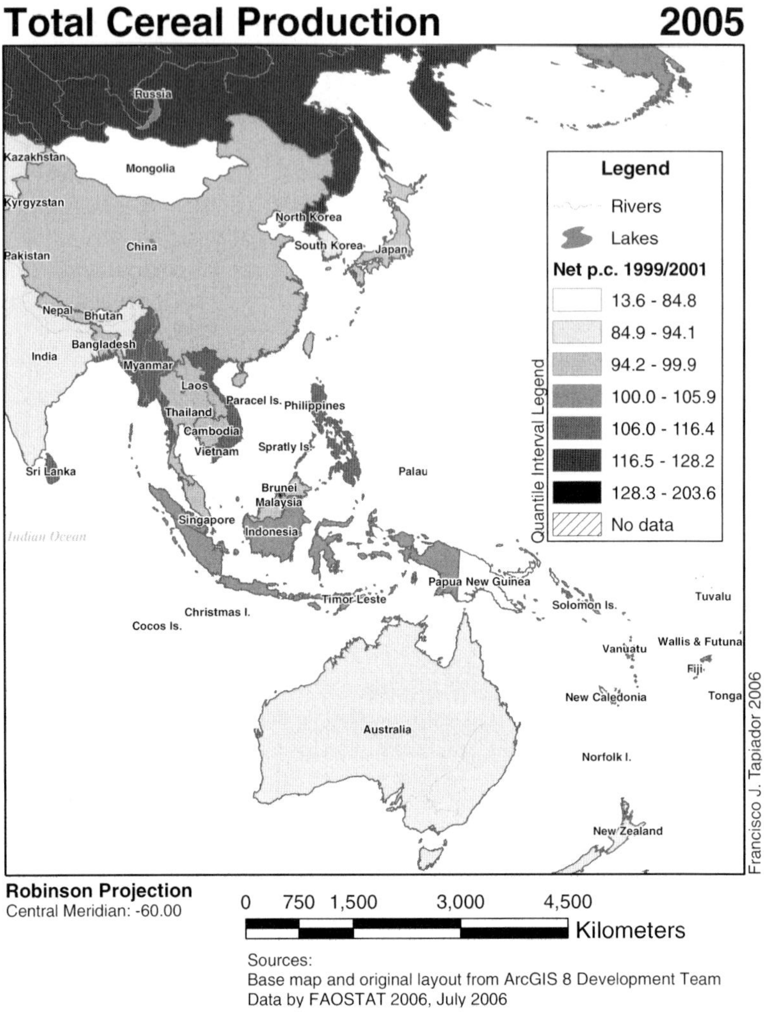

Fig. 8.37 Total cereal production in 2005 compared with 1999/2001

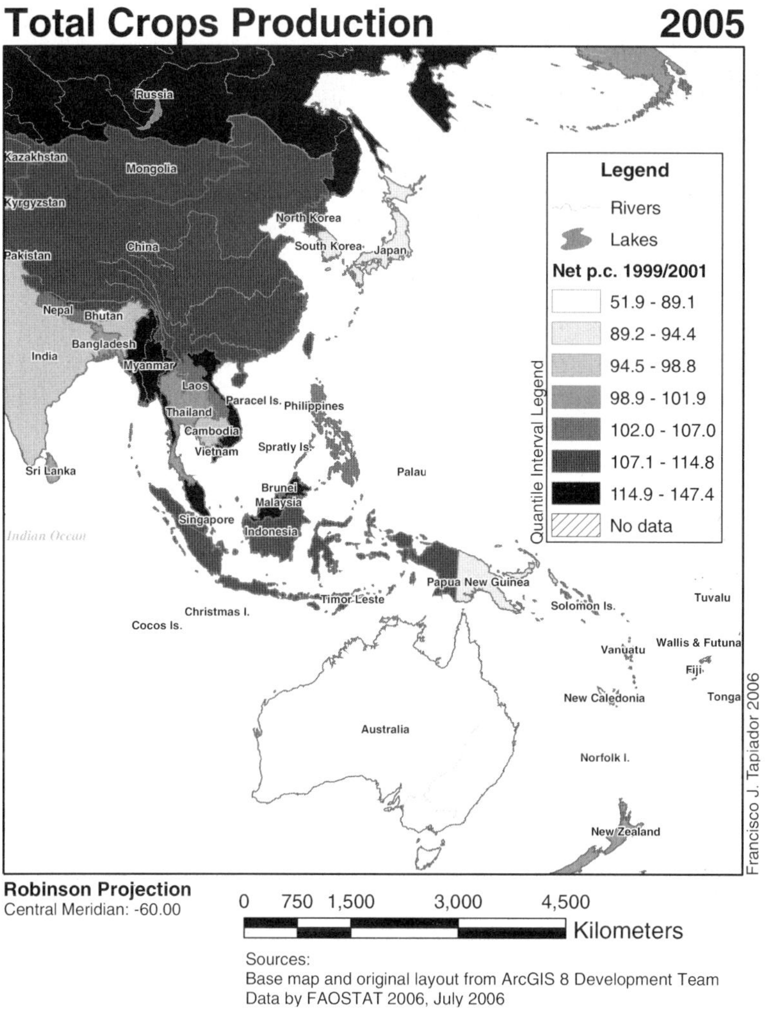

Fig. 8.38 Total crops production in 2005 compared with 1999/2001

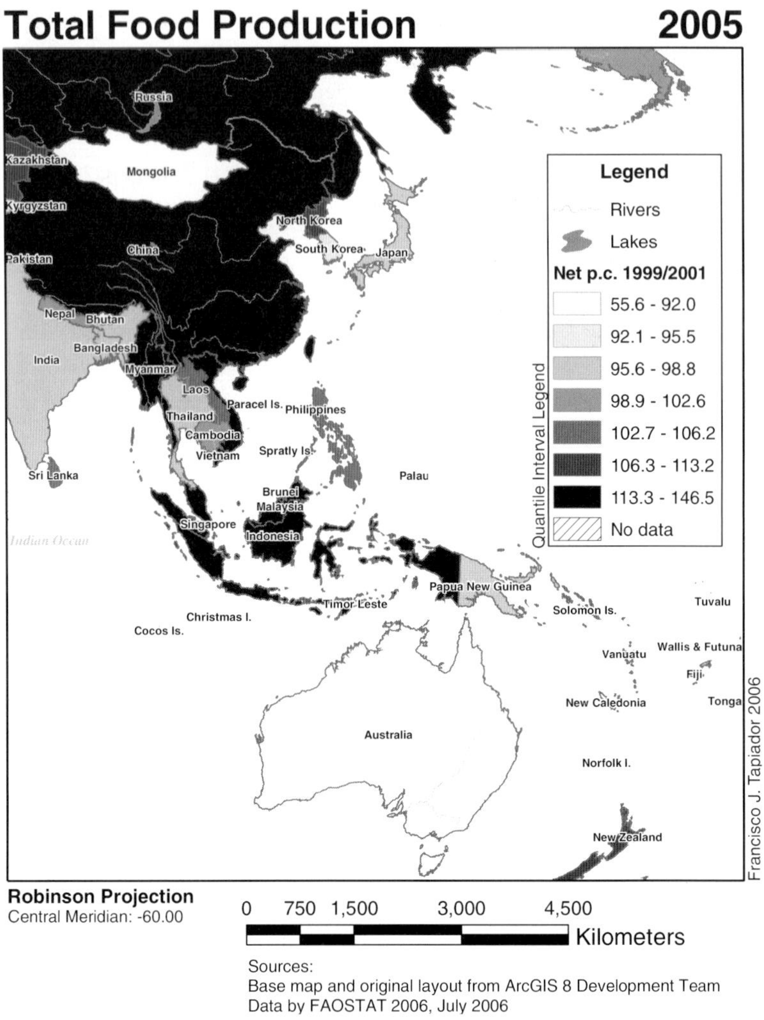

Fig. 8.39 Total food production in 2005 compared with 1999/2001

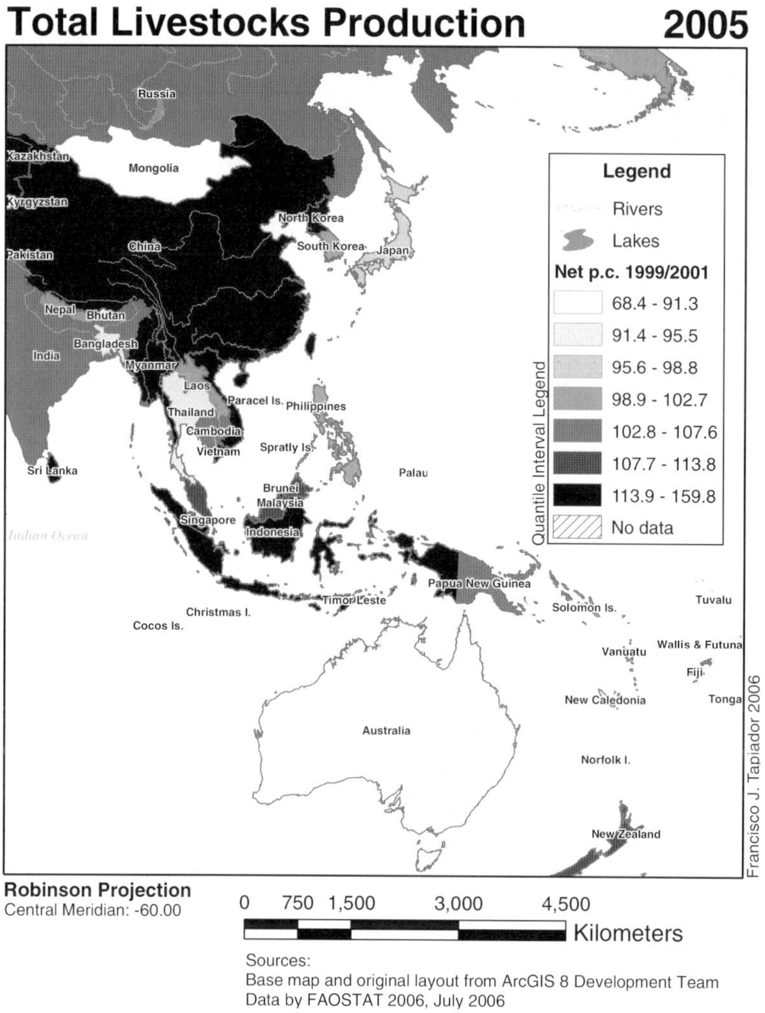

Fig. 8.40 Total livestock production in 2005 compared with 1999/2001

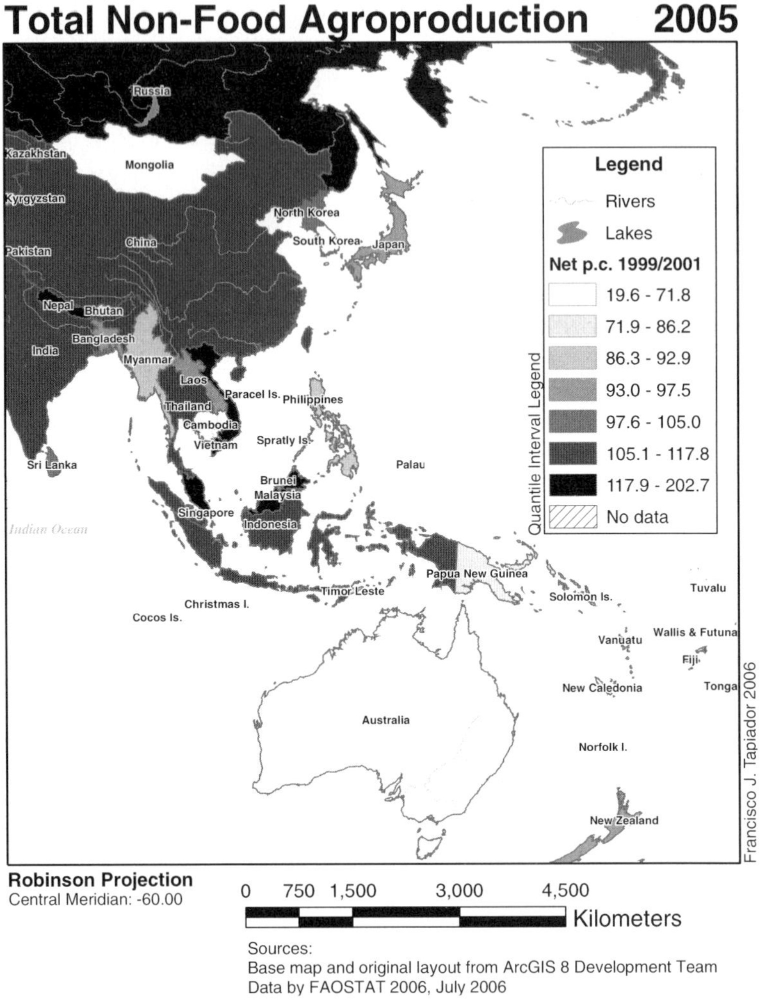

Fig. 8.41 Total non-food agroproduction in 2005 compared with 1999/2001

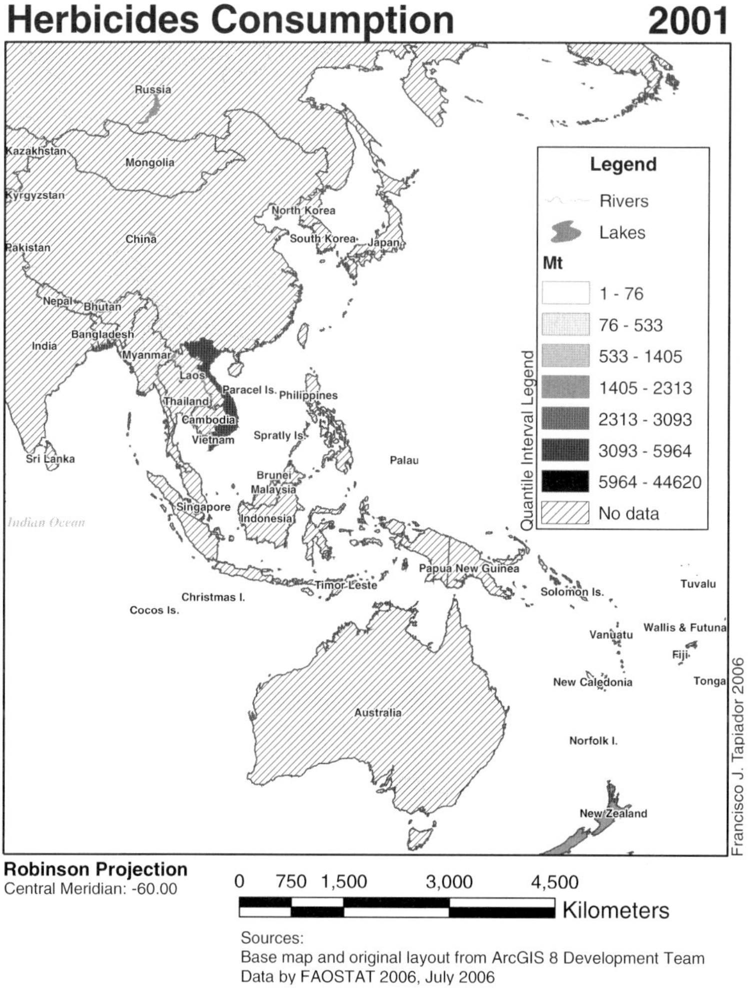

Fig. 8.42 Herbicides consumption in 2001

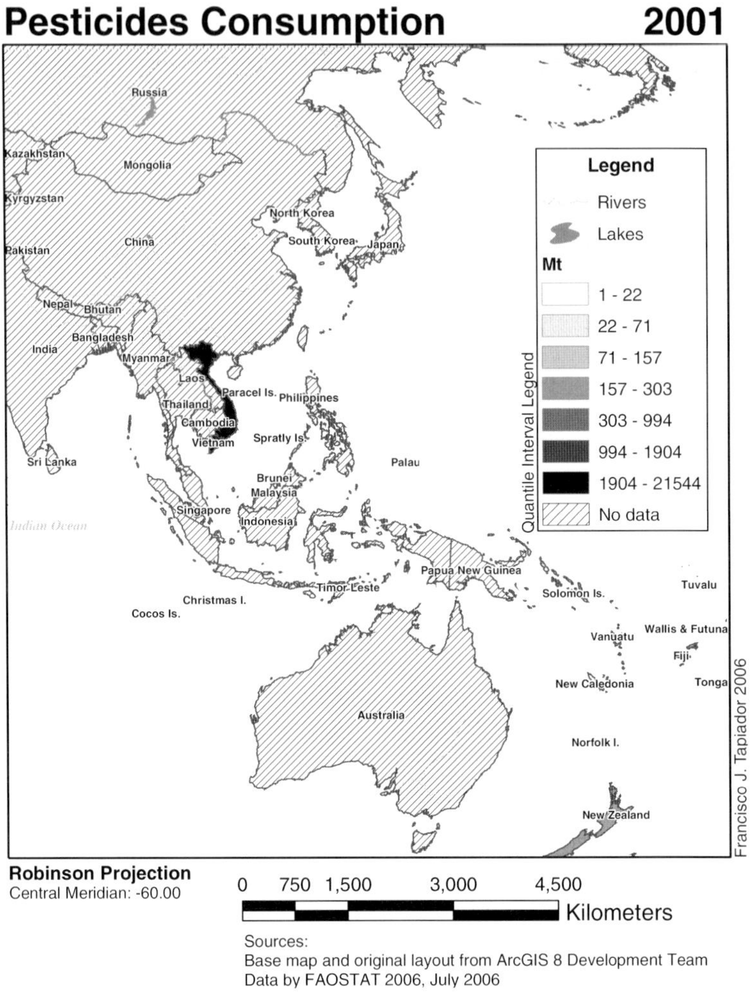

Fig. 8.43 Pesticides consumption in 2001

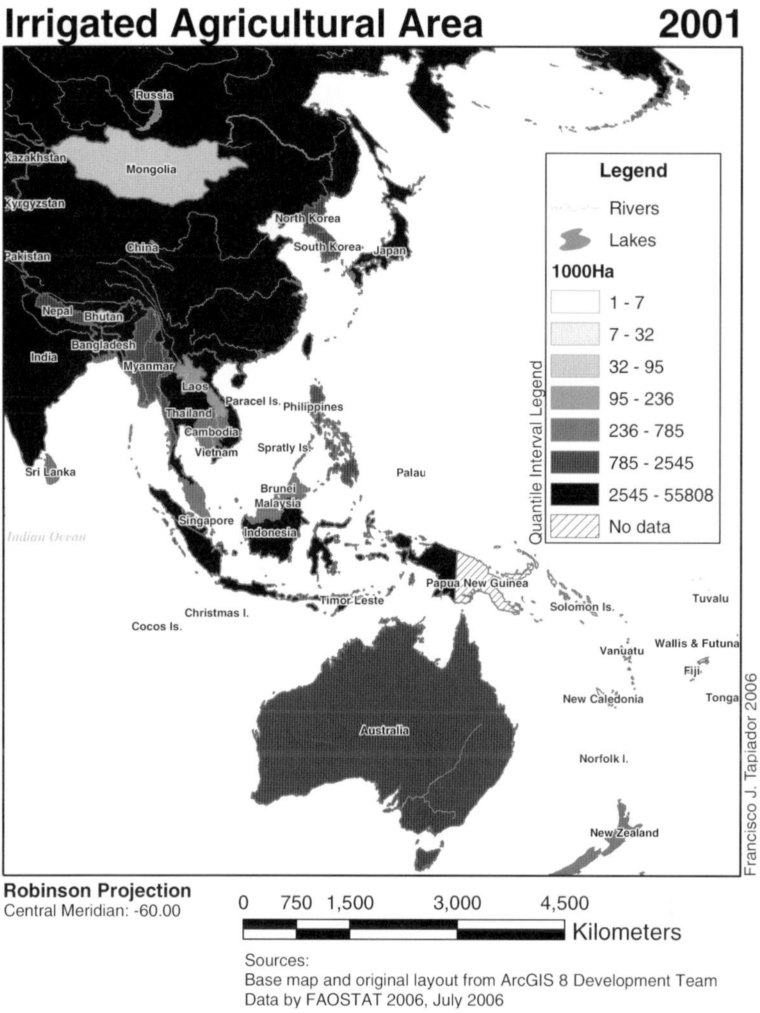

Fig. 8.44 Irrigated agricultural area in 2001

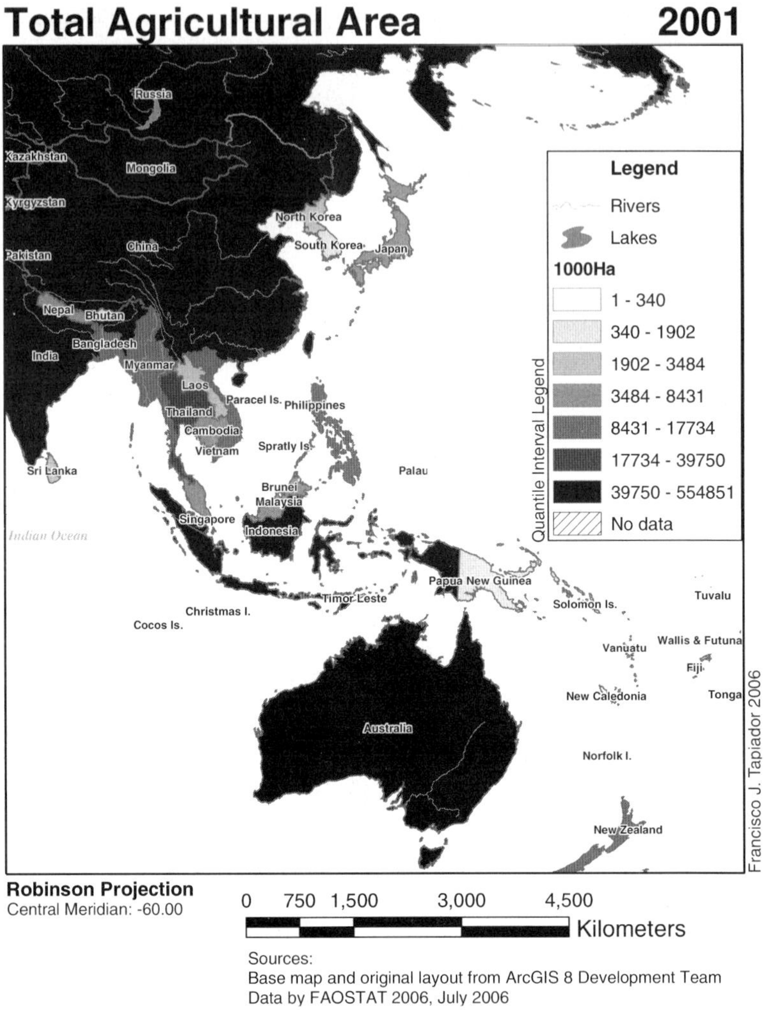

Fig. 8.45 Total agricultural area 2001

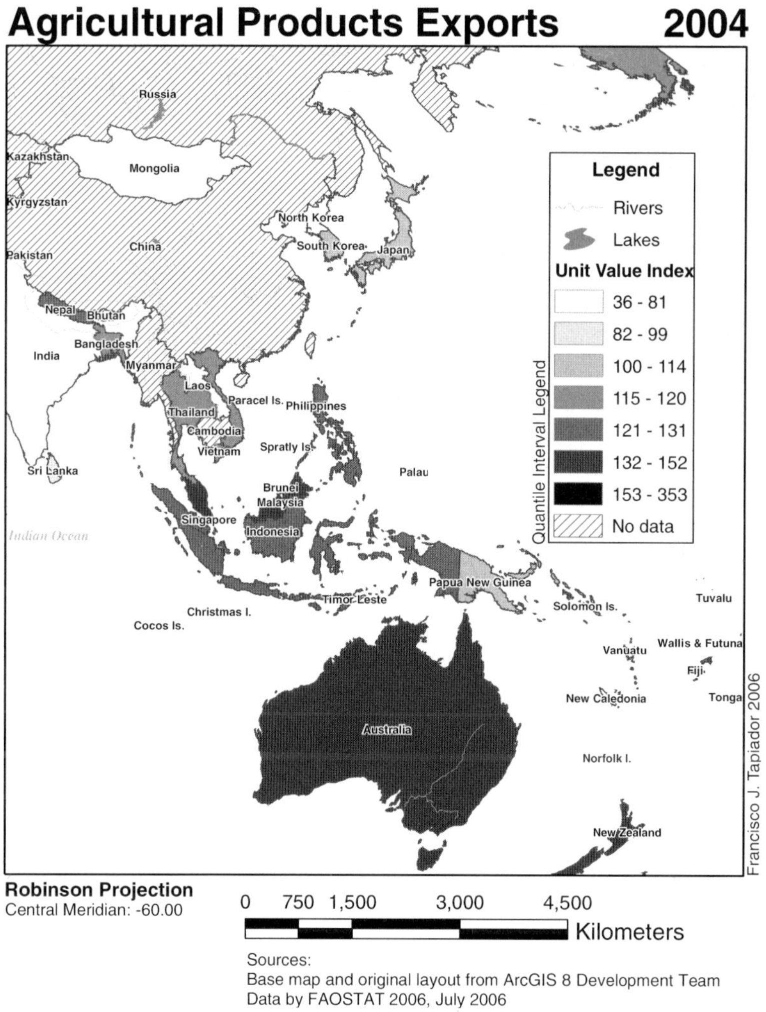

Fig. 8.46 Agricultural products exports in 2004

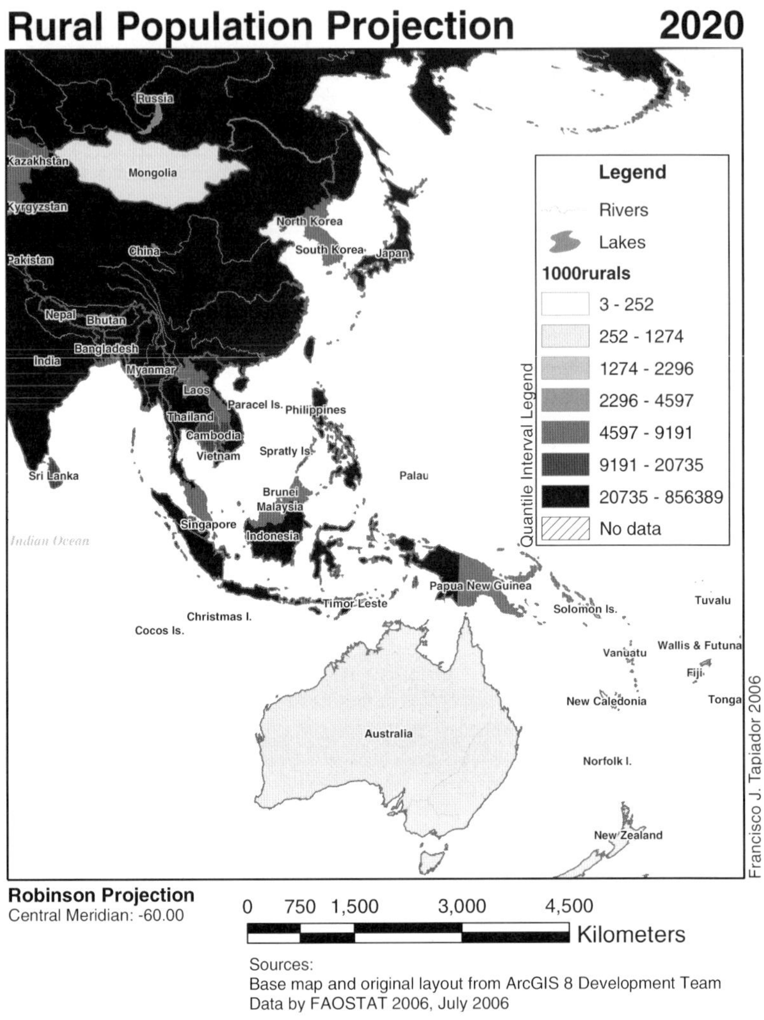

Fig. 8.47 Rural population projection in 2020

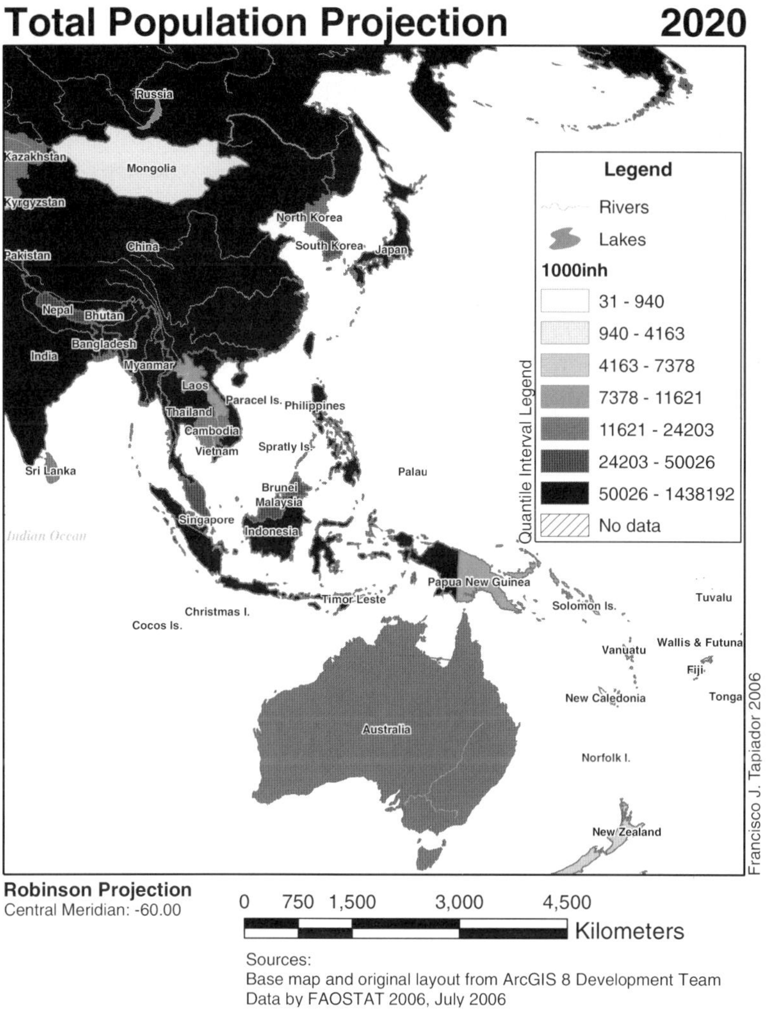

Fig. 8.48 Total population projection in 2020

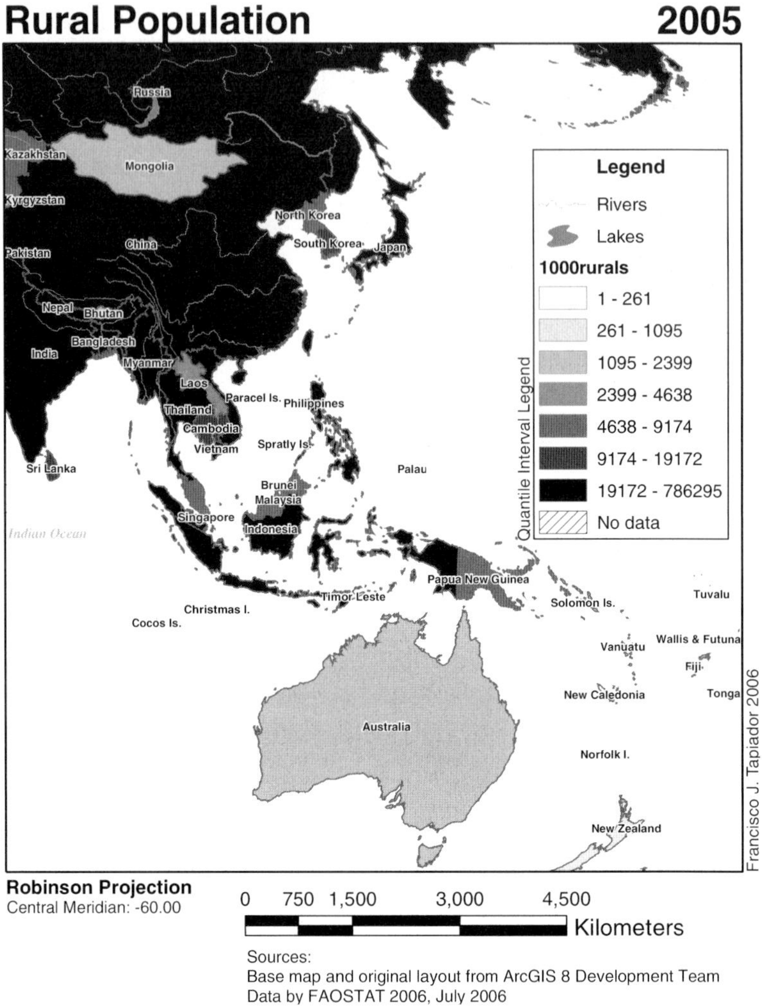

Fig. 8.49 Rural population in 2005

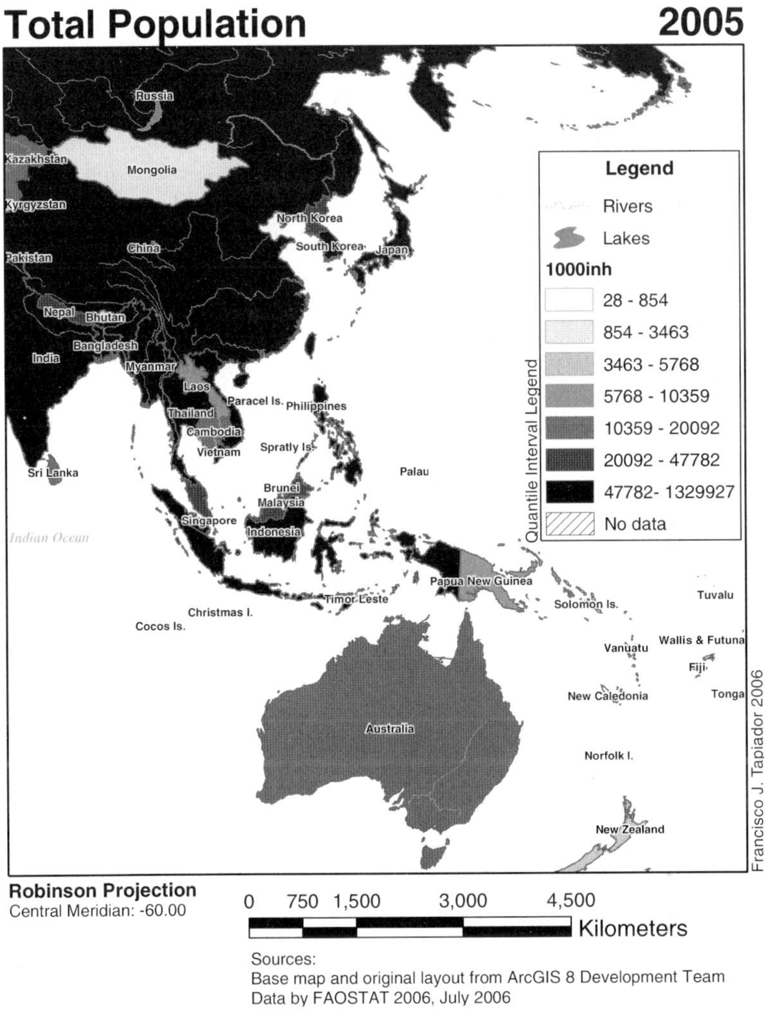

Fig. 8.50 Total population in 2005

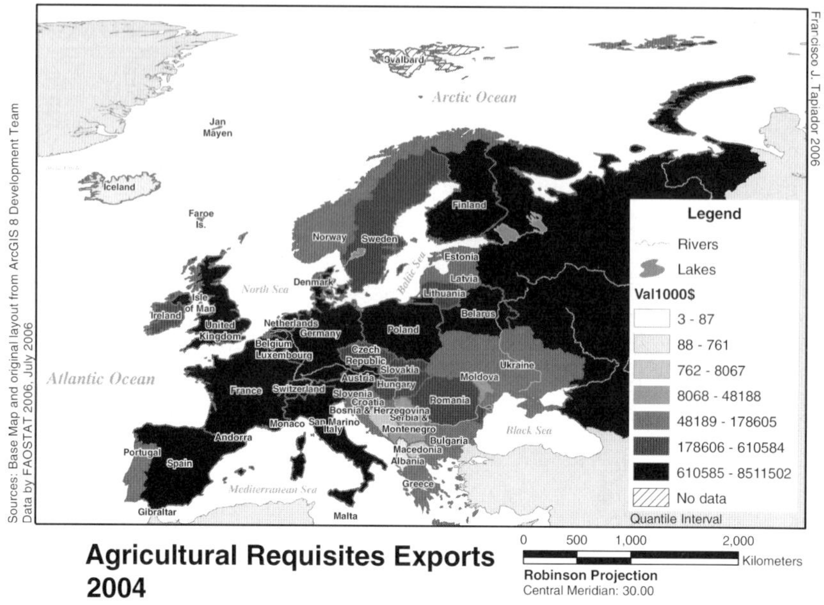

Fig. 8.51 Agricultural requisites exports in 2004

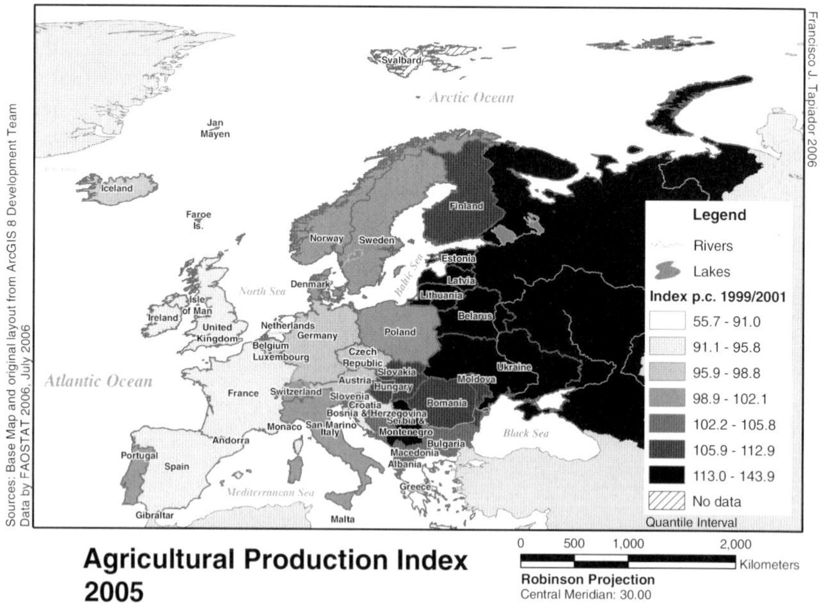

Fig. 8.52 Agricultural production index in 2005 compared with 1999/2001

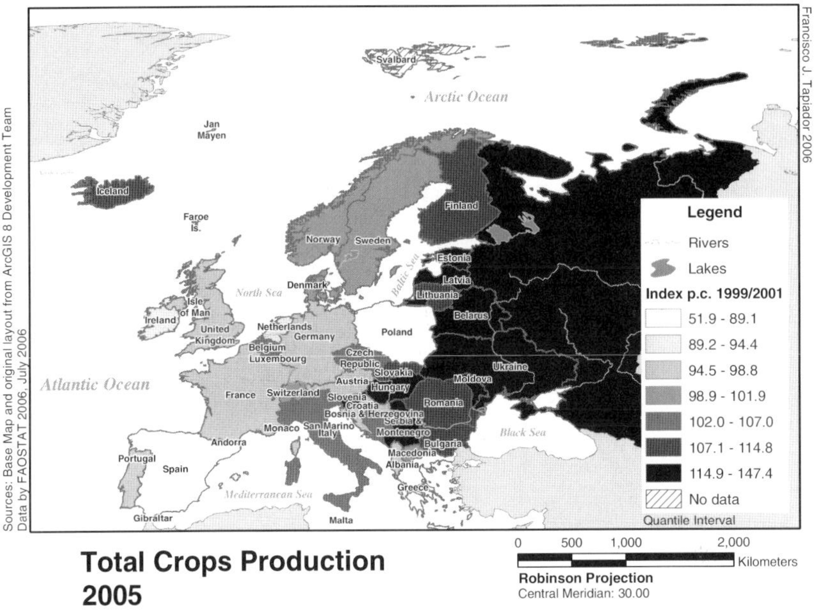

Fig. 8.53 Total crops production in 2005 compared with 1999/2001

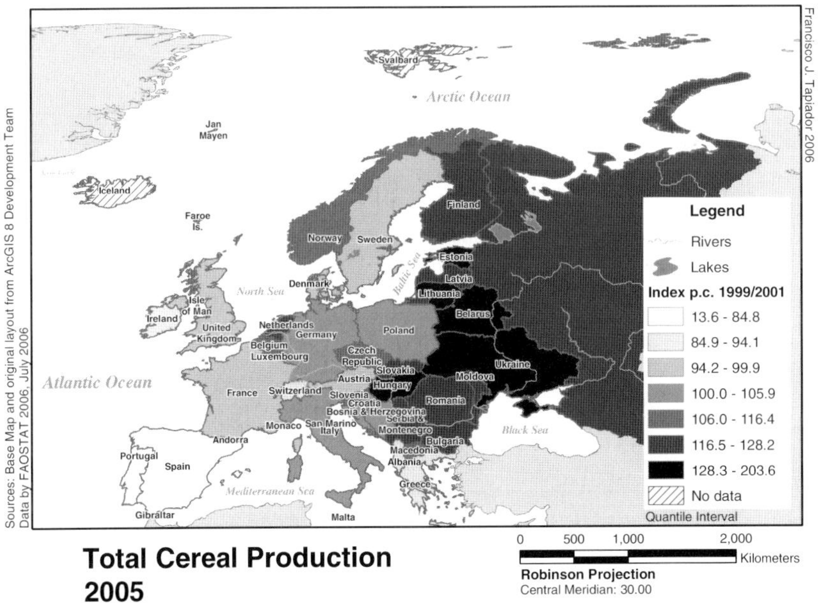

Fig. 8.54 Total cereal production in 2005 compared with 1999/2001

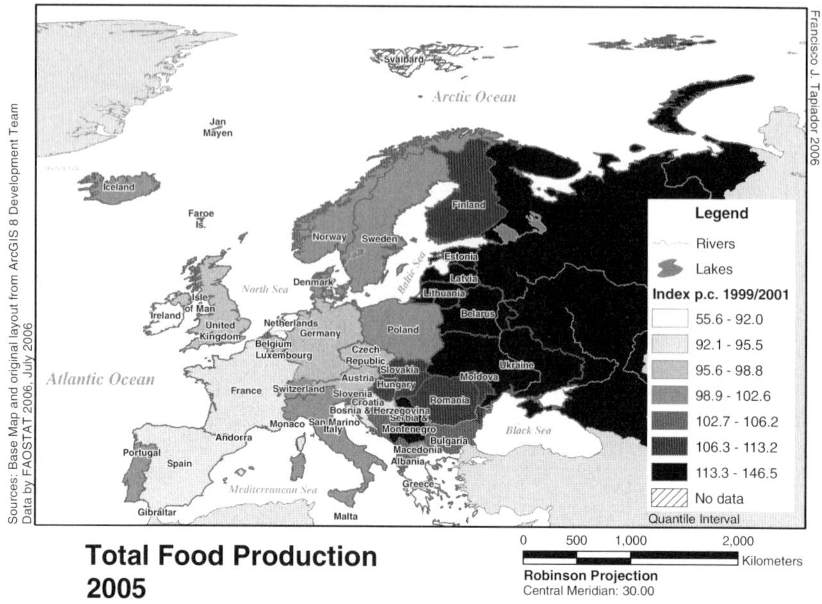

Fig. 8.55 Total food production in 2005 compared with 1999/2001

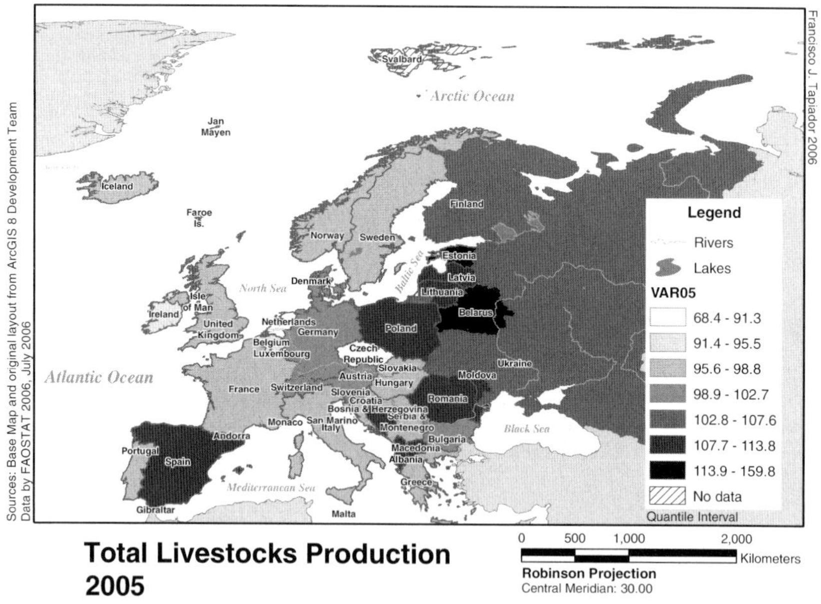

Fig. 8.56 Total livestock production in 2005 compared with 1999/2001

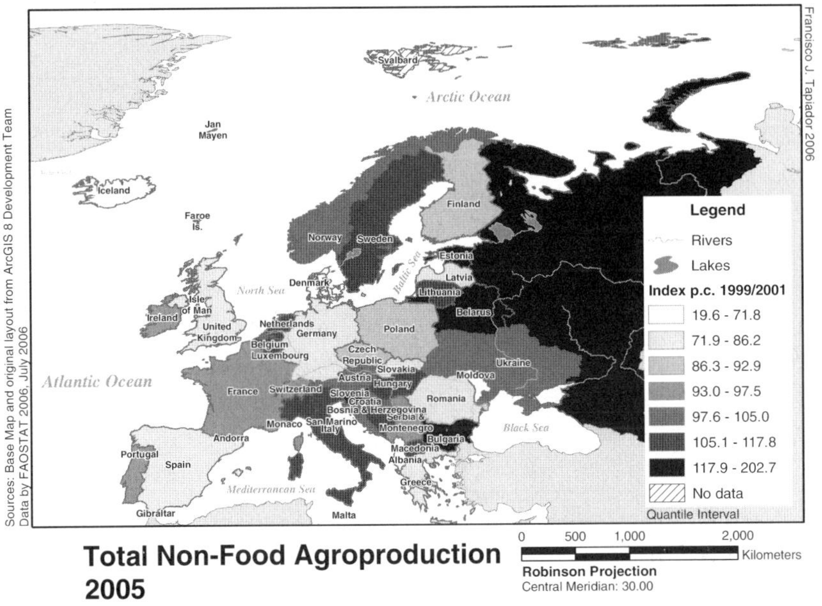

Fig. 8.57 Total non-food agroproduction in 2005 compared with 1999/2001

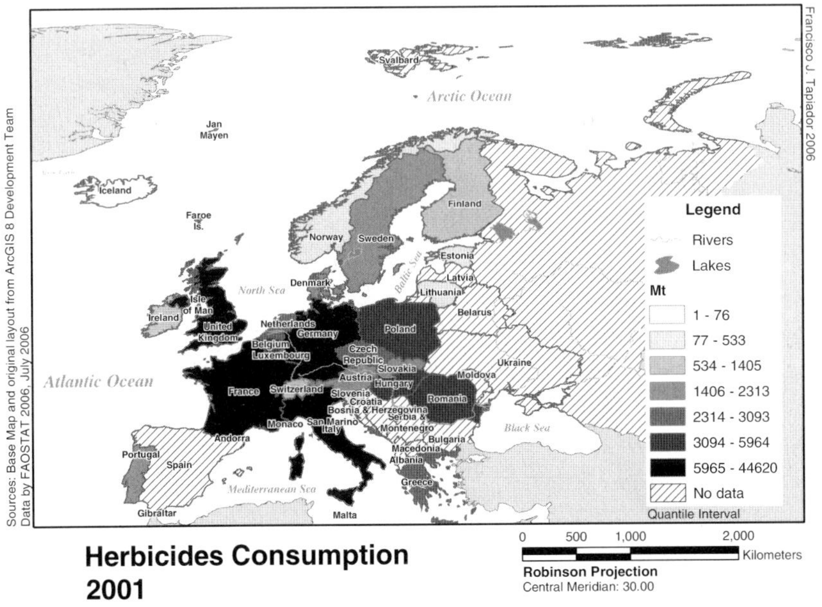

Fig. 8.58 Herbicides consumption in 2001

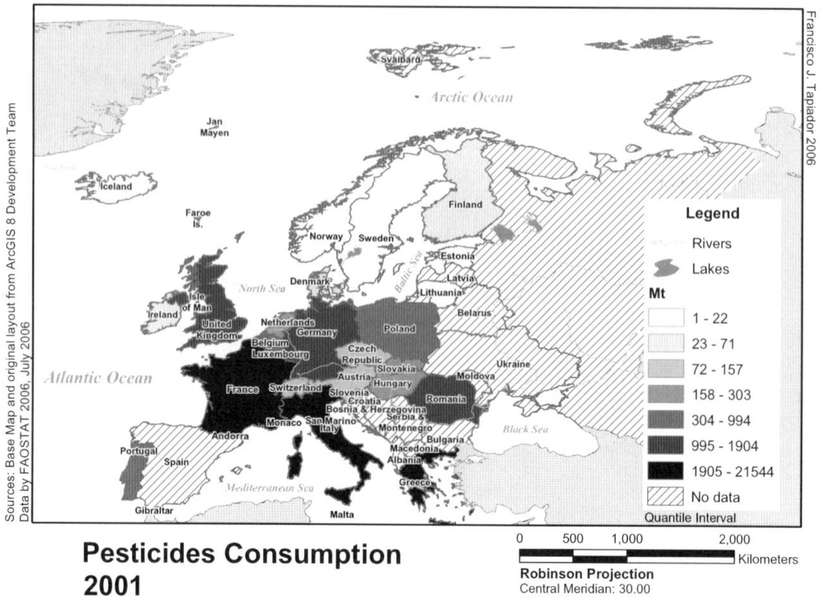

Fig. 8.59 Pesticides consumption in 2001

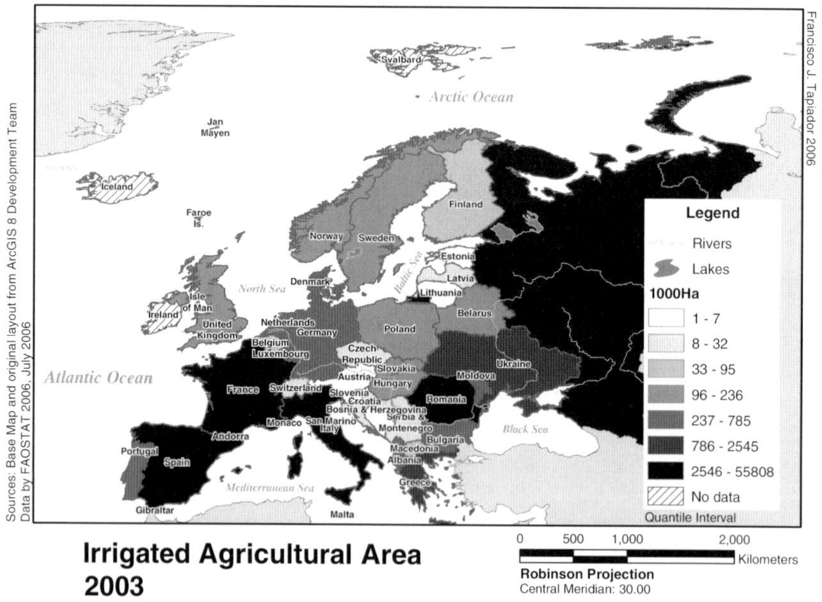

Fig. 8.60 Irrigated agricultural area in 2003

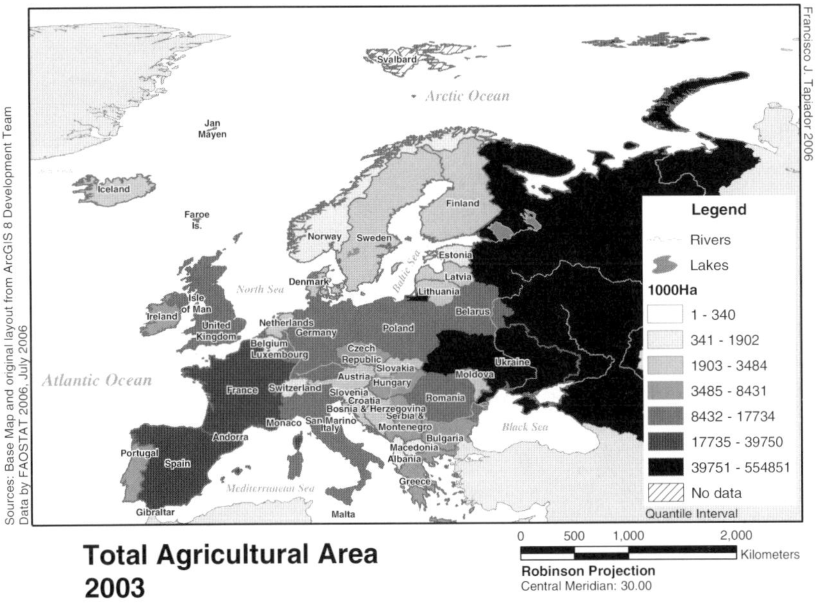

Fig. 8.61 Total agricultural area in 2003

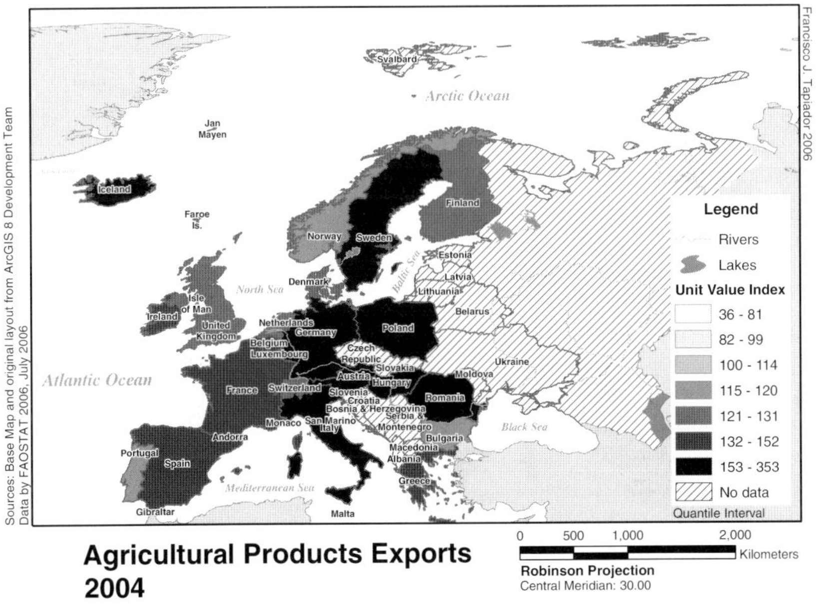

Fig. 8.62 Agricultural products exports in 2004

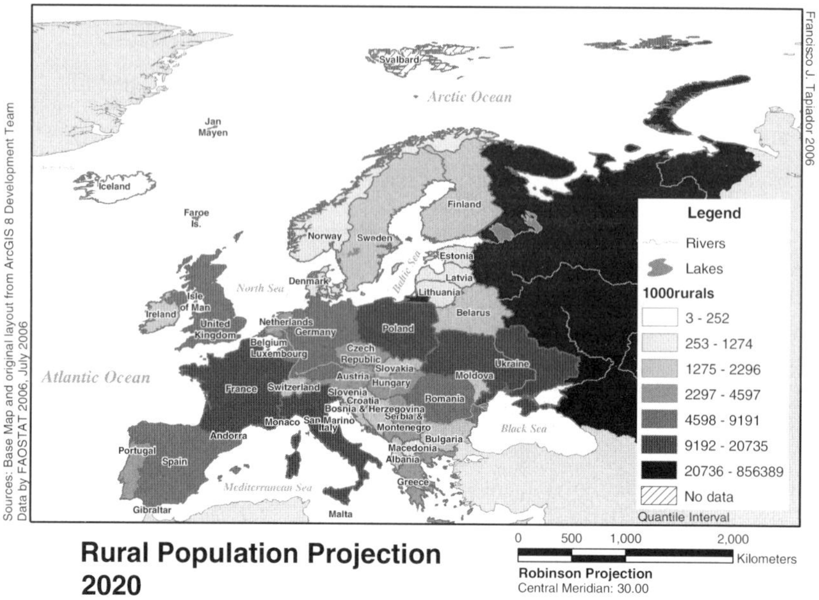

Fig. 8.63 Rural population projection in 2020

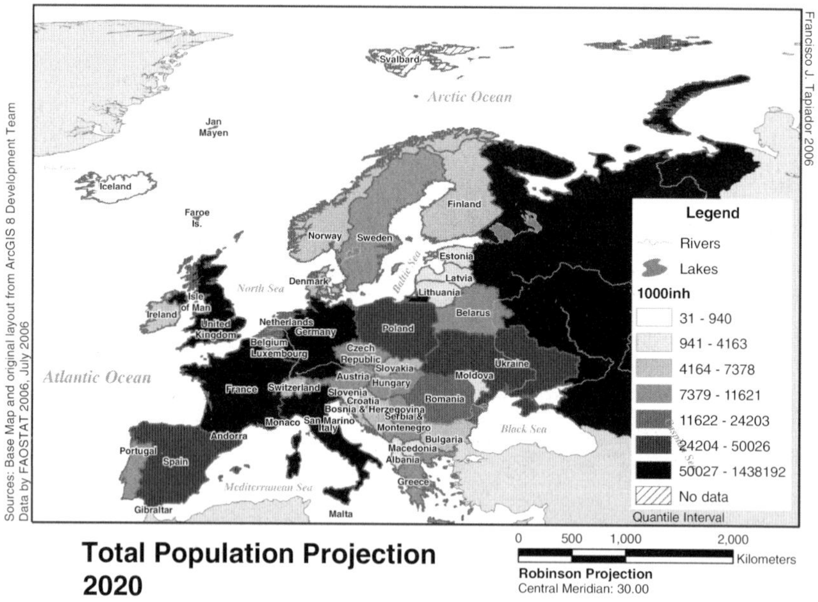

Fig. 8.64 Total population projection in 2020

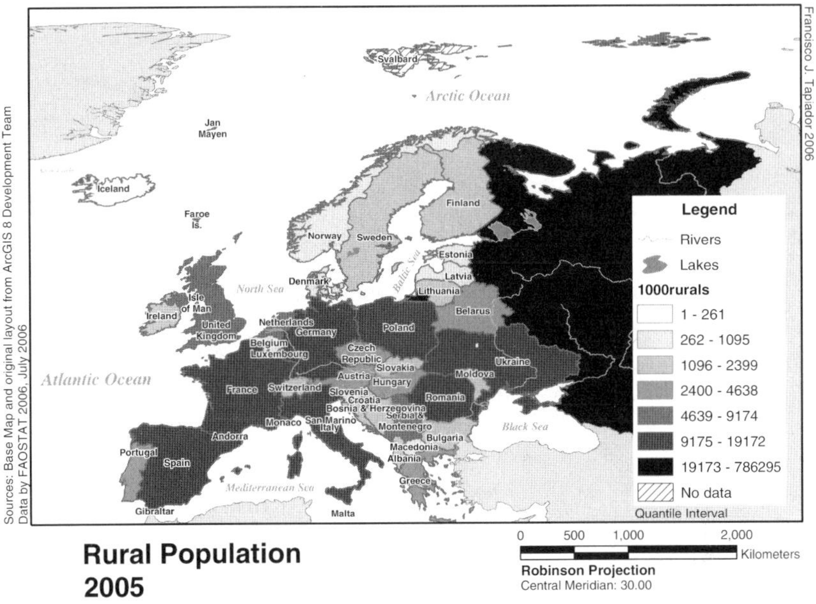

Fig. 8.65 Rural population in 2005

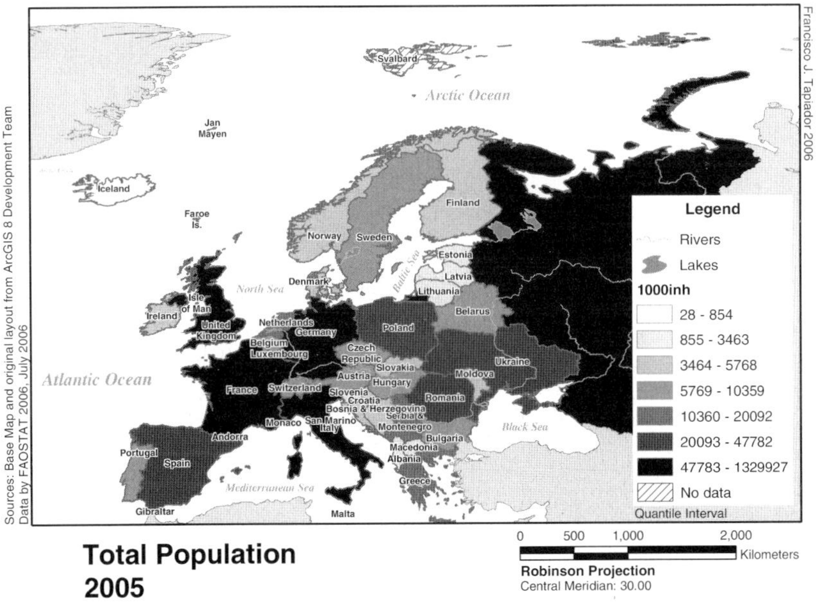

Fig. 8.66 Total population in 2005

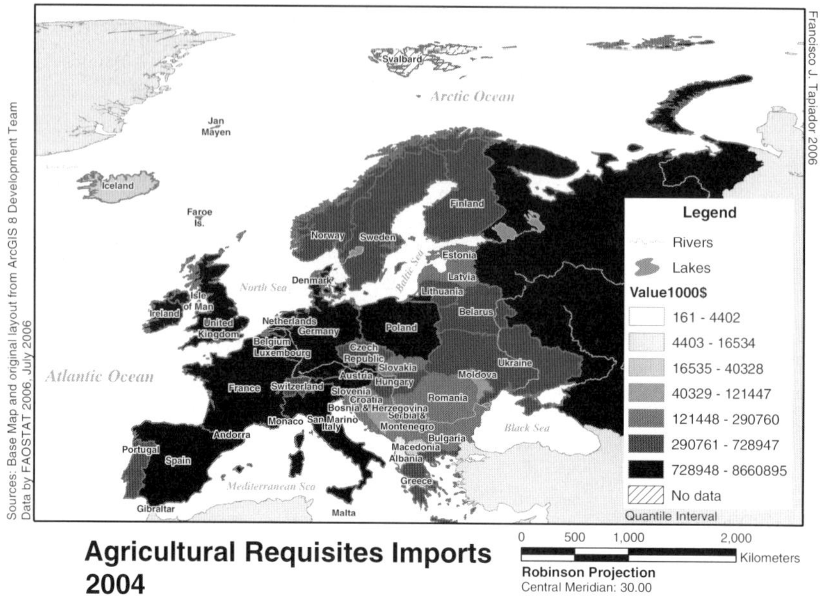

Fig. 8.67 Agricultural requisites imports in 2004

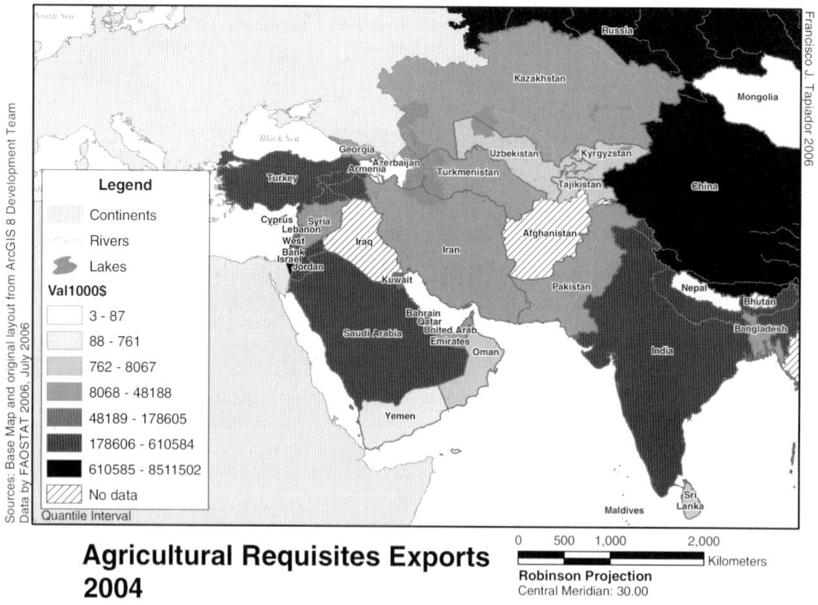

Fig. 8.68 Agricultural requisites exports in 2004

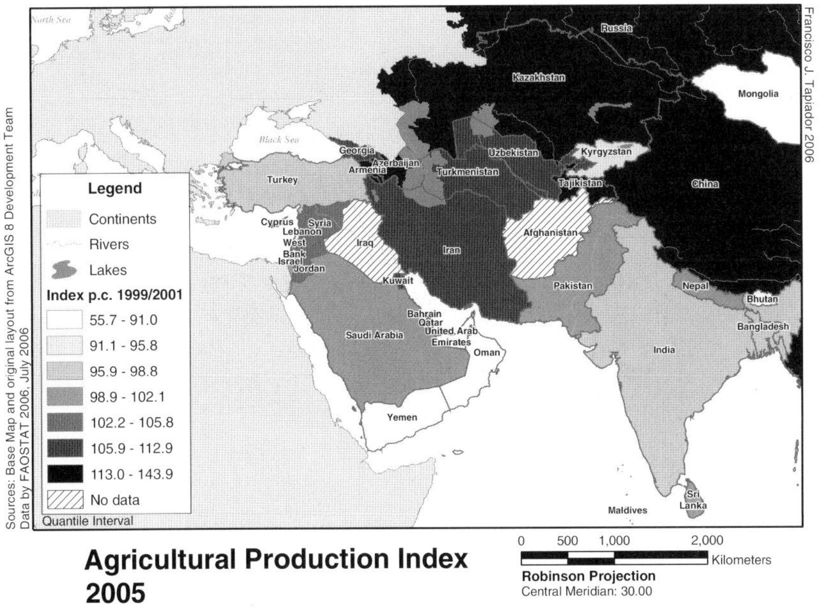

Fig. 8.69 Agricultural production index in 2005 compared with 1999/2001

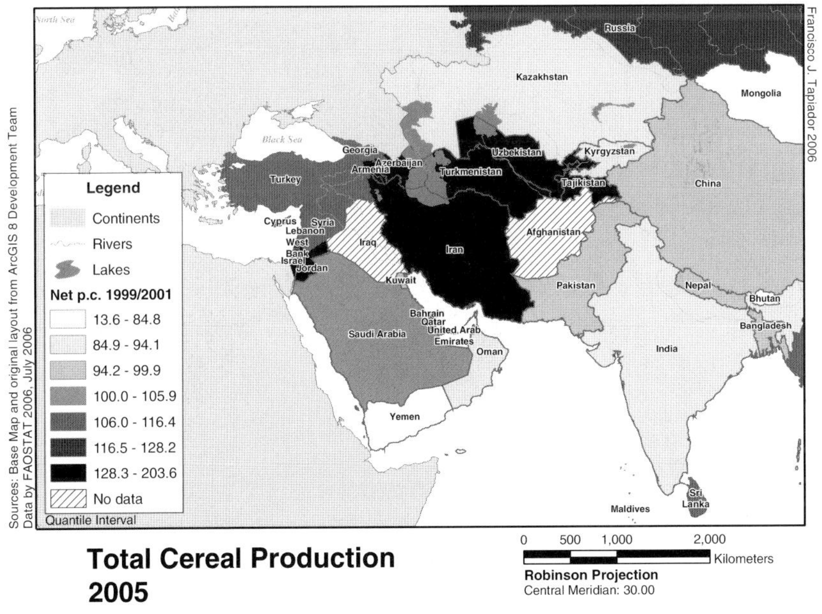

Fig. 8.70 Total cereal production in 2005 compared with 1999/2001

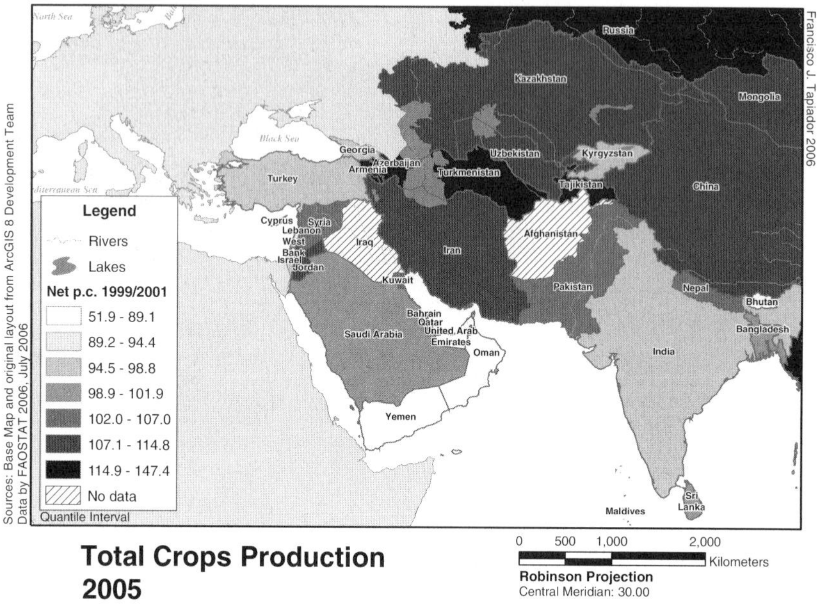

Fig. 8.71 Total crops production in 2005 compared with 1999/2001

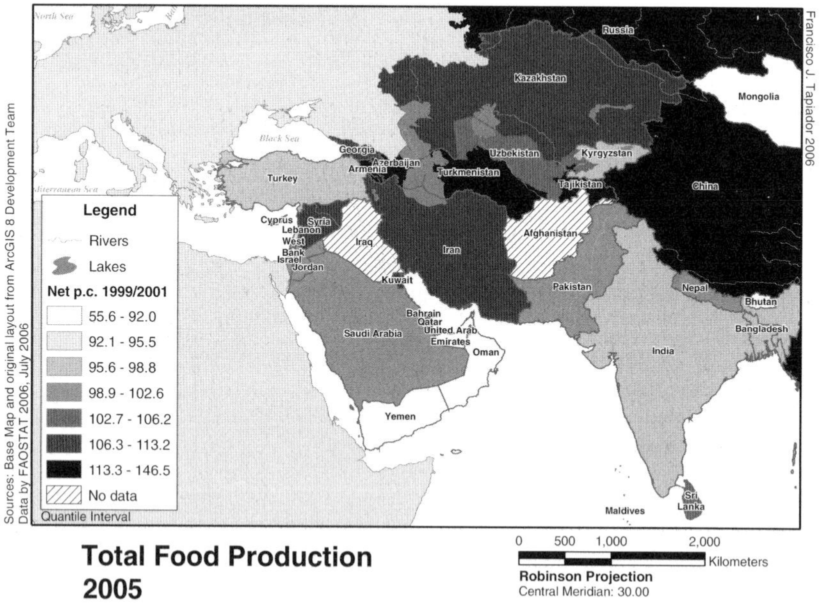

Fig. 8.72 Total food production in 2005 compared with 1999/2001

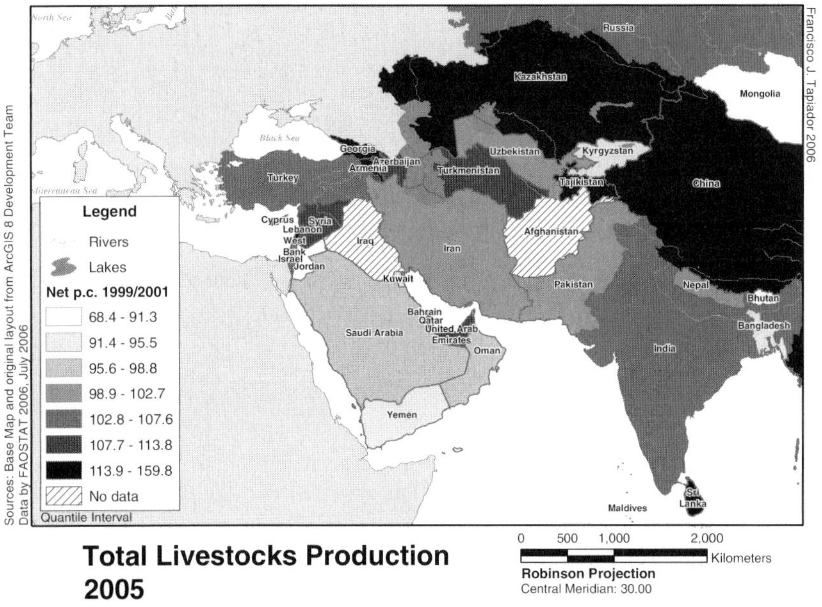

Fig. 8.73 Total livestocks production compared with 1999/2001

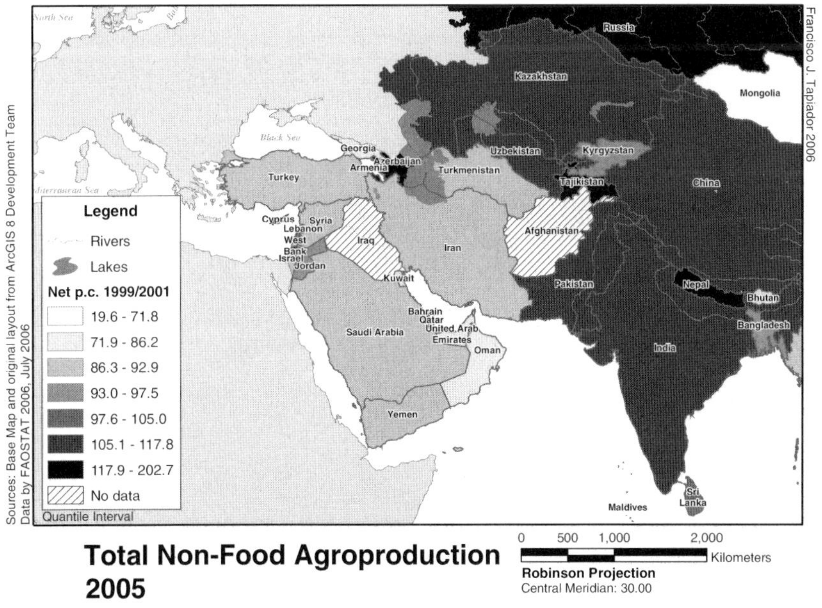

Fig. 8.74 Total non-food agroproduction in 2005 compared with 1999/2001

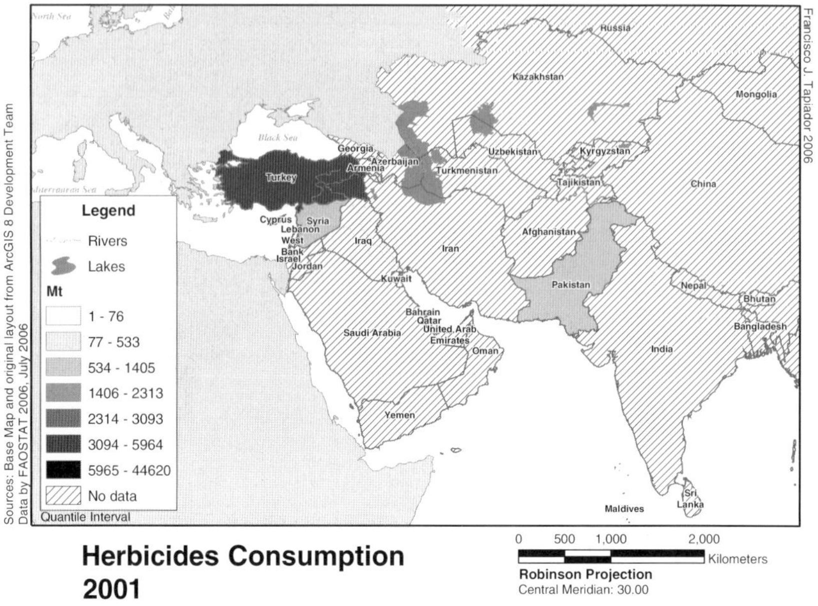

Fig. 8.75 Herbicides consumption in 2001

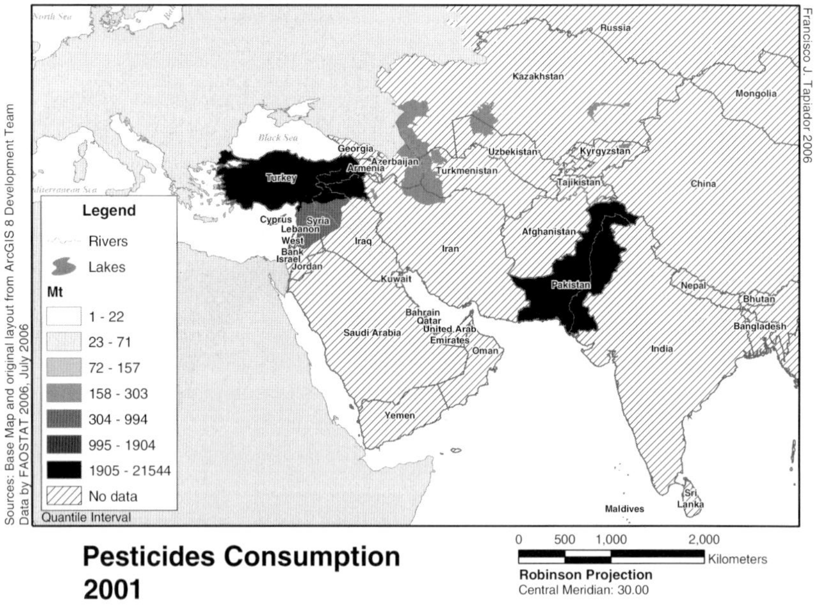

Fig. 8.76 Pesticides consumption in 2001

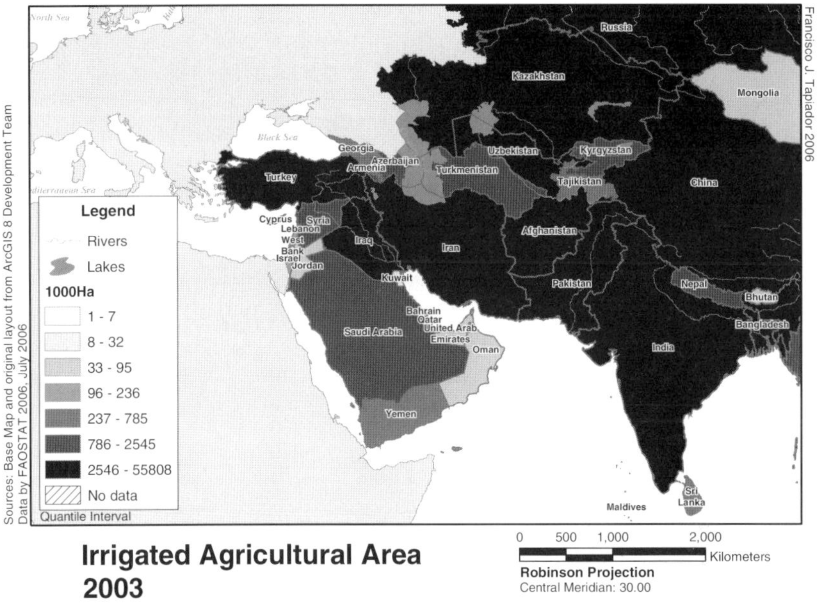

Fig. 8.77 Irrigated agricultural area in 2003

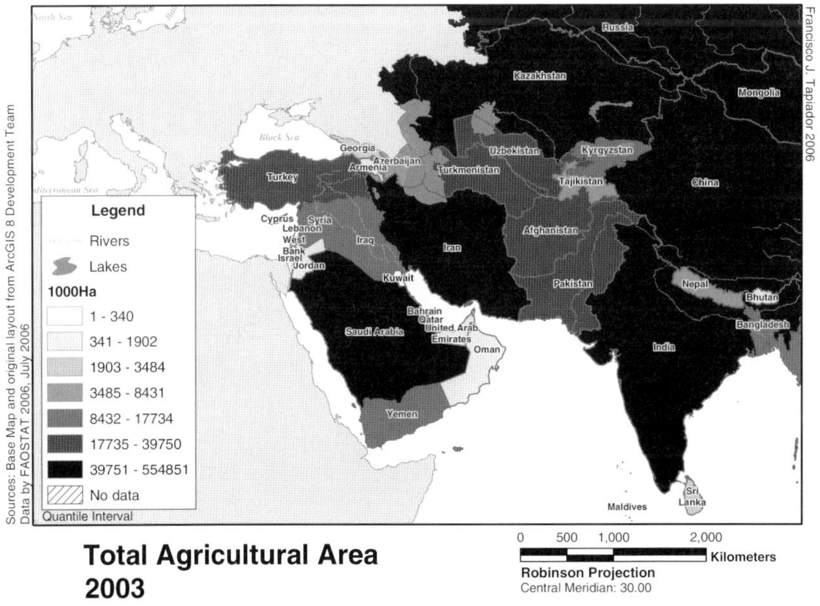

Fig. 8.78 Total agricultural area in 2003

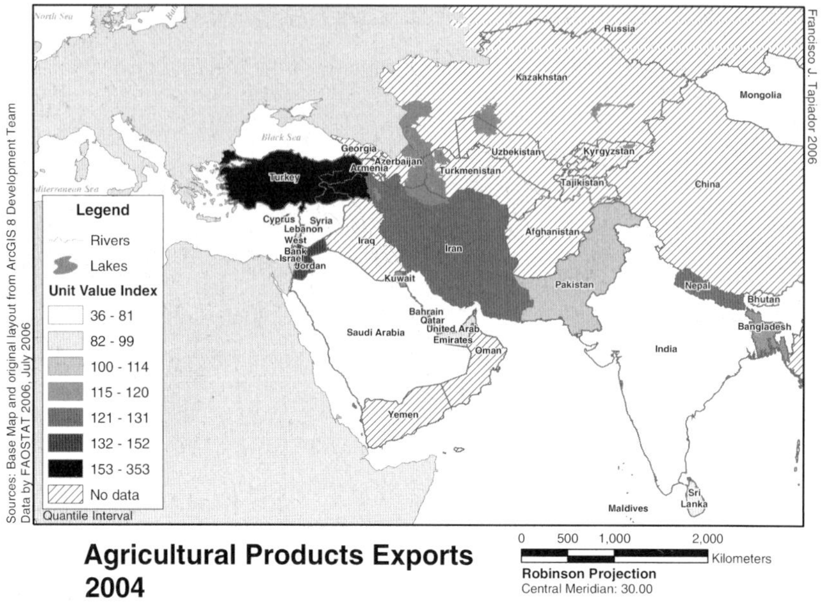

Fig. 8.79 Agricultural products exports in 2004

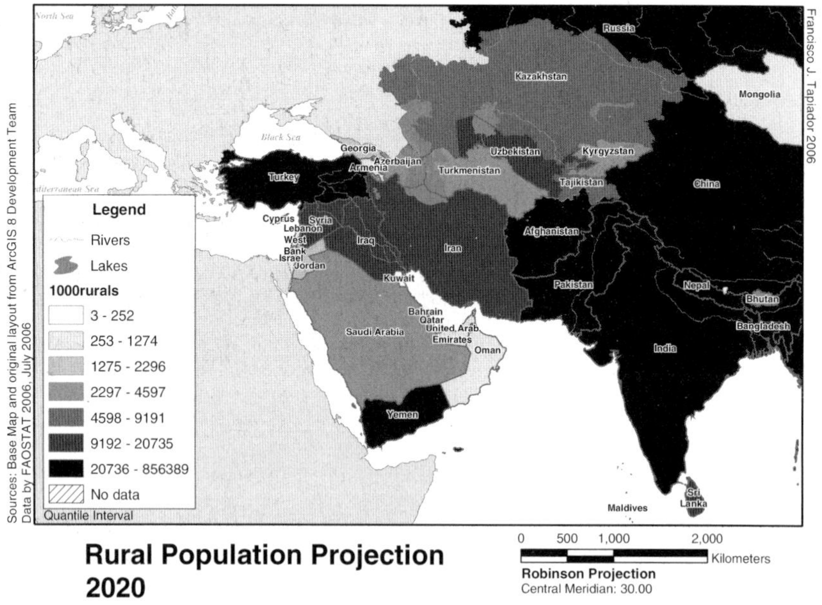

Fig. 8.80 Rural population projection in 2020

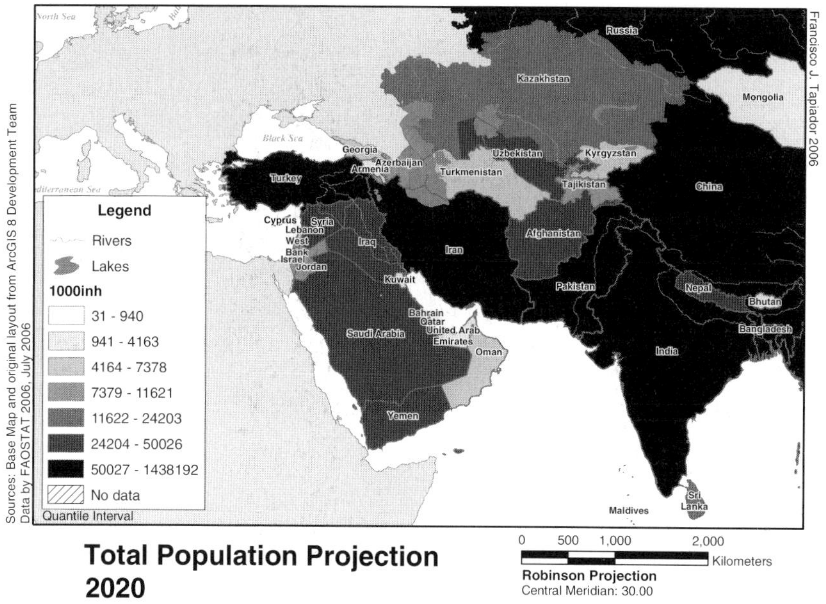

Fig. 8.81 Total population projection in 2020

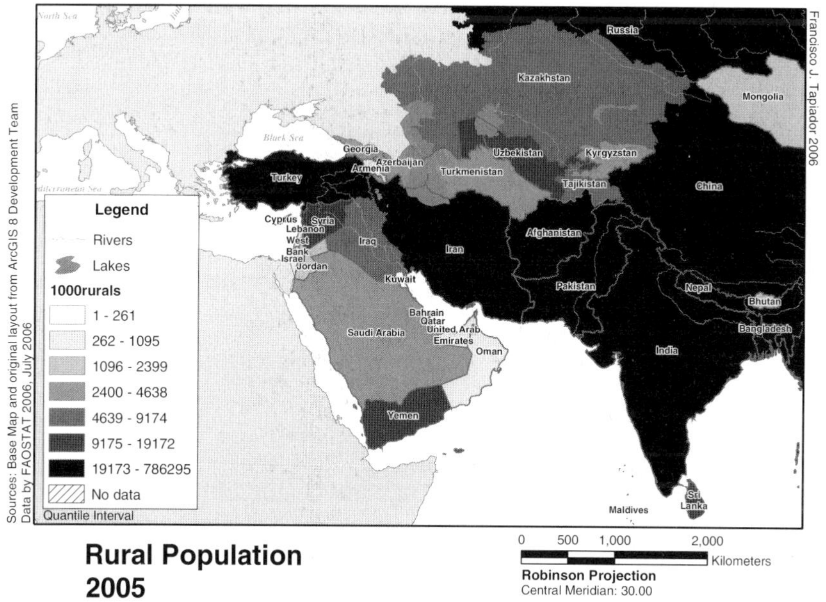

Fig. 8.82 Rural population in 2005

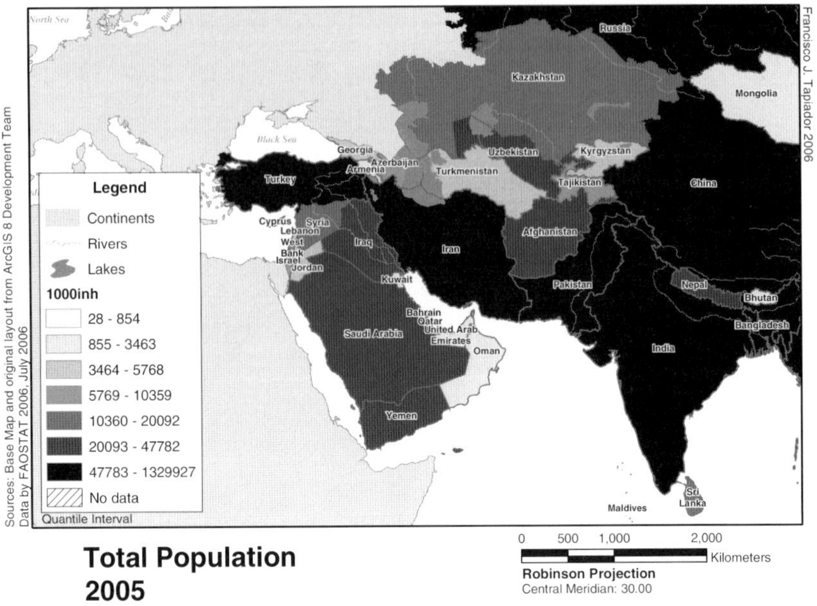

Fig. 8.83 Total population in 2005

Chapter 9
Regional Issues in the Rural Areas of the World

9.1 North America

The rural areas of North America (Canada, the US of America and the US of Mexico) present very different rural characteristics. Far from being an homogeneous area, the North part of the continent exhibits a variety of climates (from Polar to Tropical), occupation patterns, and cultures (from the Canadian melting pot to the more homogenous Mid-West of the USA) that generate marked differences in the structure of the rural areas and in the problems and opportunities they face.

Migration is currently one of the emergent issues in the developed areas of the subcontinent. Ex-urbanization (or counter-urbanization) has been fuelled by the misperception of improved quality of life in rural areas. The pace of this process has increased in the last few decades, especially in the USA. In some cases, as reported by Egan and Luloff (2000), newcomers soon affect local rural life: sudden changes in social structure and in local economy may affect old local forestry practices and policies, generating a potential conflict and imbalance in the delicate rural structure. The case of Mexico is different, with a progressive abandoning of rural areas in favour of city opportunities. The host of problems found here are similar to the areas of Central and South America. Durand (2003) analyses the migratory conflict and its consequences; such as increased gentrification of indigenous communities and farmers. Another related problem is the food supply and the need of regulatory instruments to ensure at least the self-sufficiency of the country.

The impact of new rural-related technologies cannot be disregarded and the effects should be anticipated: for example, the emergence of bio-diesel, -a potential opportunity for rural areas suited to corn cultivation- yielded a sharp increase in the price of this cereal, which constitutes the basic diet of many people. Here, the pace of industrial innovation was faster than the rural ability to take advantage of this new opportunity. Even if some rural areas did benefit from the price increase of the raw material, the problem propagated to urban areas, illustrating the complex relationships between countryside and cities.

Genetically-modified crops are another key issue here. Modern farming practices need scale economies to be competitive, and high performance varieties are thus a valuable resource. Nonetheless, a blind generalization of modern technologies may damage the delicate balance of the ecosystems. In Mexico, these problems are linked with natural resource destruction and the patenting of local ancestral knowledge on plants and animals by trans-national companies.

Rural poverty is not limited to developing countries but also to technological societies. Following Tickamyer and Duncan (1990) some of the most deprived areas in the USA were rural areas in the eighties, and that situation has not changed. As in the urban counterpart, rural poverty is related with both limited opportunities and social and economic policies.

Environmental degradation is also an issue in the rural areas of those developed countries. Inadequate resource management and agrarian practices may yield from moderated degradation to serious contamination in otherwise fertile lands. The effect on groundwater needs also to be addressed. Nitrate pollution in the USA due to overfertilization or to inadequate irrigation (Power and Schepers 1989) is an issue not to be overlooked. As it happened in the past or in other areas, as rural areas go beyond the subsistence threshold, environmental problems appear as the second immediate concern.

9.2 The European Union

The European Union (EU) faces a series of rural problems steaming from the economic development model followed in the past, and from the political nature of the Union itself. It is sometimes said that most of the EU budget is devoted to agriculture, whereas only a 3% of the population is engaged in this economic activity. Being this is a fact, it is also true that rural Europe plays a threefold role. It has to produce food and raw materials in a sustainable framework, it has to respect alleged ecological balances and host the leisure activities of urbanites.

There is both structural and temporary problems in the rural areas of Europe, such as the limited scale economies achievable by relatively small farms, and the protectionist framework in which markets operate. Being that agriculture is a subsidized activity, the incentives for improvement are somewhat limited. Some farmers adapt their production to adjust for the crop being subsidized that year, instead of following their own market strategies or even abandon farming altogether for other more productive activities.

Today, the EU can be regarded as the most environmentally concerned society on Earth. Within this framework, the massive use of fertilizers, herbicides and other chemical inputs required by current intensive farming practices has generated conflicts in some areas. Also, erosion, deforestation and land contamination are regarded as serious problems for the future development of some European rural areas. Synergetic approaches to agriculture and biodiversity have been described (Rounsevell et al. 2006), arguing for a joint approach to biodiversity/agriculture analysis in terms of vulnerability.

Current problems in the rural anthropic elements are in the form of an ageing of the rural population and its associated problems. As could be expected, more dynamical areas, such as those with recreational activities or agrarian activity, are less prone to suffer from progressive ageing of the population. At the end, economy plays a determinant role in fixing population in the countryside, and in those areas the prospects of making a life for a young population impedes a slow degradation of the habitat.

9.3 The Post-socialist Rural Areas

Privatization -after the socialist period in the countries of the former USSR- generated several challenges associated with a progressive land fragmentation and a dichotomy between those adapting for modern agrarian markets and those falling to subsistence farming practices. The variety of patterns in rural post-socialist Russia reflects the sudden changes in the policies of this country, with a growing difference between big and small landlords. Today, a large number of farmers in Eastern Europe own the land they farm but have a disparate access to technology. It has been argued that the socialist regime destroyed farmers' identity and their entrepreneurial initiative. Most analysts emphasize legal and political obstacles, and rural population resistance to land privatization, whereas others such as Wengren (2002) call attention upon economic factors. In modern economies, land ownership is not a good by itself. Machinery, fuel, fertilizers, herbicides and seeds are all expensive but required to make land cultivation a profitable business.

Land degradation is a major concern in areas that have undergone intensive exploitation in the past. Common problems include ecological habitat destruction, soil erosion and contaminated soils. Nonetheless, the pace of degradation has slowed down in the last decades due to extensive cultivation and abandoned crops, with a renewed interest in valuable land reclamation and biodiversity preservation.

Rural employment has experienced a slight improvement thanks to rural diversification. Nevertheless, there is room for an exponential improvement in the quality of life of those areas. It is foreseeable that an overall improvement in those countries' economies will permeate first to the most dynamic cities and towns, and then to the countryside. How can rural areas contribute to that development is an open research question, but a careful preservation of vulnerable natural areas should be a necessity in order to avoid the mistakes made in the past in some western rural areas. This will not only be a middle-term economic investment, but a tool for new economic activities to appear in those areas.

9.4 Central/South America and the Caribbean

Extreme poverty is a widespread element in many rural areas of Central/South America and the Caribbean. As it is well known, rural poverty is deeper and more difficult to overcome than urban poverty. Subsistence in an agrarian framework is not always guaranteed, and redistribution cannot be guaranteed everywhere.

A relevant factor of the current situation is social inequality. In those areas extreme wealth and poverty coexist back to back. Large, almost unexploited properties remain in the very same areas where some people have no land to cultivate and struggle to survive. The problem is particularly acute with the indigenous population. From the native civilizations and the European conquistadores, some people have been left behind and it is a fact that poverty strikes most harsh in such populations. Most authors agree that a development strategy should first tackle social inequalities. Agrarian reforms have been tried in this area but reducing lands speculative accumulation is one of the proposed prior measures before attempting redistribution.

An intense exploitation of the natural resources and a loss of biodiversity are also two major problems in this area. A marked reduction in forest cover to agrarian uses is apparent in areas of the Amazon and the Orinoco rivers. Problems associated with suboptimal use of the land and natural resources, with rural populations presenting limited incentives for improving management. Policies to improve farming practices and diversify crops may help to alleviate rural poverty. Other authors warn about the motivation on the rural decision-makers. The progressive collapse of natural resources in some areas, and the abandoning of crops have also motivated a growing rural exodus to cities. Instead of reducing the pressure on the land, this process has induced decline and impoverishment in many areas, where the young population has left. Another growing problem is the increase of agribusiness exploitations, introducing crops not always suited for the natural conditions of the area and completely unconnected with the alimentary needs of the local population.

9.5 South-East Asia

Rural poverty spans across large areas of South-East Asia. Countries such as Myanmar, Laos and Nepal are regarded as spaces with the largest structural agrarian problems in the World. Agriculture impedes extreme poverty, but productions are too slim to generate trade. Also, uncertainties in production and several risk factors impede overcoming the poverty threshold. As Warr (2000) analyses, amongst the factors of poverty in this area structural factors and crisis coexist. Recovery from crisis (such as the 1997 crisis) is hindered by structural weaknesses. On the one hand, environmental degradation affects large areas. Diffuse pollution, soil erosion and land use changes have adversely affected the potentialities of many landscapes. Deforestation, an unordered expansion of the agriculture and a changing pattern of food and timber in international markets are factors to be considered in understanding the problems facing this area. Political instability is another factor to be taken into account when analysing this area. Also, it is here where natural disaster risks are amongst the largest on the planet, with the rural areas being the most severely affected.

Problems related with poultry and cattle are similar to developed countries. Massive and disordered expansion to intensive husbandry systems have brought soil and groundwater contamination, including heavy metals. In a developing scenario, those problems are usually considered minor nuisances, but an intelligent and comprehensive strategy would encompass rural development with natural and environmental care. The cost associated with a proper strategy does not necessarily exceed that of a rushed development.

The role of infrastructures is crucial for this area. A crucial problem was highlighted by *The Expert Group Meeting on Integration of Energy and Rural Policies and Programmes, 2003*. Namely that energy in Asia is seldom considered in rural planning, being a crucial element for alleviating poverty. On the other hand, accessibility has been proven to affect development (Castella et al. 2005). Improvements in the transportation network should also be aware of remote rural areas, aiming to improve connectivity and thus access to resources, services and markets.

9.6 China

It is easy to forget when analysing China that this country represents more than one fifth of the World population, and the current situation of the country does not adequately reflect the role of this country in human history. China is not just a country, but the result of a civilization. The western perception of the country is tainted by an educational veil in which history is told as the history of the western civilization. Being the world's fastest growing economy, it is almost certain that China will become the first world's economic power in the early 21st century.

Rural problems in China today are related with a limited supply of cultivated land for a growing population (1,313 million people in 2006). Ecological disorders are related to an explosive growth and need of raw materials, the expansion of urbanization, rural migration, and a startling disparity in wages and opportunities between rural and urban areas.

China will continue to grow in the near future. Landscapes are being altered to accommodate for rapid industrialization and a large mass of population is changing their way of life at an unprecedented rate. Private ownership was made legal in early 2007, a change that may benefit rural areas by encouraging individuals to increase productivity and think of new ways to improve their quality of life. In a country in which the GDP is half of the World average (2003 data), subsistence agriculture is still widespread, but new developments in biotechnology, renewable energies, and water policies are being applied with noticeable success.

The potential for rural planning is immense in China. Promoting efficiency in how the land resources are exploited, and in programs to improve rural energy supplies, and an overall intelligent rural planning that accounts for as many sectors and areas as possible could greatly benefit the country as a whole. The large scale of the plans, the challenge of public participation, and the enormous resources available make rural planning in China of primary interest for anyone interested in rural planning. A two-way exchange between rural researchers and practitioners in China and abroad could be mutually beneficial, providing each side acknowledges the particularities and history of the other side.

9.7 India

As in China's case, when approaching the problems in India it is advisable to change the scale. India has a population of 1,129 million people, which is about one sixth of the population of the World (6,525 million people).

Poverty marks rural India. In a country where a 26–44% of the population is below the poverty line (Borooah 2006), rural areas are those suffering the most with large differences between the Indian states.

A large part of the Indian economy has long been agriculture, with agrarian sufficiency in a pre-eminent place. After getting rid of colonial patterns in rural planning, India was one the first countries involved in the Green Revolution, but globalization is starting to generate unwelcome effects (Shiva 2004). Rural monopolies are back, people are dying of hunger, and at the same time, food is affected by chemical pollution of agribusiness. Infections are also a major concern. Almost 90% of HIV-infected people in Asia live in India (Abeysenaa and Janaka de Silva 2005). This represents a major concern not only in the cities but also in rural areas. Any development strategy should account with these constraints in mind.

One of the challenges of rural planning is to make the good health of the macro indicators (India is one of the world's first producers of wheat and has its own space program) compatible with a limited local development.

9.8 Japan

Rural problems of an opulent rural society, such as the Japanese countryside, are quite different from the problems faced in India. The improvement of socioeconomic conditions in rural Japan has generated an increase in anthropized landscapes, and a loss of biodiversity to the point of some species being in danger of extinction (Iiyama et al. 2005). The conservation of traditional landscapes is one means to avoid further problems. The development of new management systems considering ecological premises is a possible course of action.

Another issue of concern is crop-induced erosion. Expansion of cultivated land and mechanization can reduce soil infiltration, allowing wind and runoff to erode. Landscape and stream restoration, remediation, slope contention and reforestation are practices that can help to tackle such problems. The problem of soil contamination is also present here. Nitrogen excess in fertilizers causes problems not only to crop productivity but also to water supplies.

On the anthropic elements side, the ageing rural population is a serious concern to the point of new imaginative plans of rural habitation being proposed (Ogawa 2004). The role of public care and how to provide recreation activities for such population are also current discussion issues. As in other opulent societies, the major problems facing rural planning here are orientated to an improvement in the quality of life of the rural population and not to survival strategies.

9.9 Africa

Africa is a continent with a high proportion of people under extreme deprivation. Sub-Saharan rural Africa is probably the poorest region on Earth. Both endogenous and external causes are to blame for this situation (Dorward et al. 2004). Rural poverty in Africa is linked with both agrarian shortcomings and the absence of other alternatives. Hunger is a daily problem and bad management, regional conflicts, climate change, AIDS and international trade agreements are deepening the problem. Sub-Saharan Africa has been systematically ignored in the globalization scheme as it does not play a major role in the international market share in spite of its size. Other major problems in Africa are malaria and HVI/AIDS, which is propagating at an impressive pace leaving behind entire generations.

Mitigation policies are unsuccessful due to limited resources, inversions and public management. Corruption is a serious drain in improving life and in optimizing resources. The way this part of Africa exploits its resources is crucial to the future of the continent, with China having a growing investment role. Examples of good practices in forest management (Shackleton et al. 2007) shows the potential of natural resources to avoid extreme poverty, but also shows the fragility and dependence of those rural societies.

It is debatable if rural planning here has to be focused on leaving the locals to follow their own projects or in proposing foreign development strategies. It is clear that part of the responsibility of the current situation can be traced back to *western* society's intervention in the past, but reality shows that local impulse is lacking, and that external help is needed. How to combine an aseptic non-intervention policy with thousands of human beings dying every day of hunger requires a clear perspective on what is acceptable not only in planning, but on ethical grounds.

9.10 Australia and Oceania

Almost 10% of the Australian population lives in rural areas, and in some States economy is highly dependent on agriculture. Since the fifties, drought has been a major concern in Australia, so irrigation and water policies are the focus. Ecological problems add up to crop production, reducing the GDP up to 1.6% for this single cause (Horridge et al. 2005 data for the 2002–2003 droughts). This also permeates other sectors of the economy, increasing unemployment and the exterior trade.

Other problems are related with water are eutrophization, forest degradation and loss of biodiversity. Forest fires are also a major concern, affecting not only property but life itself (Malcom 2005). An integrated approach to palliate its effects is an integral part of rural development strategies. In a framework of rural depopulation and low population density, the need to coordinate policies and pool national, regional and local resources is instrumental in solving rural problems with national effects, and to improve the quality of life of the rural population.

9.11 The Middle East

Water availability is also a major problem in the Middle East. Hydrological resources are scarce and future demands are not guaranteed. Droughts in the nineties clearly showed the problems the area might be facing in the near future in a global warming scenario. Water in the Arabian Gulf is in the form of groundwater, and increased urbanization and industrialization are starting to press on even fossil waters. Sea water desalination could be an alternative, but this is a costly strategy. Also, there are ecological concerns on the impact of those installations on the environment. Nonetheless, any rural planning strategy should balance different alternatives in an objective way.

Poverty is not absent in those oil-rich fields and political instability does not help to promote a more equal society. Migration to cities is a common process. In the case of Israel though, migration is being done in the opposite direction (Sofer and Applebaum 2006). The *moshav* communities are regaining their appeal, in part due to political and personal safety reasons. Changes in local communities brought on by the newcomers have been described. Israel also shows how technology can help to overcome natural constraints and limitations.

References and Bibliography Used

Areiqat, A.Y., Mohamed, K.A. (2005): Optimization of the negative impact of power and desalination plants on the ecosystem. Desalination, Vol. 185, Issues 1–3, 1 November 2005, pp. 95–103

Avni, Y., Porat, N., Plakht y Avni, G. (2006): Geomorphic changes leading to natural desertification versus anthropogenic land conservation in an arid environment, the Negev Highlands, Israel. Geomorphology, Vol. 82, Issues 3–4, 15 December 2006, pp. 177–200

Bryceson, D.F. y Fonseca, J. (2006): Risking death for survival: Peasant responses to hunger and HIV/AIDS in Malawi. World Development. Vol. 34, Issue 9, September 2006, pp. 1654–1666

Castella, J.C., Hung Manh, P., Pheng Kam, S., Villano, L. y Tronche N.R. (2005): Analysis of village accessibility and its impact on land use dynamics in a mountainous province of northern Vietnam. Applied Geography, Vol. 25, Issue 4, October 2005, pp. 308–326

Chaplin, H., Davidova. S. y Gorton, M. (2004): Agricultural adjustment and the diversification of farm households and corporate farms in Central Europe. Journal of Rural Studies. Vol. 20, Issue 1, January 2004, pp. 61–77

Deininger, K., Zegarra, E. y Lavadenz, I. (2003): Determinants and Impacts of Rural Land Market Activity: Evidence from Nicaragua. World Development. Vol. 31, Issue 8, August 2003, pp. 1385–1404

Díaz Gacitúa, M. (2000): El desarrollo rural y el medio ambiente en las últimas décadas: dos pasos atrás y uno adelante. Pontificia universidad Javeriana. Seminario internacional 'La nueva ruralidad en América Latina'. Bogotá, Colombia. Agosto 2000

Dorward, A., Kydd, J., Morrison, J. y Urey, I. (2004): A Policy Agenda for Pro-Poor Agricultural Growth. World Development. Vol. 32, Issue 1, January 2004, pp. 73–89

Durand Alcántara, C. H. (2003): La cuestión agraria: un análisis de coyuntura (el caso mexicano). Agroalim, jun 2003, Vol. 8, n° 16, pp. 55–63

Egan, A. y Luloff, A. (2002): The Exurbanization of America's Forests: Research in Rural Social Science. Journal of Forestry, Vol. 98, Number 3, 1 March 2000, pp. 26–30(5)

Ellis, F. y Mdoe, N. (2003): Livelihoods and Rural Poverty Reduction in Tanzania. World Development. Vol. 31, Issue 8, August 2003, pp. 1367–1384

Emdad Haque, C. (2005): Perspectives of Natural Disasters in East and South Asia, and the Pacific Island States: Socio-economic Correlates and Needs Assessment Natural Hazards. Vol. 29, Number 3 / July, 2003

Gerber, P., Chilonda, P., Franceschini, G, y Menzi, H. (2005): Geographical determinants and environmental implications of livestock production intensification in Asia. Bioresource Technology, Vol. 96, Issue 2, January 2005, pp. 263–276

Hadad, M. Y Lindner (2001): Sustainable water demand management versus developing new and additional water in the Middle East: a critical review. Water Policy, Vol. 3, Issue 2, June 2001, pp. 143–163

Hölzela, N. et al. (2002): The return of the steppe large-scale restoration of degraded land in southern Russia during the post-Soviet era. Journal for Nature Conservation. Vol. 10, Issue 2, 2002, pp. 75–85

K. M. Homewood, 'Policy, environment and development in African rangelands', Environmental Science & Policy 7 (2004), 125–143

Horridge, M., Madden, J. y Wittwer, G. (2005): The impact of the 2002–2003 drought on Australia. Journal of Policy Modeling, Vol. 27, Issue 3, April 2005, pp. 285–308

Iiyama, N., Kamada, M., y Nakagoshi, N., (2005): Ecological and social evaluation of landscape in a rural area with terraced paddies in southwestern Japan. Landscape and Urban Planning Vol. 73, Issue 1, 15 August 2005, pp. 60–71

Jurskis, V. (2005): Eucalypt decline in Australia, and a general concept of tree decline and dieback. Forest Ecology and Management, Vol. 215, Issues 1–3, 25 August 2005, pp. 1–20

Kay, C. (2005): Reflections on Rural Poverty in Latin America. The European Journal of Development Research. Vol. 17, Number 2/June, 2005.

Kay, C. (2006): Rural Poverty and Development Strategies in Latin America. Journal of Agrarian Change 6 (4), 455–508

Lin , G. C. S. y Ho S. P. S. (2003): China's land resources and land-use change: insights from the 1996 land survey . Land Use Policy. Vol. 20, Issue 2, April 2003, pp. 87–107

Lofgren, H. y Richards, A (2003): Food security, poverty, and economic policy in the Middle East and North Africa. Research in Middle East Economics, Vol. 5, 2003, pp. 1–31

Malcom Gill, A. (2005): Landscape fires as social disasters: An overview of 'the bushfire problem'. Global Environmental Change Part B: Environmental Hazards. Vol. 6, Issue 2 , 2005, pp. 65–80

Masoud, A.A. y Koike, K. (2006) Arid land salinization detected by remotely-sensed landcover changes: A case study in the Siwa region, NW Egypt. Journal of Arid Environments, Vol. 66, Issue 1, July 2006, pp. 151–167

Mensah, O.A. y Kumaranayake, L.(2004): Malaria incidence in rural Benin: does economics matter in endemic area?. Health Policy. Vol. 68, Issue 1, April 2004, pp. 93–102

Nagasaka, A., Yanai, S., Sato, H. y Hasegawa, S. (2005): Soil erosion and gully growth associated with cultivation in southwestern Hokkaido, Japan. Ecological Engineering, Vol. 24, Issue 5, 30 May 2005, pp. 503–508

Ogawa, T. 2004 Japanese rural aging in transition: Evaluation of the model plan of habitation for older persons Geriatrics & Gerontology International 4 (s1), S63–S64

Oh, K., Kato, T., Li, Z.-P. and Li F.Y. (2006): Environmental Problems From Tea Cultivation in Japan and a Control Measure Using Calcium Cyanamide. Pedosphere, Vol. 16, Issue 6, December 2006, pp. 770–777

Ortuño, S.F. y Martín A.J.(2006): Forest externalities, demography and rural development in inland Spain. Forest Policy and Economics. Vol. 8, Issue 2 , March 2006, pp. 109–122

Pallot, J. Y Nefedova T. (2003): Trajectories in people's farming in Moscow oblast during the post-socialist transformation. Journal of Rural Studies. Vol. 19, Issue 3, July 2003, pp. 345–362

Power, JFy Schepers, JS (1989): Nitrate Contamination of Groundwater in North America. Agriculture. Ecosystems and Environment AEENDO Vol. 26, No. 3/4, pp. 165–187, October 1989. 2 fig, 5 tab, 30 ref.

Rounsevell, M.D.A., Berry, P.M. y Harrison, P.A. (2006): Future environmental change impacts on rural land use and biodiversity: a synthesis of the ACCELERATES project. Environmental Science & Policy. Vol. 9, Issue 2 , April 2006, pp. 93–100

Shackleton, C.M., Shackleton, S.E., Buiten, E. y Bird, N. (2007): The importance of dry woodlands and forests in rural livelihoods and poverty alleviation in South Africa. Forest Policy and Economics. Vol. 9, Issue 5 , January 2007, pp. 558–577

Smailes, P.J., Argent, N. y Griffin, T.L.C. (2002) Rural population density: its impact on social and demographic aspects of rural communities. Journal of Rural Studies, Vol. 18, Issue 4, October 2002, pp. 385–404

Sofer, M. y Applebaum, L. (2006): The rural space in Israel in search of renewed identity: The case of the moshav. Journal of Rural Studies, Vol. 22, Issue 3, July 2006, pp. 323–336

Sthiannopkao, S., Takizawa, S., Homewong, J. Y Wirojanagud, W.(2007): Soil erosion and its impacts on water treatment in the northeastern provinces of Thailand Environment International, In Press, Corrected Proof, Available online 1 February 2007

Swinton, S.M., Escobar, G. y Reardon, T. (2003): Poverty and Environment in Latin America: Concepts, Evidence and Policy Implications. World Development. Vol. 31, Issue 11, November 2003, pp. 1865–1872

T.H.S., Watanabe, T., Ogino Y. Y Tanji, K.K. (2000): Soil salinization in the Nile Delta and related policy issues in Egypt. Agricultural Water Management, Vol. 43, Issue 2, March 2000, pp. 239–261

Tickamyer, A.R. y Duncan, C.M. (1990): Poverty and Opportunity Structure in Rural America. Annual Review of Sociology, Vol. 16, 1990 (1990), pp. 67–86

Tonmanee, N. y Kanchanakool, N. (1999): Agricultural diffuse pollution in Thailand. Water Science and Technology, Vol. 39, Issue 3, February 1999, pp. 61–66

Upadhyay, T.P., Solberga, B. y Sankhayana, P.L. (2006): Use of models to analyse land-use changes, forest/soil degradation and carbon sequestration with special reference to Himalayan. Forest Policy and Economics Vol. 9, Issue 4, 31 December 2006, pp. 349–371

VVAA (2003): Discusión: Sustainable formula sought for rural energy development in Asia. Refocus, Vol. 4, Issue 4, July 2003, p. 14

Warr, P.G. (2000): Poverty incidence and economic growth in Southeast Asia. Journal of Asian Economics, Vol. 11, Issue 4, Autumn 2000, pp. 431–441

Wegren, S.K. (2002): The rise, fall, and transformation of the rural social contract in Russia. Communist and Post-Communist Studies. Vol. 36, Issue 1, March 2003, pp. 1–27

Wezel, A., Steinmüller, N. y Friederichsen, J.R. (2002): Slope position effects on soil fertility and crop productivity and implications for soil conservation in upland northwest Vietnam. Agriculture, Ecosystems & Environment, Vol. 91, Issues 1–3, September 2002, pp. 113–126

Wu, W., Viisainen, K. y Hemminki, E. (2006): Determinants of High Sex Ratio among Newborns: A Cohort Study from Rural Anhui Province, China. Reproductive Health Matters. Vol. 14, Issue 27, May 2006, pp. 172–18

Xu, W. y Tan, K. C. (2002): Impact of reform and economic restructuring on rural systems in China: a case study of Yuhang, Zhejiang. Journal of Rural Studies. Vol. 18, Issue 1, January 2002, pp. 65–81

Zammit, Silvapalan, M., Kelsey, P. y Viney N.R. (2005): Modelling the effects of land-use modifications to control nutrient loads from an agricultural catchment in Western Australia. Ecological Modelling. Vol. 187, Issue 1, 10 September 2005, pp. 60–70

Zhang, Y. y Goza, F.W. (2006): Who will care for the elderly in China? A review of the problems caused by China's one-child policy and their potential solutions. Journal of Aging Studies. Vol. 20, Issue 2, April 2006, pp. 151–164

Part II
Rural Management – The Applied Approach

Rural management can be understood in at least two different ways: it may be a follow-up of a previous rural analysis, or an independent activity that may benefit from a rural analysis study. In the first case, a rural analysis report can be the starting point of a rural management or rural development program. This may be in a project aimed to improve the socioeconomic structure of the rural area, or in a development strategy. Therefore the analysis of a rural area can be used to improve the living conditions of its inhabitants by reinforcing the strengths and avoiding the observed weaknesses; by linking the area to a wider economic integration area, or by devising new development strategies. The necessity of a rural analysis prior to any management attempt is clear, since a developed knowledge on the different aspects of the rural area is required to take full advantage of its potentialities, and only a quality rural analysis can provide such an insight.

In this part, we will review some issues relevant to the management of rural areas. We will discuss some topics that are equally relevant for the management as the prior analysis itself, such as how to convey research result to the agents and how to define policy scenarios. Rural management and rural development plans are participative processes. Albeit expert advice is always welcomed, local knowledge is fundamental to achieve first-class, permanent results. The actors of the management process are the people inhabiting the rural areas, and no long-term plan can be pursued without their willingness, collaboration and participation. Since the whole point of a rural plan is to benefit the local population, it is instrumental to convey to them the proposed actions and how these will increase their quality of life. Moreover, it is fundamental to develop such plans directed at their own interests and perceptions, no matter how objective they might be. It is sometimes preferable to under perform when applying an optimization process at the expense of increasing the public reception of the process and to carry out an objectively optimum task with a degree of opposition.

Chapter 10
Approaches to Management

We use here the term 'management' to describe the process of decision-making based upon the best scientific advice, which in turn must be based on the best available information. Thus, a 'rural analysis' is a scientifically-oriented activity, whereas 'rural management' aims to provide the best solution within a set of constraints. The managerial approach also differs from the analysis in that management is a continuous activity, while analyses are point estimate appraisals of the current state of a rural area.

10.1 From Problem Solving to Resource Optimization

There are several approaches to management depending on the current situation of a rural area. A rural area can be suffering from severe economic depression and social problems, or can be facing a new regional challenge, for example, a modification in the regional economic specialization within the national economy, or a massive tourist influx. Alternatively the area can be in a relatively stable situation with numerous opportunities to improve the quality of life of the rural population. Any of these possibilities requires a different managerial approach and different management tools, ranging from continuous assessment and monitoring of the area to a surgical action to serve as a catalyst for further improvement.

A rural analysis is usually triggered by a problem, and seldom by a top-down interest in local planning. Even less frequently, the impulse for analysis comes from an endogenous interest in self-improvement when no problem is present. The difference, in terms of management, between solving an existing problem and optimizing rural resources is crucial.

Problem solving is the process of identifying the difference between the current and desired state of affairs and then taking action to resolve this difference. Following several authors, a full problem solving strategy involves the following steps:

- Identify, define and parameterize the problem
- Determine the alternatives
- Determine a metric to evaluate the alternative
- Evaluate the alternatives

- Choose an alternative
- Implement the alternative
- Evaluate the results to determine whether a satisfactory result has been obtained.

These may seem to be common-sense steps, but a careful examination would illustrate that this method presents both a quantitative approach and a feed back loop that is critical for problem solving. Other more qualitative approaches might engender serious methodological problems from the start: the absence of any metrics precludes the appraisal of the success of planning alternatives, and an incorrect definition of the problem may generate an even worse situation when palliative measurements are put into practice.

Optimization is another possible approach to management. In this case, the goal is to improve the current situation to extract the best of the available resources, or to take advantage of new opportunities. This approach is not to be confused with a quest for pure economic benefits: almost every rural area in the world can be optimized to yield better performances in terms of improved quality of life, lower environmental impact or increased social harmony.

To decide whether management would be a problem or an optimization –oriented task, the first step is to perform a SWOT appraisal of the situation (SWOT stands for Strengths, Weaknesses, Opportunities, and Threats). This can only be properly done after an analysis phase has been carried out, that is once the relevant data have been gathered and the dynamics of the area modelled. Attempts of a SWOT analysis without the analysis step produce naïve analyses at best, which are of little use and can introduce some confusion in the whole management process.

10.2 A SWOT Approach

SWOT is a methodology used in industry to evaluate the situation of a company when confronted with a challenge, or to propose strategies for increasing productivity. This method can also be used in a rural analysis to help identify an overall management strategy for the area. It is useful as a first approximation when the analysis impetus comes not from a well-identified problem but from a desire of developing a rural area and to optimize the available resources.

A SWOT analysis comprises the exhaustive listing of the four components. First, a compilation of the strengths is made in the light of the analysis previously done. Potential strengths could be an advantageous situation: proximity to a dynamic centre, to a special natural beauty spot, a dynamic social structure, or a functional specialization in the markets. The possibilities are endless and should be identified by both the analysts and the agents involved.

It should be noticed that elements can be bipolar: both a strength and a weakness. This could be a location close to a natural park: it can be considered a strength in terms of having a natural resource near by, but it may also constrain the development of industrial activities deemed as suitable for the area being analyzed. Bipolar elements are interesting because they provide clues on *'orthogonal planning alternatives'*, that is, alternatives that are independent from the others as possible. The change in the main characteristic of the area is linked to the selection of one orthogonal alternative or another. The so-called vocation of the area is not always as clear as desired, and if one of the aims of the planning is to actually decide which alternative is the most appropriate for the rural area, it is a critical choice to make.

Weakness can be a result of environmental, historical or social factors. The most relevant for an analysis are those that are difficult, if not impossible, to overcome, such as a location in a remote high mountain area or a harsh climate. However, one of the interests of management strategies is to find the way to transform a weakness into a neutral or even positive advantage point. This could be in thinking of marshes as wildlife sanctuaries capable of attracting tourists and proliferating money, or converting otherwise unhealthy areas into recreational spaces for locals. In other cases, the analyst cannot be so creative: severely contaminated lands, badlands or frequently flooded areas are serious weaknesses not easily translatable into neutral elements.

Opportunities have a temporal component. An opportunity should be defined within a window of opportunity open for a short time only. For example, Eastern European countries currently have a window of opportunity in accessing EU funds for rural development. The same window is about to close to other countries such as Spain, Portugal and Greece. Other opportunities such as temporal appeal for the area based upon a new trend or fashion, or a new sensitivity for its values, can be less obvious to identify. Being capable of catching the zeitgeist of societies is a valuable asset for the rural analyst.

In relation to the threats, external threats (competitive disadvantages, new regional policies, invasive infrastructures) are more easily recognized than internal threats such as demographic decline or social discomfort. The relationship between strengths and weaknesses is applicable to the relationship between opportunities and threats: a threat can be transmuted into an opportunity to remove old structures, to advance development and to catch the so-called 'pace of the times'. The energy to overcome psychological or social barriers usually comes from vividly perceived threats (or for catastrophic events, which are in fact realized threats) and those critical moments might be seen as opportunities for change.

This is often exemplified in the inspirational sentence stating that 'what the caterpillar perceives as the end, it is the beginning for the butterfly'. Strategies can be proposed to neutralize threats, but if they are unavoidable it is better to face them and work out a mitigation strategy, or even better, find a way to take advantage of the energy they provide. Because apathy is one of the most prevalent problems in rural development, any energetic surplus is sometimes welcomed to be judiciously reoriented to move energy forward. The opportunities that European rural areas have had to preserve their traditional character by housing highly-qualified technicians capable of carrying out tele-working is an example of how a threat (the end of the rural manufacturing industries in advanced countries) can be used as an incentive to change, and even to improve, rural economies in decline. Especially now that unskilled manufacturing can be cheaply done in emerging countries.

In practice, SWOT analysis is a complex task that requires a great deal of careful and objective consideration of the rural reality. The cases shown above are just educational examples to familiarize the reader with the meaning given to the four SWOT components. Even when it should be possible to systematize many SWOT lists suitable for rural analysis, the casuistic is too wide to be useful. Moreover, a fresh SWOT analysis is always desirable since it prevents the need for applying recipes or preconceptions from other cases.

SWOT analysis can be applied independently to several natural and anthropic elements. Therefore, a SWOT analysis should be done once the analytical part is completed (such as the one presented in the first part of this book is done) since SWOT analysis must be based upon scientific, quantifiable evidence and not on a naïve appraisal of the reality. One of the leitmotivs of this text is the emphasis placed on backing any decision or recommendation on empirical evidence. This empirical data should be then modelled using repeatable and objective methodologies, and then, and only then, would it be possible to offer meaningful advice as an analyst.

Bad practices in SWOT analysis includes the use of clichés or statements based on unsubstantiated evidence, a superficial analysis considering only lineal relationships, or the transposition of similar cases in different geographical areas. Common malpractices include analyses based on the expectations of the client, the more or less explicit requests of some lobbies, biased analysis due to prejudices, or a simple copy-and-paste of void statements.

10.3 Management Agents

Just as 'how' to manage a project on rural planning is important, so is 'who' is going to manage the project. One key aspect is 'decentralised management': the principle of delegating policy-making into the lowest effective local levels of public authority. This concept has been promoted as a result of the UN Commission on Sustainable Development Agenda 21, and is being taken up increasingly around the world.

If either the driving force of the project is an external body, such as regional or national administration following mandatory regulations in regional planning, or is a committed rural group aiming to propose developing strategies for its area, the agents have to be identified.

The success of a project lies in the competence and willingness of the local people to carry out the required actions; the reception of the actions by the local communities; and the overall situation in which the rural area is inserted, including the circumstances of the national economies. The driving force of the management process however, corresponds to a reduced number of people that can be either locals or people who are indirectly linked with the area. External expertise depends on a number of factors, including funding availability, cultural issues, and the extent and complexity of the rural management required. In some cases, it would be necessary to hire a team to carry out a complicated, time-consuming process. In other cases, the locals may carry out the process by themselves. The choice of 'who' is going to be the agents of the rural management process is usually done in an advanced stage of the analysis, where a vision on the rural area is built.

References and Bibliography Used

N. Bamfo, 'A grassroots developmental strategy for Africa: Towns as agents of growth through financial credit', Policy Studies Journal 29 (2001), 308–318.

K. Brown and F. Ekoko, 'Forest encounters: Synergy among agents of forest change in Southern Cameroon', Society & Natural Resources 14 (2001), 269–290.

M. Fandino, C. J. Alvarez, R. Ramos and M. F. Marey, 'Agricultural cooperatives as transforming agents in rural development - The case of Galicia', Outlook on Agriculture 35 (2006), 191–197.

M. Kajanus, J. Kangas and M. Kurttila, 'The use of value focused thinking and the SWOT hybrid method in tourism management', Tourism Management 25 (2004), 499.

M. K. Masozera, J. R. R. Alavalapati, S. K. Jacobson and R. K. Shrestha, 'Assessing the suitability of community-based management for the Nyungwe Forest Reserve, Rwanda', Forest Policy and Economics 8 (2006), 206.

T. R. Wojan and A. Rupasingha, 'Crisis as opportunity: Local context, adaptive agents and the possibilities of rural development', Regional Studies 35 (2001), 141–152.

Chapter 11
Management Tools: Geographical Information Systems (GIS) and Expert Systems

This chapter describes some of the tools that can be used for rural management. First, the role of databases is discussed. It has to be remembered that a rural management project should refer to objective data, and that this data should always be stored in a database. The secondary tools are a GIS, used to manage the rural area from a single technological platform. A third instrument is represented by Expert Systems that are used to analyze alternatives or to develop scenarios. Finally, Management Systems can be appropriated for a more comprehensive integrated management in areas with complex problems.

11.1 Databases

In a low-level description, a database is comprised of fields and registers (columns and rows in spreadsheet parlance). Databases are not only tools for storing large volumes of information on the individual elements of the analysis, but also the empirical evidence required for a critical assessment of the analysis by other researchers. Databases should be regarded as the repository of the best information available upon which the analysis has been performed. If a piece of information is not in the database, either it is not relevant or cannot be scientifically used in the analysis.

A database is seldom a closed instrument. Newly available information can be integrated into the database and then re-analyzed, in the same way as meteorologists use the information they have to provide a weather forecast and then improve this forecast when new information becomes available.

This process illustrates the need of automatic and flexible analytical tools to respond to the rapidly-growing availability of new data. It is the case of processing near-real-time information such as those provided by satellites. Conceding (for the sake of the argument) that a qualitative analysis could provide useful insight into rural problems, such analyses would soon become obsolete if this approach is unable to quickly integrate new information. It is usually the case that this kind of analyses are ready many years after the processes under study were acting, making the research only a matter of academic curiosity.

Order, planning and flexibility are three key elements in the database design task. A database needs to be planned well before any programming task begins. It is often said that each hour of prior planning saves three hours of programming, and probably tens of hours of analysis. A poorly designed database would give the impression of a quick advance in the first stages of the rural analysis and management process, but it usually damages the whole process by generating many back steps, redesigning of the analytical tools and unbalanced analyses.

The process of database building is closely linked to the data model definition. In the first part of this book the elements of the rural analysis have been described. Each of these elements has many quantifiable features. Some of them may have numerical values, while others may consist of logical or categorical values. Data model building is about defining the relevant elements, their measurable characteristics and their links.

A database should be as complete and gapless as possible. Many missing registers usually means a poor selection of the field, since the available information should be as comprehensive as possible. Some tens of samples in several hundreds of registers can render the field useless regardless of the interest of the data itself. Nonetheless, an intelligent selection of specific rare cases can be advisable.

Deciding what is relevant is an ability that is acquired with a scientific or technical education, and that is refined through experience, either in training studies or in real projects. There is no such thing as the perfect rural analysis. All the analyst can expect is to make a sensible project based upon professional work and an honest approach to his task.

The task of database creation is an interdisciplinary assignment typically involving computer engineers, archivists, accountants, rural analysts, spatial analysis experts, technical staff, clerks, and local and regional agents. Ideally, the database of the project would also be useful for the daily basis management, making it crucial to involve right from the beginning as many future customers and users as possible. This (a) helps to build a sense of interest in the project, (b) saves many hours of training and improves the familiarity with the system, and (c) improves the actual database creation by including privileged information from the people who have first hand information of the problem. Some crucial elements that could have otherwise been neglected by a plain application of *template schemes* may then be included if the analysis team shares its needs with the appropriate informers. This could be the case with complex ownership issues regarding agrarian land, sentimental links with some natural areas, unexpected local laws, the behavioural patterns of the rural inhabitants, or the real or imagined place of the area in relation with its neighbours. It might be relevant in terms of database design, to map unusual aspects such as religious festivals or sportive preferences. Memorial sites, places of recent war events or irrational misgivings toward other areas should be carefully considered since it could play an even more relevant role in the area than purely economical activities.

There are other technical elements to be considered in database design that should be considered by the team, such as security procedures to impede database tamping or to prevent unwanted access to sensitive information. These aspects are relevant in that they contribute to a robust and reliable database. The aim of the analyst is to extract better performances out of a well-oiled machine, and not to deal with the mechanical problems that could arise. In that sense competent technical staff is necessary.

This handbook is not intended to delve deep into the technicalities of database design and exploitation but to show the most common elements required for a successful design of a database for a rural analysis. A crude definition of a database is a computer program in an information system that stores information. Many commercial solutions exist in almost all the Operative Systems, from simple but efficient Access from Microsoft, to the expensive but comprehensive Oracle system. Linux users have at their disposal many SQL solutions. The selection of one of these systems depends on many factors such as:

- The formal organization of the project. If the project is done by a consulting firm using their own information systems it could be preferable to stick to existing technologies. This could be a good policy if the problem is small enough that it does not require a great deal of involvement from the community. On the other hand, some rural areas require a permanent system for management and then this factor should also be considered bearing in mind the personnel who will finally deal with the system.
- The temporal and spatial scale of the project.
- The availability of specialized technical staff to future maintenance and management tasks.
- The budget of the project.
- The balance between analytical capabilities, costs, and how friendly the system is for new users.
- Regional and national policies. Some regions prefer to use Linux systems for geopolitical or strategic reasons while proprietary software could be more extended in others.

A good analyst should provide a wide range of options to the client, depending on the client's needs and not on the personal preferences or work habits of the analyst.

11.2 GIS = CAD+DB+Analysis

Rural information needs to be geographically-located (georeferenced in GIS lingo). This means that every element needs to be given a set of coordinates to be able to relate one element to others. The further establishment of spatial relationships is one of the key points in rural analysis, being also an issue to observe in management. The elements that are to be georeferenced are both the natural and anthropic elements. All their attributes has to be defined beforehand, in a process termed as data model definition, so to assign the relevant information to a set of coordinates. As we saw, how this task is accomplished depends on the prior knowledge on the area, being therefore dependent on the analytical stage of rural management. It is only from in-deep knowledge of the relationships and hierarchies of the region that a useful data model can be defined.

The physical part of a Geographical Information System (GIS) comprises three elements: A system to georeference and visualize the information (CAD module), a database (DB) to store the attributes of the elements, and a good toolbox of analytical capabilities. The combination of these three elements is what makes the GIS valuable. It goes well beyond the scope of this handbook to discuss GIS potentialities. The reader is referred to the general works cited in the reference section of this chapter.

11.3 Expert Systems

Expert systems represent a step forward over GIS. Expert systems embed most of the information and knowledge stored into a GIS, but includes some artificial intelligence or reasoning-based methods to generate additional results to the mapping and modeling capabilities. They can be directly used to simulate scenarios, or to develop strategies for real time events. A typical case is expert systems for forest fires management: real time information using GPS or satellite data on fire spread, land use, orography, weather, or fire brigades location is ingested by the system, which is programmed to find the optimum routes for the brigades, safe keeping the population or provide escape routes (Fig. 11.1). All this information is usually supported by a GIS. Actually, an expert system can be a GIS with additional tailored functions for the problem and capable of being used as a planning or as a management tool.

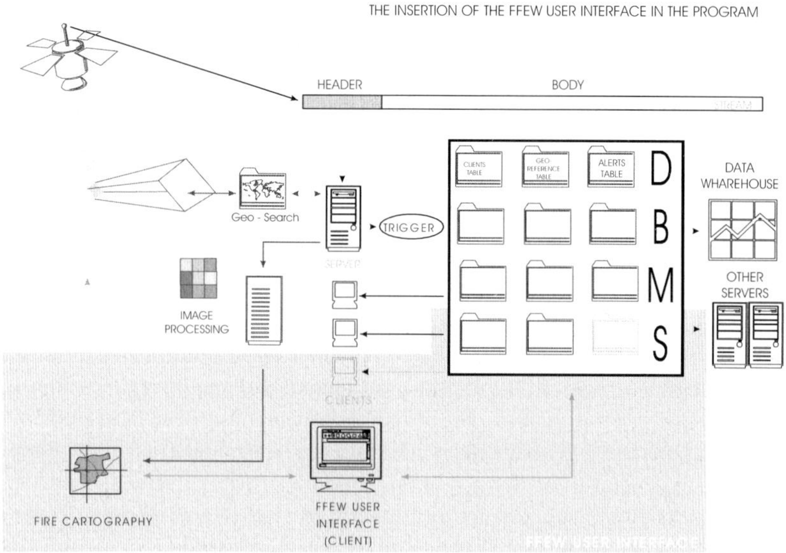

Fig. 11.1 An expert system example: expert system for forest fires early warning. A system of this kind can automatically calculate contingency plans and emergency routes for the population, and are an excellent support for decision-making (see Color Plate on following p. 333)

11.4 Management Systems

Management systems are a step forward over expert systems. They are comprehensive systems for the planning and management or rural areas. In some cases, a rural management system can be build using a suite of management programs. In other cases, management systems are specifically oriented to cadastral purposes. Thus, the term Land Information Systems (LIS) refers to 'a tool for legal, administrative and economic decision-making and an aid for planning and development. A land information system consists, on the one hand, of a database containing spatially referenced land-related data for a defined area and, on the other, of procedures and techniques for the systematic collection, updating, processing and distribution of the data. The base of a land information system is a uniform spatial referencing system, which also simplifies the linking of data within the system with other land-related data'. (UNECE, 1996). Therefore, a LIS is a computer system to manage spatial data within a government or management context. It can be also be describe as a Geographical Information System (GIS) used for land information, as the software and hardware part of most of the LIS are a commercial GIS tuned for storing, analysing and plotting cadastral information.

Closely linked with LIS, Cadastral Management Information Systems also exist. The cadastre is defined as 'a parcel based and up-to-date land information system containing a record of interests in land (i.e. rights, restrictions and responsibilities). It usually includes a geometric description of land parcels linked to other records describing the nature of the interests, and ownership or control of those interests, and often the value of the parcel and its improvements. It may be established for fiscal purposes (e.g. valuation and equitable taxation), legal purposes (conveyancing), to assist in the management of land and land use (e.g. for planning and other administrative purposes), and enables sustainable development and environmental protection.' (International Federation of Surveyors, 1991)

Chukwudoize et al. (1996) have identified the following elements as typical tenure information data that may be found in a western land tenure-based LIS:

- ownership (name, postal address, street address);
- occupier (name, postal address, street address);
- full private rights (title reference, dealings, transfer documents);
- parcel boundaries and location (legal description, e.g. plan and lot number, cadastral boundary data, map reference);
- private restrictions (mortgage, covenants, caveats);
- public restrictions (zoning);
- limited rights (leases, water);
- land value (assessed, market);
- other ownership attributes (historic descriptions);
- other parcel attributes (land-use, cultural features, e.g. buildings, roads).

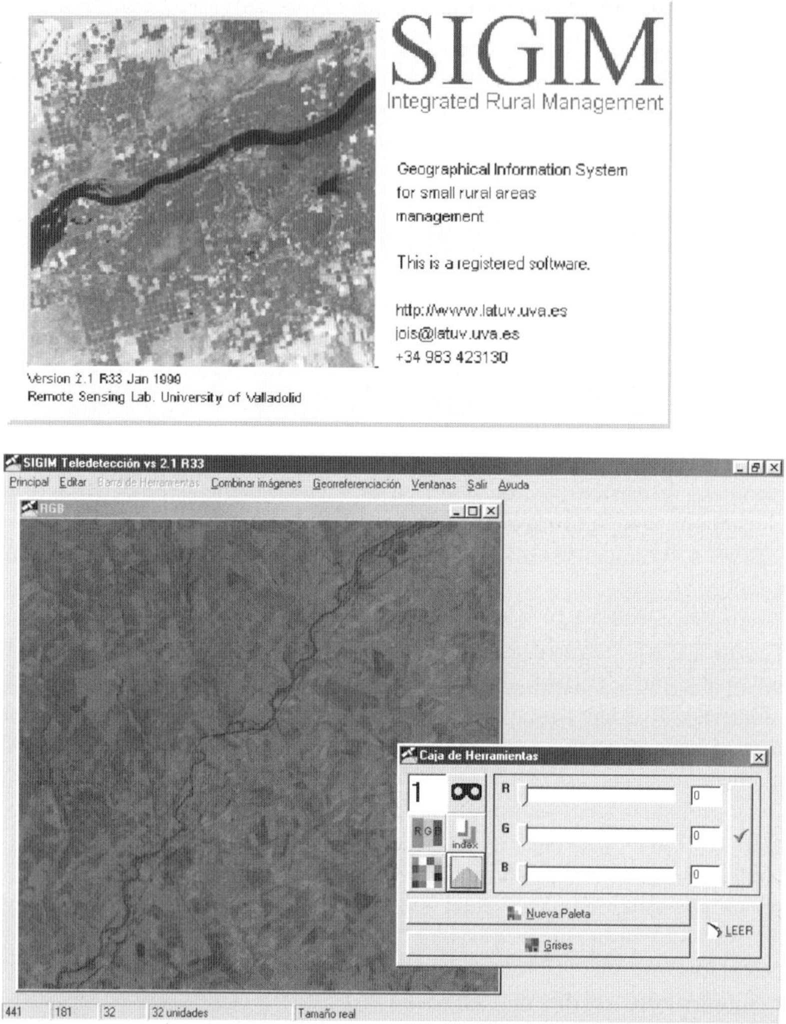

Fig. 11.2 SIGIM welcome pop-up screen (translated to English; original version in Spanish) and the remote sensing module, featuring masking, band combination, georeferencing, vegetation indexes calculations and thresholds tools over a LANDSAT image (see Color Plate on following p. 333)

Those authors recognize that the LIS concept has been developed primarily to serve the needs of countries with a western-style land market where individual land rights are the norm. They notice that many countries where customary landholdings exist may also be interested in establishing LISs to manage their land resources better, demonstrating the importance of analytical tools in rural management.

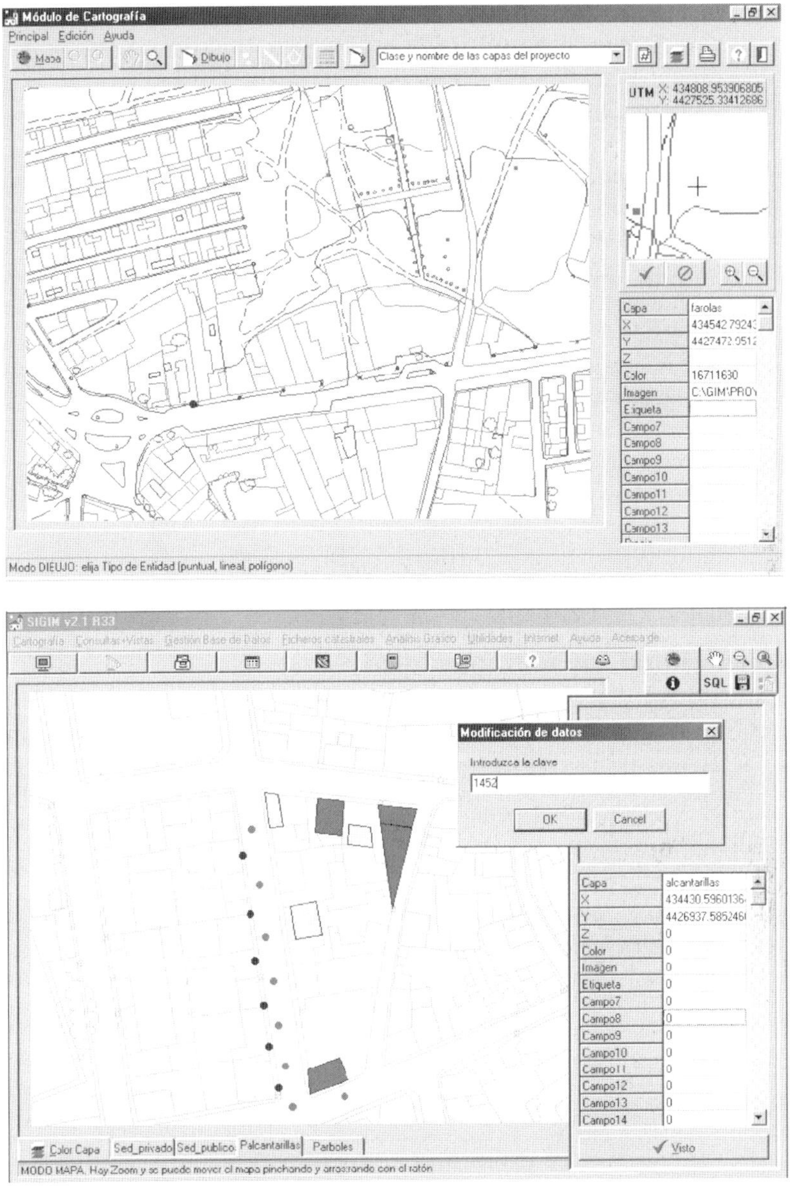

Fig. 11.3 SIGIM cartographical module, with the basic cartography layout and the georeferencing tools (see Color Plate on following p. 333)

An Example of a Management System

It is worth to analyze the components of a real rural-oriented management system. The SIGIM was a simple integrated rural management tool for small villages (Figs. 11.2, 11.3). It was developed in the University of Valladolid, Spain, at the beginning of the century by the author. The software part of this GIS was generated in C++, allowing Object Oriented Programming (OOP) features to be used in the definition of the environmental and anthropic elements.

After a number of successful implementations SIGIM was finally discontinued, making it a suitable choice for a neutral, non-commercial analysis of rural management systems capabilities.

SIGIM comprised several modules. A remote sensing layer dealt with agricultural information including vegetation indexes calculation, irrigation practices, and ownership issues. The cartographical segment was used to include new information over basic cartography. A third part generated certificates and official documents for the users.

More importantly in terms of rural management, SIGIM was not a closed system. The suite was tuned to adapt to the needs of each village. In some cases, additional analytical capabilities were required to for instance, water quality assessment analysis. In some other places, reducing the tools to a minimum set of features increased the usability of the system by reducing the qualifications required to operate the system.

References and Bibliography Used

J. Abbott, 'The use of GIS in informal settlement upgrading: its role and impact on the community and on local government', Habitat International 27 (2003), 575–593.

S. Anuchiracheeva, H. Demaine, G. P. Shivakoti and K. Ruddle, 'Systematizing local knowledge using GIS: fisheries management in Bang Saphan Bay, Thailand', Ocean & Coastal Management 46 (2003), 1049.

J. Ball, 'Towards a methodology for mapping "regions for sustainability" using PPGIS', Progress in Planning 58 (2002), 81–140.

D. Belshaw, 'Strategising poverty reduction in sub-Saharan Africa: The role of small-scale agriculture', Journal Of Agricultural Economics 53 (2002), 161–193.

G. Bocco, F. Rosete, P. Bettinger and A. Velazquez, 'Developing a GIS program in rural Mexico - Community participation = success', Journal Of Forestry 99 (2001), 14–19.

C. Bussink, 'GIS as a tool in participatory natural resource management - Examples from the Peruvian Andes', Mountain Research And Development 23 (2003), 320–323.

S. Dragicevic and D. J. Marceau, 'An application of fuzzy logic reasoning for GIS temporal modeling of dynamic processes', Fuzzy Sets And Systems 113 (2000), 69–80.

S. Dragicevic and D. J. Marceau, 'A fuzzy set approach for modeling time in GIS', International Journal Of Geographical Information Science 14 (2000), 225–245.

E. A. Ellis, P. K. R. Nair, P. E. Linehan, H. W. Beck and C. A. Blanche, 'A GIS-based database management application for agroforestry planning and tree selection', Computers and Electronics in Agriculture 27 (2000), 41.

I. A. El-Magd and T. Tanton, 'Remote sensing and GIS for estimation of irrigation crop water demand', International Journal of Remote Sensing 26 (2005), 2359–2370.

E. Feoli, L. G. Vuerich and W. Zerihun, 'Evaluation of environmental degradation in northern Ethiopia using GIS to integrate vegetation, geomorphological, erosion and socio-economic factors', Agriculture, Ecosystems & Environment 91 (2002), 313.

A. Gobin, P. Campling and J. Feyen, 'Logistic modelling to derive agricultural land use determinants: a case study from southeastern Nigeria', Agriculture Ecosystems & Environment 89 (2002), 213–228.

R. M. Gonzalez, 'Joint learning with GIS: multi-actor resource management', Agricultural Systems 73 (2002), 99.
W. R. Gordon and N. M. Soubra, 'Geographical Information-Systems and Planning in the USA - Selected Municipal Adoption Trends and Educational Concerns', International Journal of Geographical Information Systems 6 (1992), 267–278.
R. Grabaum and B. C. Meyer, 'Multicriteria optimization of landscapes using GIS-based functional assessments', Landscape and Urban Planning 43 (1998), 21–34.
I. Gunatilleke, C. V. S. Gunatilleke and P. Abeygunawardena, 'Interdisciplinary Research Towards Management Of Non-Timber Forest Resources In Lowland Rain-Forests Of Sri-Lanka', Economic Botany 47 (1993), 282–290.
E. M. Hamin, 'Legislating growth management - Power, politics, and planning', Journal Of The American Planning Association 69 (2003), 368–380.
R. K. Jaiswal, S. Mukherjee, J. Krishnamurthy and R. Saxena, 'Role of remote sensing and GIS techniques for generation of groundwater prospect zones towards rural development - an approach', International Journal Of Remote Sensing 24 (2003), 993–1008.
H. G. P. Jansen, B. A. M. Bouman, R. A. Schipper, H. Hengsdijk and A. Nieuwenhuyse, 'An interdisciplinary approach to regional land use analysis using GIS, with applications to the Atlantic Zone of Costa Rica', Agricultural Economics 32 (2005), 87–104.
B. D. Jokisch and C. G. Boone, 'Teaching and tragedy: Lessons from a population and GIS workshop in Cambodia', Journal Of Geography 102 (2003), 13–20.
V. Keller, K. Fox, H. G. Rees and A. R. Young, 'Estimating population served by sewage treatment works from readily available GIS data', Science of The Total Environment 360 (2006), 319.
Y. Koumpouros, B. D. Mahaman, M. Maliappis, H. C. Passam, A. B. Sideridis and V. Zorkadis, 'Image processing for distance diagnosis in pest management', Computers and Electronics in Agriculture 44 (2004), 121–131.
G. Krause, M. Bock, S. Weiers and G. Braun, 'Mapping land-cover and mangrove structures with remote sensing techniques: A contribution to a synoptic GIS in support of coastal management in North Brazil', Environmental Management 34 (2004), 429–440.
P. A. K. Kyem, 'Of intractable conflicts and participatory GIS applications: The search for consensus amidst competing claims and institutional demands', Annals Of The Association Of American Geographers 94 (2004), 37–57.
G. Lien, 'Assisting whole-farm decision-making through stochastic budgeting', Agricultural Systems 76 (2003), 399–413.
R. K. Mallick and J. K. Routray, 'Identification and accessibility analysis of rural service centers in Kendrapara District, Orissa, India: a GIS-based application', International Journal of Applied Earth Observation and Geoinformation 3 (2001), 99.
D. J. Marceau, L. Guindon, M. Bruel and C. Marois, 'Building temporal topology in a GIS database to study the land-use changes in a rural-urban environment', Professional Geographer 53 (2001), 546–558.
O. Masera, A. Ghilardi, R. Drigo and M. Angel Trossero, 'WISDOM: A GIS-based supply demand mapping tool for woodfuel management', Biomass and Bioenergy 30 (2006), 618.
S. Mtetwa, S. Kusangaya and C. F. Schutte, 'The application of geographic information systems (GIS) in the analysis of nutrient loadings from an agro-rural catchment', Water Sa 29 (2003), 189–193.
W. R. Mukabana, K. Kannady, G. M. Kiama, J. N. Ijumba, E. M. Mathenge, I. Kiche, G. Nkwengulila, L. Mboera, D. Mtasiwa, Y. Yamagata, I. van Schayk, B. G. J. Knols, S. W. Lindsay, M. C. de Castro, H. Mshinda, M. Tanner, U. Fillinger and G. F. Killeen, 'Ecologists can enable communities to implement malaria vector control in Africa', Malaria Journal 5 (2006).
L. Palmeri and M. Trepel, 'A GIS-based score system for siting and sizing of created or restored wetlands: Two case studies', Water Resources Management 16 (2002), 307–328.
R. Pendall, R. M. Wolanski and D. McGovern, 'Property rights in state legislatures: rural-urban differences in support for state anti-taking bills', Journal of Rural Studies 18 (2002), 19–33.
T. V. Ramachandra and B. V. Shruthi, 'Wind energy potential mapping in Karnataka, India, using GIS', Energy Conversion and Management 46 (2005), 1561–1578.

G. R. Robinson and K. E. Kapo, 'A GIS analysis of suitability for construction aggregate recycling sites using regional transportation network and population density features', Resources Conservation and Recycling 42 (2004), 351–365.

S. Sabesan and K. H. K. Raju, 'GIS for rural health and sustainable development in India, with special reference to vector-borne diseases', Current Science 88 (2005), 1749–1752.

J. Sanyal and X. X. Lu, 'Remote sensing and GIS-based flood vulnerability assessment of human settlements: a case study of Gangetic West Bengal, India', Hydrological Processes 19 (2005), 3699–3716.

J. Sanyal and X. X. Lu, 'GIS-based flood hazard mapping at different administrative scales: A case study in Gangetic West Bengal, India', Singapore Journal of Tropical Geography 27 (2006), 207–220.

S. K. Srivastava, C. B. S. Dutt, R. Nagaraja, S. Bandyopadhayay, H. C. M. Rani, V. S. Hegde and V. Jayaraman, 'Strategies for rural poverty alleviation in India: A perspective based on remote sensing and GIS-based nationwide wasteland mapping', Current Science 87 (2004), 954–959.

J. Tourino, J. Parapar, R. Doallo, M. Boullon, F. F. Rivera, J. D. Bruguera, X. P. Gonzalez, R. Crecente and C. Alvarez, 'A GIS-embedded system to support land consolidation plans in Galicia', International Journal Of Geographical Information Science 17 (2003), 377–396.

E. L. Webb and Thiha, 'Integrating social preference in GIS-aided planning for forestry and conservation activities: A case study from rural SE Asia', Environmental Management 30 (2002), 183–198.

C. D. Yue and S. S. Wang, 'GIS-based evaluation of multifarious local renewable energy sources: a case study of the Chigu area of southwestern Taiwan', Energy Policy 34 (2006), 730–742.

S. X. Zeng, C. M. Tam, V. W. Y. Tam and Z. M. Deng, 'Towards implementation of ISO 14001 environmental management systems in selected industries in China', Journal of Cleaner Production 13 (2005), 645–656.

Chapter 12
The Concept of Rural Quality of Life

The concept of quality of life has historically depended on income and education. People in modern western societies present a different attitude on what they deemed necessary for their well-being than people in the nineteen century, or than people in remote areas. Cultural factors play a role in what is acceptable as living standards in a given period, but economic factors are of greater importance. As income increases, people tend to valuate other elements apart from material wealth.

Several studies have proposed a number of lists of the elements comprising quality of life. Most of the researchers conclude that quality of life is a weighted average of basic needs such as food, water, shelter, safety and a healthy condition, plus other needs such as freedom of movement, speech, beliefs, leisure, services availability and living in a congenial environment. It could be argued however that most if not all of those elements are linked with material wealth. Money can buy from food to safer environments, and in a sense quality of life is a euphemism in opulent societies for increased money availability beyond basic needs.

Even in developed countries, some rural areas may appear has having a low quality of life because they do not generate enough material wealth. Paradoxically, rural areas in developing countries can more easily reach the basic limit of subsistence, which is the quality of life bottom line. Agriculture may provide food, and housing is less expensive in the rural environment than in the urban areas. The next stage of increased wealth, however, can be harder to obtain in such rural areas. Since agrarian products have a lower added value than manufactures or services, they cannot be easily converted into money. Rural services others than agrarian services lack of scale economies, and accumulation for expensive activities such as holidays—an activity that many people in developed countries might consider as essential for their well-being-may not be an easy task.

This discussion should not mask the fact that some rural areas are even incapable of generating resources to cover the basic needs of life. Three quarters of the humans live below the subsistence threshold, and most of them are rural population.

12.1 Quality of Life Indicators

The differences between urban and rural quality of life are so marked, that the indicators have to be different between rural and urban areas when an assessment exercise is carried out. In this context, the First European Quality of Life Survey (EQLS) identified the following working hypotheses from the literature review:

- In late modernity, there are no essential differences between urban and rural areas; urban–rural differences will therefore emerge to a greater degree in poorer countries where the process of modernisation is more uneven.
- Urban–rural differences in income and levels of deprivation are greater in poorer countries.
- Household self-provision is greater in rural areas and in poorer countries.
- Housing problems are more severe in urban areas and in poorer countries.
- Education levels are higher in urban areas, with no gender differences.
- Internet usage is higher in urban areas.
- Unemployment is higher in rural areas and in poorer countries.
- Rural areas have witnessed a general shift to a service-based economy.
- Quality of employment is higher in urban areas, especially for women.
- Work–life balance is more favourable in rural areas and in richer countries.
- Access to work, school, family, friends and services is more difficult in rural areas, especially for women.
- Rural areas are perceived as offering a better quality of life and, in particular, a better quality of family life.
- People living in urban areas are more optimistic about their future.
  (European Foundation for the Improvement of Living and Working Conditions First European Quality of Life Survey: Urban–rural differences Luxembourg: Office for Official Publications of the European Communities 2006, ISBN 92-897-0960-X)

This list could be useful as a general orientation, but for specific studies a more detailed and case-tailored appraisal should be developed. Finding what the local think on their quality of life, and combine those perceptions with objective indicators is also part of the process of improving living conditions. Otherwise, the danger is imposing extraneous standards to the population.

12.2 The Urban Concept of the Rural

The idyllic, romantic vision of the countryside is an historical product of the nineteen century that propagated by mass-media in the next century, and percolated into mainstream culture. The idea of rural life that most urbanites have is the kind of life lived by the aristocratic elite of preindustrial societies. Most people today would no identify rural life with the ploughman or the yeoman lives, but with the manor masters'. Industrial cinema and marketing have contributed to this distorted vision of what was (and still is in many areas) the core of rural life: an area with fewer opportunities for social promotion or to wealth increase.

In a rural management project such realities have to be taken into account when dealing with initiatives such attracting urban population to deprived areas, or to disseminate the problems of the area in rural forums. Being depopulation a shared problem of the rural, initiatives to attract new inhabitants have flourished in the last decade. In some cases, housing and a work is provided by the local authorities to those interested in move to grow a family in their municipality. Rural development officials are well used however to deal with urbanites looking to a change being soon dismayed by poor accessibility in the winter season, a closed society, and a limited choice of products and relationships. Adequate information is the key to avoid disappointments. Rural officials should invite applicants to think twice about setting down in remote areas which look blissful in spring season but are *six feet under* (of snow) in the winter; and should recommend them to come back in the harsh months of the year. Parents should also be advised to think about having their progeny wandering around with just a handful of children of their own age during a decade. Also, quite usually children born and breed in cities do no share the enthusiasm of their parents to embrace *the good life* in nature.

12.3 Rural Areas for Rural People?

The rural areas have a structural problem in the continuing fight with the urban areas. They represent a small fraction of the total population and in democracy more people means increased power. The interests of those dwelling in the cities are usually opposed to the people in the rural areas in terms of industrial development, free access to natural resources or regional planning initiatives so conflicts are frequent in regional planning.

It has been argued that rural planning should first satisfy the needs of the rural population, and only after the needs of those than only occasionally have access to the countryside. While it is true that the 'common good' is a tenet of any regional planning exercise, the 'right of place' should not be forgotten. In any regional initiative, the actual inhabitants of the area have to be the main beneficiaries of the changes. To that end, consultation and a formal public participation framework for regional or national project should be a part of the rural managing process. If the tools, instruments and mechanisms to conflict-solving are already set up when new problems arise, it will be easier to respond to them in a coordinate way.

References and Bibliography Used

C. J. Andrews, 'Analyzing quality-of-place', Environment and Planning B-Planning & Design 28 (2001), 201–217.

J. Little and P. Austin, 'Women and the rural idyll', Journal of Rural Studies 12 (1996), 101–111.

H. Matthews, M. Taylor, K. Sherwood, F. Tucker and M. Limb, 'Growing-up in the countryside: children and the rural idyll', Journal of Rural Studies 16 (2000), 141–153.

B. Mclaughlin, 'Rural Policy in the 1980S - the Revival of the Rural Idyll', Journal of Rural Studies 2 (1986), 81–90.

G. Valentine, 'A safe place to grow up? Parenting, perceptions of children's safety and the rural idyll', Journal of Rural Studies 13 (1997), 137–148.

F. van Dam, S. Heins and B. S. Elbersen, 'Lay discourses of the rural and stated and revealed preferences for rural living. Some evidence of the existence of a rural idyll in the Netherlands', Journal of Rural Studies 18 (2002), 461.

M. Whitby, 'The Rural Idyll - Mingay, Ge', Journal of Rural Studies 7 (1991), 131–133.

H. P. M. Winchester and M. W. Rofe, 'Christmas in the "Valley of Praise": Intersections of the rural idyll, heritage and community in Lobethal, South Australia', Journal of Rural Studies 21 (2005), 265–279.

Chapter 13
The Rural Areas in the Structure of the Territory

This section addresses how rural areas are integrated into a larger space which is the region. It is argued here that an integrate approach is required to maximize the benefits of planning at several scales, and that rural management is a part of that objective. In this context, an especially relevant process, namely the counter urbanization in developed countries, has to be analyzed with more detail.

Most of the topics here may seem to apply only to developed countries where planning practices are severely enforced by law. They are, however, compelling reasons to consider planning in developing countries as an opportunity to initiate growth and development from the lowest scale. Hence, rural planning appears as a down-top approach to development, opposite to the centralized top-down way in most developed countries. The possibility of creating a better society improving first some carefully-selected rural areas and then spreading development to the rest of the territory is a real possibility.

13.1 Towards an Integrated Approach

The integrated approach of managing a territory involves a careful consideration of the multiscale effects of the policies. Large civil works impact at several levels, from the national to the local, and the effects must be independently considered for each scale. Rural development plans are a multidepartamental task at each administrative level, from the national ministries to the local civil organization.

Funding allocated at regional level should be directed to maximize the regional benefit, and in that respect, the share of the several rural areas comprising the region will depend on the ability of each sector to contribute most to the total. Sensible policies are directed to maximize the social benefit of the community but resources are generally scarce so an intelligent manager will try to fund the most profitable plans. It is up to the local authorities to create the conditions to attract such investments and to direct the money from the regional and national level to the best interest of the area. This will involve developing the recommendations or guidelines suggested in a rural analysis exercise, but also to keep tuned to supralocal plans and policies to adapt to funding opportunities in terms of for example attracting new economic activities to the area or to increase the number of visitors.

A certain degree of competition between local areas always exists, but each village has their own particularities and potentialities. Taking advantage of the competitive advantages involves a comparative with neighboring areas which should be included in either a rural analysis report or in a managing strategy for the rural area. Indeed, competition can be also view in terms of re-enforcing mutual strengths and avoid weaknesses, creating synergism between neighbors. Within this collaborative scheme, regional planning should aim to find how to optimize the potential of every rural area in the region.

13.2 Rural Management in Regional Planning

While Chap. 16 provides additional guidance on this topic, it is worth to discuss here the role of rural management in regional planning here in order to frame it into the structure of the territory discussion.

The official levels of planning vary nationally, but they can be conceptualized into national legislation, regional guidelines, rural guidelines, local plans, and special actions. How rural guidelines, local plans and special actions cast into regional planning also depend on the local legislation but the relevant thing in terms of management is the ability of rural planners to influence high-level planning from the local scale. This can be more easily accomplished if the interests of several local communities are coincidental and they are all capable of acting together.

Rural areas might directly benefit from regional planning by asking a role in the regional planning making. This task is easier if the locals are highly organized, having a permanent rural management office, which has not to be confused with a part of the local government or policies. A rural management office (also known as a rural development office in some countries) does not necessarily depend on the local government. It can be run by another public office or being privately owned to provide a paid service to the community. This office can also be set up as a part of a trade-off with a private company for locating into the community (a large supermarket chain locating in the area, a mining or quarrying corporation, a timber business, etc.). In any case, the primary information source of a rural management office is a rural analysis, which ideally will be repeated from time to time to adapt to new needs and realities. One critical part of the work of a rural management office is to be able of make sure that the area is listen in other forums, and that any decision that might affect the village is known beforehand in order to be capable of acting if necessary. A good deal of networking and negotiation is required to that end.

Strategic knowledge of the neighboring areas is also a part of a proper rural management program. In coordinated actions such as European rural development actions (PRODER in Spain, POMO in Finland, EQUAL), ex-post information memories are a wealth of comparative information between areas with similar problems and challenges (Figs. 13.1 to 13.5). The term ex-post refers to an assessment of the development and results.

Fig. 13.1 The regionalization of the study area. Ex-post evaluation of a PRODER project in Catalonia, Spain. 2006

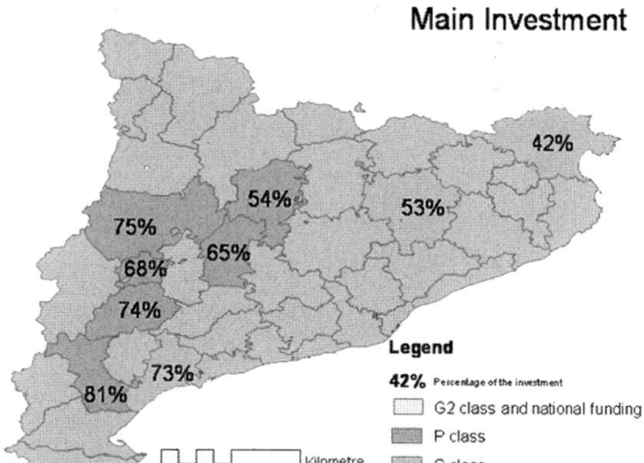

Fig. 13.2 Main investments in the area, by classes of funding. Ex-post evaluation of a PRODER project in Catalonia, Spain. 2006

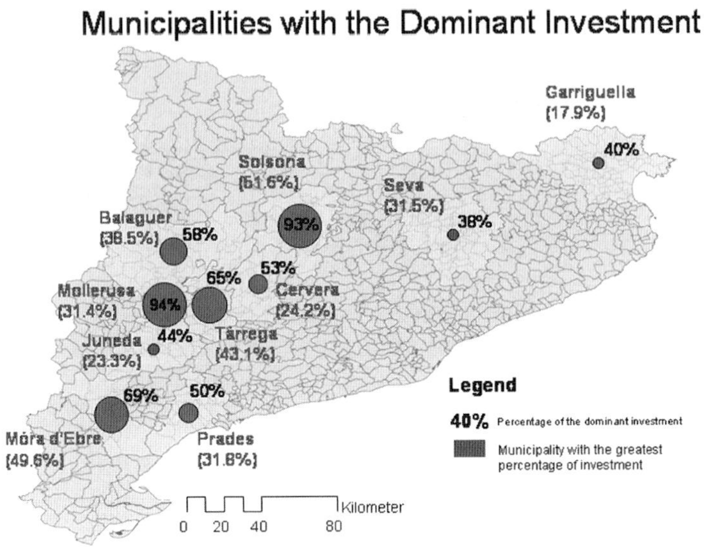

Fig. 13.3 Municipalities with the dominant investment. Ex-post evaluation of a PRODER project in Catalonia, Spain. 2006

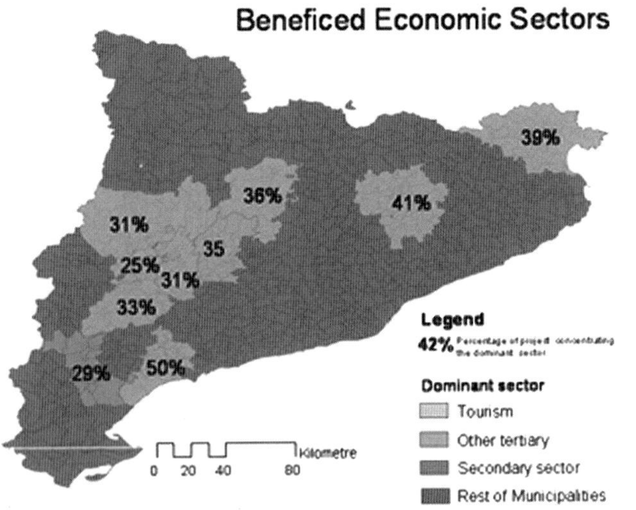

Fig. 13.4 Beneficed economic sectors in each municipality. Ex-post evaluation of a PRODER project in Catalonia, Spain. 2006

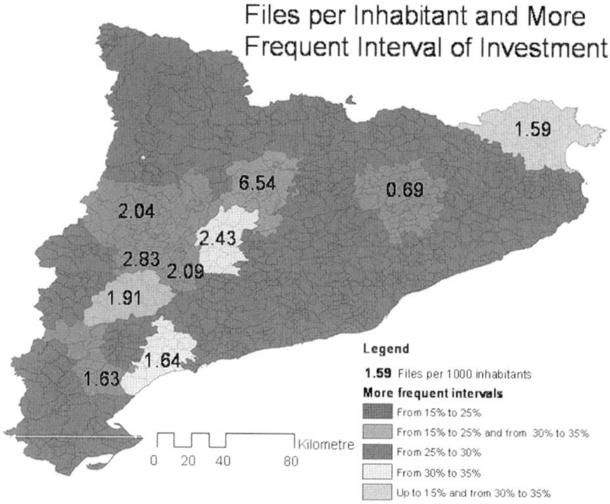

Fig. 13.5 Administrative files per inhabitant. Ex-post evaluation of a PRODER project in Catalonia, Spain. 2006

13.3 Counter-urbanization Processes in Developed Countries

If coordinate policies with other rural areas are relevant, the relationship with the closest urban areas is fundamental. Many elements are involved in under that topic, but counter-urbanization is one of the most relevant, thus requiring a separate discussion.

Counter-urbanization is a process almost limited to developed countries. It refers to the movement of people originally from the cities to the countryside, or to people who went to the cities in the past and then came back to the rural areas (usually after retirement). While the urban push and pull has been a constant in modern history, it is the scale of this process what is new. Poor urban quality of life, the lack of spaces, crowding or environmental factors partly explain this tendency. Also, changes in the work system (telework enabled by information technologies, flexible work terms prompted by tertiarisation, specialization, and technification in modern economies), and increased mobility have made possible a new migration to some rural areas. Psychological and cultural factors also exist as it was mentioned before. In the West, the bucolic vision of the countryside spans several centuries (from Horatio's *Beatus Ille*), but the concept it is not limited to this culture: then same idea can be traced back in Zen-influenced areas such as China and Japan. Regardless of objective factors, urbanites are keen to accept that rural areas present a simpler, quieter, safer life than those in cities.

Overcrowded megalopolises are the natural places in which the 'Paradise Lost' sensation thrives. Nonetheless, even in medium or small cities the perception of a land-based life has permeated into popular culture. A non negligible sector of the middle classes who grew in cities has experienced a double process. First, some of them moved to the urban fringe in search of a rural idyll not too far from the urban facilities (suburban rail or buses, motorways), and at commuter distance of the cities. After a while, those areas that became more and more urban by the own demands of the newcomers became less rural and then far from their original dreamed landscape. For some moving to the outer circle where rurality presents their pristine characters, rural life soon showed the face or limited and steady leisure facilities, poor accessibility, limited health and educational services, social rigidity and a limited choice for those unconsciously used to a variety of choice. After a few years, many families come back to the cities with rural homes usually transformed into second residences or sold to the next wave of urbanites.

In terms of planning and management, it is necessary to quantify those processes. There are some rural areas which will clearly become suburbia loosing their rural character altogether, whereas remote areas will be suitable to accommodate the few looking for a life far away from the urban influence. In the middle, a large part of the rural land will be affected by either an increasing pressure to accommodate and provide services to a limited number of newcomers, albeit in some other cases the area may be simply subject to endogenous transformations at its own pace.

New residents come with their own standards, and sometimes conflicts appear when the for instance traditional farming practices are perceived as nuisances by the new neighbors. The pressure over the environment is also an issue to be considered. The bucolic vision requires attractive sites such as woods, national park fringes or valuable natural spaces in general. Natural habitats fragmentation is also a problem, and it is always advisable to consider limiting the spread of new settlements if the rural nucleus is still capable of holding new inhabitants.

It is therefore worthy to identify the place of the rural area being studied within these processes. This task can be understood either as a part of the rural analysis or as a management issues once the analysis has identified this as a relevant element. It widely depends on the scope and orientation given to the rural plan.

References and Bibliography Used

S. O. Akatch, 'Rural planning in regional development: The Kenyan experience', Discovery And Innovation 13 (2001), 123–131.

O. Bastian, 'Landscape classification in Saxony (Germany) - a tool for holistic regional planning', Landscape and Urban Planning 50 (2000), 145–155.

T. K. Buxton and D. M. Dunham, 'Some Aspects of Regional Planning and Rural Development Policy in Ghana', Development and Change 2 (1970), 54–60.

D. Conyers, 'Regional-Development and the Local-Community - Planning, Politics and Social-Context - Weaver, C', Journal of Rural Studies 1 (1985), 290–291.

F. Dahms and J. McComb, 'Counterurbanization, interaction and functional change in a rural amenity area - a Canadian example', Journal of Rural Studies 15 (1999), 129–146.

F. A. Dahms, 'Dying Villages, Counterurbanization and the Urban Field - a Canadian Perspective', Journal of Rural Studies 11 (1995), 21–33.

T. Gunton, 'Megaprojects and regional development: Pathologies in project planning', Regional Studies 37 (2003), 505–519.

T. Jackson and J. Curry, 'Regional development and land use planning in rural British Columbia: Peace in the woods?' Regional Studies 36 (2002), 439–443.

H. U. W. Jones, 'Counterurbanization - Case Studies of Urban to Rural Movement - Perry,R, Dean, Pk, Brown, B', Journal of Rural Studies 3 (1987), 373–373.

R. Kanbur and X. B. Zhang, 'Fifty years of regional inequality in China: a journey through central planning, reform, and openness', Review Of Development Economics 9 (2005), 87–106.

G. Kohlhepp, 'Amazonia 2000: An evaluation of three decades of regional planning and development programmes in the Brazilian Amazon region', Amazoniana-Limnologia Et Oecologia Regionalis Systemae Fluminis Amazonas 16 (2001), 363–395.

C. J. A. Mitchell, 'Making sense of counterurbanization', Journal of Rural Studies 20 (2004), 15–34.

T. K. Moulik, B. H. Dholakia, R. H. Dholakia, K. V. Ramani and P. R. Shukla, 'Energy Planning in India - the Relevance of Regional-Planning for National Policy', Energy Policy 20 (1992), 836–846.

M. Poppe, 'Decentralised regional development planning in Indonesia - Implications for training', Third World Planning Review 20 (1998), 331–350.

A. M. Soliman, 'Regional planning scenarios in South Lebanon: the challenge of rural-urban interactions in the era of liberation and globalization', Habitat International 28 (2004), 385–408.

J. Sterkenburg, 'Rural development and the evolution of regional planning in Botswana: the implementation of national policy in Ngamiland district', Journal of Rural Studies 6 (1990), 217.

P. Vartiainen, 'The End of Drastic Depopulation in Rural Finland - Evidence of Counterurbanization', Journal of Rural Studies 5 (1989), 123–136.

C. A. Vogt and R. W. Marans, 'Natural resources and open space in the residential decision process: a study of recent movers to fringe counties in southeast Michigan', Landscape and Urban Planning 69 (2004), 255–269.

Chapter 14
Rural Planning and Sustainability

The concept of sustainability has been defined by the UN as 'a particular course of progress that meets the needs of the present without compromising the ability of future generations to meet their needs'. This definition has been criticized from several perspectives. First, it is not clear whether or not we are capable of understanding, let alone predict, our future needs and capabilities. Also, it is not clear whether or not we could even have a glimpse of our future interests, morals and philosophies. Our choices today may bring foreseeable problems in the future, but they may also bear new means to tackle those problems.

There are also other possible challenges that we cannot even imagine. The global catastrophe scenario is a good example: if in a particular moment in the future, life on Earth would be threatened by an imminent asteroid collision, having an orbital nuclear missile system ready to fire would be judged as a really good idea of the ancestors. Many of those ancestors, however, surely considered at the time that aerospatial development was a waste of resources in a world of widespread hunger, a potential menace to the environment, and a danger to the future of forthcoming generations.

Another similar argument can be said about the role of inventions. We do not know what future will bring, but we can foresee some hints. A final success of nuclear fusion technology would completely change the economic structure of the world, providing inexhaustible and ubiquitous clean energy for those who have the technology. In that scenario, having chosen to use most of the oil available in the past would then be deemed as not too serious, and could even be considered as beneficial since it boosted technology and new chemical techniques were developed.

In criticizing the sustainability concept, other researchers have pointed out that it is not up to us to decide what is in the best interest of the future inhabitants of the planet. The future might see our capitalistic economic structure in the same light we hold today the Roman slave-based system, thus considering our sustainable development concept as a mean of perpetuating a particularly efficient system of human exploitation. The application of the sustainability concept to the western Middle Ages mentality would have prevented the development of new economic practices such as industry, just as the blind application of sustainable development could today hinder material progress in some remote rural areas requiring a shift, as some might find that is not in the best spiritual interest of their children leaving traditional subsistence agriculture and welcoming advanced, global market-orientated practices.

Regardless of the more academic than practical criticisms, the sustainable development concept has the virtue of providing an easy-to-grasp idea that most cultures would agree upon: that a sensible use of Earth's resources involves avoiding complete depletion, since we might need those resources in the future.

It is however difficult to ascertain the consequences of our decisions. As we have no idea of what the future might bring, all we can do is use the best-available knowledge to act rationally. To that end, perspective is required. Earth is a dynamical system in which any equilibrium is actually a transient state. Whereas this is clear at a geological scale, the average person's timescale is very short and we tend to consider our environment and society as a closed and stable entity. If sustainability is focused on maintaining the status quo, evolution can be hindered and thus the ability of some animals to adapt could be affected. On the other hand, not all the animals and plants that benefit humans have a high tolerance to sudden changes or to invasive human activities. One of the problems of ongoing global change is the unprecedented pace of the changes, leaving no time for species to adapt to the new reality. Development can act as a powerful selection mechanism, but can result in favouring unwelcome species that might otherwise be controlled by a natural balance. Examples exist, such as some rats and cockroaches have evolved in the sense of being attached to human supplies, making them independent of their original food fluctuations.

Rural planning has to take into account sustainability issues on from a more local perspective. The effects of policies have to be planned and the consequences of the proposal taken into account. This involves the use of modeling techniques in order to fully quantify the changes.

14.1 Planning with Maps

Some of the first rural plans were developed using McHart's overlay mapping approach. This planner advocated the use of several layers of information to find suitable places for specific uses by using simple logical operations (AND, OR, XOR) with the layers. He uses actual translucent sheets of paper, one for each use, infrastructure or constraint. By combining them, he could find the most appropriated location for buildings, industrial sites or other activities. As we have discussed in Chap. 21, modern GIS systems work in a similar (albeit more complex) way.

14.2 Scenario Definition

Scenarios are plausible future situations. Defining a scenario involves common sense and a good knowledge of the processes involved. Each and every simulation or forecast implicitly accepts what is called a 'pink future', that is, a future in which no major tragedies have occurred. Under those circumstances, the initial hypotheses still hold and the simulation or forecast is valid. The difference between a forecast and a simulation is that a simulation is not intended to provide a definite state of the system being simulated, but a statistical description of the dynamics.

The importance of objective information has been stressed throughout this handbook. Information is used to develop models and test hypotheses. In terms of planning, 'what if' techniques can be integrated into mapping technology to test scenarios: once the system has been parameterized and a model can be established. It is then time to suggest sustainability-related scenarios, such as steady increase in tourist input, or a progressive tertiarisation of the economy. The usefulness of these quantitative approaches depends on how complex the model is and its ability to capture the relevant processes.

14.3 The Strategy of Development

Strategic evaluation is the cornerstone of management in a sustainability framework. A feasibility plan has to be initiated for any proposal. This activity has many things is common with a company or business plan. It has to include (among many other topics) an itemized guide to achieving the proposed objectives. Assessing any feasibility plan includes an appraisal on how realistic the hypotheses are, and whether or not the resources will permit the accomplishment of the objectives.

Additionally, the feasibility plan should be analyzed from other perspectives. Integration with other planning figures, at every level, has to be considered. If the plan collides with the aims of high-level legislation, it is unlikely to succeed. Also, if regional or national plans will be affecting the area, synergisms may appear and money can be saved. The sustainability of the proposal must also be assessed using one the widely available methodologies.

The societal value of the plan is another element to be considered. The strategy should explicitly state who the final beneficiaries of the plan are. Finally, it would be worth carrying out a DAFO analysis of the most likely scenario providing the strategic plan was effectively realized.

References and Bibliography Used

P. Allanson, J. Murdoch, G. Garrod and P. Lowe, 'Sustainability and the rural economy: An evolutionary perspective', Environment And Planning A 27 (1995), 1797–1814.

A. Bebbington, 'Social capital and rural intensification: local organizations and islands of sustainability in the rural Andes', Geographical Journal 163 (1997), 189–197.

A. Bosshard, 'A methodology and terminology of sustainability assessment and its perspectives for rural planning', Agriculture Ecosystems & Environment 77 (2000), 29–41.

C. Bryant, 'Sustainability and the civil commons: Rural communities in the age of globalization', Journal of Rural Studies 22 (2006), 500–501.

H. H. Carey, 'The guiding star of ecological and rural sustainability', Journal Of Forestry 97 (1999), 42–43.

M. Clow, 'Sustainability and international rural development.' Canadian Review of Sociology and Anthropology-Revue Canadienne De Sociologie Et D Anthropologie 34 (1997), 438–439.
A. K. Copus and J. R. Crabtree, 'Indicators of socio-economic sustainability: An application to remote rural Scotland', Journal of Rural Studies 12 (1996), 41–54.
J. Davies and E. Young, 'Sustainability, Development And Rural Aboriginal Communities', Australian Geographer 26 (1995), 150–155.
D. B. K. Dovie, 'Rural economy and livelihoods from the non-timber forest products trade. Compromising sustainability in southern Africa?' International Journal Of Sustainable Development And World Ecology 10 (2003), 247–262.
R. N. Eliason, 'Towards sustainability in village health care in rural Cameroon', Health Promotion International 14 (1999), 301–306.
A. Farrow and M. Winograd, 'Land use modelling at the regional scale: an input to rural sustainability indicators for Central America', Agriculture, Ecosystems & Environment 85 (2001), 249.
J. Gutberlet, 'Rural development and Social Exclusion: a case study of sustainability and distributive issues in Brazil', Australian Geographer 30 (1999), 221–237.
Y. Komatsu, A. Tsunekawa and H. Ju, 'Evaluation of agricultural sustainability based on human carrying capacity in drylands - a case study in rural villages in Inner Mongolia, China', Agriculture, Ecosystems & Environment 108 (2005), 29.
N. T. Krogman, 'Sustainability and international rural development - Bakker, JIH', Rural Sociology 61 (1996), 396–398.
H. Kulkarni, P. S. V. Shankar, S. B. Deolankar and M. Shah, 'Groundwater demand management at local scale in rural areas of India: a strategy to ensure water well sustainability based on aquifer diffusivity and community participation', Hydrogeology Journal 12 (2004), 184–196.
M. Kurtz, 'The condition of rural sustainability', Journal of Rural Studies 20 (2004), 257–258.
E. M. Makhanya, 'Demographic dynamics and land degradation at Ratau, Lesotho, in the context of rural sustainability', Land Degradation & Development 15 (2004), 257–269.
R. Panelli, 'Taking stands: Gender and the sustainability of rural communities', Journal of Rural Studies 22 (2006), 493–496.
F. Pauwels and H. Gulinck, 'Changing minor rural road networks in relation to landscape sustainability and farming practices in West Europe', Agriculture, Ecosystems & Environment 77 (2000), 95.
L. H. Qiao, Z. B. Yang and H. P. Benwang, 'A Computer-Aided Process Planning Methodology', Computers in Industry 25 (1994), 83–94.
K. Scott, J. Park and C. Cocklin, 'From "sustainable rural communities" to "social sustainability": giving voice to diversity in Mangakahia Valley, New Zealand', Journal of Rural Studies 16 (2000), 433–446.
L. Tacconi and C. Tisdell, 'Rural-Development Projects in LDCs - Appraisal, Participation and Sustainability', Public Administration and Development 12 (1992), 267–278.
H. Tovey, 'Sustainability and international rural development - Bakker, JI', Sociologia Ruralis 37 (1997), 162–163.
T. Unwin, 'Tourist development in Estonia - Images, sustainability, and integrated rural development', Tourism Management 17 (1996), 265–276.
R. Willis, 'Sustainability and change in rural Australia', Journal of Rural Studies 21 (2005), 487–488.
R. C. Wimberley, 'Policy Perspectives On Social, Agricultural, And Rural Sustainability', Rural Sociology 58 (1993), 1–29.

Chapter 15
Involving People in the Process

Local people can help both in the analytical and in the management part. They are the actual actors of the process, but they are also privileged sources of first-hand information. In the analysis part, meetings provide a means to identify key people, groups or associations capable of illuminating specific aspects of the rural area or of the rural needs, interests and hopes. In the management process, the locals are the key to the successful application of directives. But to get people involved, tools and participation mechanisms have to be created.

15.1 Management Actors

The management actors are both the rural people and the planners. The rural people affected by a rural project can range from proactive to openly hostile to any interference with their way of life. Nonetheless, politicians are the key side of any project. The role of the politician in a democracy is to be the expression of the popular will, so they are actually representing the people and it is them to whom the rural analyst should respond.

Thus, the role of the planners and managers is to provide the best objective evidence in order that the politicians take the appropriate decisions. In some cases, the final outcome made by the local authorities may be at odds with the best option as identified by the professionals. It is up to the analysts' ability to foresee and integrate any potential bias or bad practice into the analysis. In terms of management, rural management cannot be divorced from the political power from which it emanates.

15.2 Tools for Participation

Public access is a significant issue in relation to rural local planning and management, as the information provided by the locals can form a very significant part of the decision-making process. Access to information in digital form is an area of rapid development as computer and internet technology increases the capabilities of selecting data. Web services, information talks, and a permanent office make easy the access to all non-personal available documentation. Setting up the appropriate channels to the locals to participate is a crucial task at this stage.

References and Bibliography Used

S. Abram, J. Murdoch and T. Marsden, 'The social construction of "Middle England": The politics of participation in forward planning', Journal of Rural Studies 12 (1996), 353–364.

B. Adam, 'Spatial policies for metropolitan regions - Identity, participation and integration', European Planning Studies 11 (2003), 739–747.

B. Agarwal, 'Environmental action, gender equity and women's participation', Development and Change 28 (1997), 1–44.

A. Agrawal and K. Gupta, 'Decentralization and participation: The governance of common pool resources in Nepal's Terai', World Development 33 (2005), 1101–1114.

R. Amin, A. U. Ahmed, J. Chowdhury and M. Ahmed, 'Poor Womens Participation In Income-Generating Projects And Their Fertility Regulation In Rural Bangladesh - Evidence From A Recent Survey', World Development 22 (1994), 555–565.

D. G. Anderson, 'Approaches to Participation in Rural-Development - Oakley,P, Marsden,D', Journal of Rural Studies 2 (1986), 255-256.

N. E. Awa, 'Participation and Indigenous Knowledge in Rural-Development', Knowledge-Creation Diffusion Utilization 10 (1989), 304–316.

J. Beaumont, 'Community participation in the establishment and management of marine protected areas: A review of selected international experience', South African Journal Of Marine Science-Suid-Afrikaanse Tydskrif Vir Seewetenskap 18 (1997), 333–340.

S. Berman, S. Wasserman and S. Grimm, 'Participation Of Colorado Pediatricians And Family Physicians In The Medicaid Program', Western Journal Of Medicine 155 (1991), 649–652.

A. W. Black and I. Reeve, 'Participation in Landcare Groups: the Relative Importance of Attitudinal and Situational Factors', Journal of Environmental Management 39 (1993), 51.

L. Boer, 'Approaches to Participation in Rural-Development - Oakley, R, Marsden, D', Sociologia Ruralis 24 (1984), 279–279.

E. A. Brett, 'Participation and accountability in development management', Journal Of Development Studies 40 (2003), 1–29.

S. D. Brody, 'Measuring the effects of stakeholder participation on the quality of local plans based on the principles of collaborative ecosystem management', Journal of Planning Education and Research 22 (2003), 407–419.

I. Brotherton, 'Farmer Participation in Voluntary Land Diversion Schemes - Some Observations from Theory', Journal of Rural Studies 5 (1989), 299–304.

G. S. Child, 'People's participation in wildlife utilization in Africa', Landscape and Urban Planning 20 (1991), 159.

C. Clarke, 'Decentralization and rural development in Mexico: Community participation in Oaxaca's municipal funds program', Journal of Latin American Studies 30 (1998), 439–440.

J. M. Cohen and N. T. Uphoff, 'Participations Place in Rural-Development - Seeking Clarity through Specificity', World Development 8 (1980), 213–235.

A. Cornwall and K. Brock, 'What do Buzzwords do for Development Policy? A critical look at "participation", "empowerment" and "poverty reduction" ', Third World Quarterly 26 (2005), 1043–1060.

S. J. Cronin, M. G. Petterson, P. W. Taylor and R. Biliki, 'Maximising multi-stakeholder participation in government and community volcanic hazard management programs; A case study from Savo, Solomon Islands', Natural Hazards 33 (2004), 105–136.

A. Curtis, J. Birckhead and T. Delacy, 'Community Participation In Landcare Policy In Australia - The Victorian Experience With Regional Landcare Plans', Society & Natural Resources 8 (1995), 415–430.

A. Curtis and M. Lockwood, 'Landcare and catchment management in Australia: Lessons for state-sponsored community participation', Society & Natural Resources 13 (2000), 61–73.

D. Curtis, 'Approaches to Participation in Rural-Development - Oakley, P, Marsden, D', Community Development Journal 21 (1986), 236–237.

S. K. Datta and K. J. Virgo, 'Towards sustainable watershed development through people's participation: Lessons from the lesser Himalaya, Uttar Pradesh, India', Mountain Research And Development 18 (1998), 213–233.

A. K. Dev, 'Beneficiary Participation - His Optimum Choice under Uncertainty of Income Generated from Rural-Development Schemes', Journal of Rural Development 7 (1988), 461–465.

S. D. Edwards, P. J. S. Jones and D. E. Nowell, 'Participation in coastal zone management initiatives: a review and analysis of examples from the UK', Ocean & Coastal Management 36 (1997), 143.

G. Elliott, B. Wiltshire, I. A. Manan and S. Wismer, 'Community participation in marine protected area management: Wakatobi National Park, Sulawesi, Indonesia', Coastal Management 29 (2001), 295–316.

P. Ellis, 'Equipping Women for Fuller Participation in the Process of Rural-Development', Community Development Journal 22 (1987), 135–140.

R. Eyre and R. Gauld, 'Community participation in a rural community health trust: the case of Lawrence, New Zealand', Health Promotion International 18 (2003), 189–197.

K. Falconer, 'Farm-level constraints on agri-environmental scheme participation: a transactional perspective', Journal of Rural Studies 16 (2000), 379–394.

A. Fleischer and L. Applebaum, 'Spatial Differences in the Labor-Force Participation of Married-Women - the Case of Israels Peripheral Areas', Journal of Rural Studies 8 (1992), 293–302.

D. Forbes, 'Popular Participation in Planning for Basic Needs - Concepts, Methods and Practices - Lisk, F', Journal of Rural Studies 4 (1988), 305–306.

F. C. Fowler, 'Participation in Ohio's interdistrict open enrollment option: Exploring the supply-side of choice', Educational Policy 10 (1996), 518–536.

C. Francis, 'Rural-Development, Peoples Participation and the Role of NGOs', Journal of Rural Development 12 (1993), 205–210.

J. Friedmann, 'Approaches to Participation in Rural-Development - Oakley, P, Marsden, D', Journal of the American Planning Association 51 (1985), 380–383.

T. A. Ghebreyesus, T. Alemayehu, A. Bosman, K. H. Witten and A. Teklehaimanot, 'Community participation in malaria control in Tigray region Ethiopia', Acta Tropica 61 (1996), 145–156.

C. M. Gichuki, 'Community participation in the protection of Kenya's wetlands', Ostrich 71 (2000), 122–125.

A. A. Gonzalez and R. Nigh, 'Smallholder participation and certification of organic farm products in Mexico', Journal of Rural Studies 21 (2005), 449–460.

P. Goodwin, '"Hired hands" or "local voice": understandings and experience of local participation in conservation', Transactions Of The Institute Of British Geographers 23 (1998), 481–499.

D. D. Gow and J. Vansant, 'Beyond the Rhetoric of Rural-Development Participation - How Can It Be Done', World Development 11 (1983), 427–446.

S. Gregson, N. Terceira, P. Mushati, C. Nyamukapa and C. Campbell, 'Community group participation: Can it help young women to avoid HIV? An exploratory study of social capital and school education in rural Zimbabwe', Social Science & Medicine 58 (2004), 2119–2132.

K. A. Greiner, C. Li, I. Kawachi, D. C. Hunt and J. S. Ahluwalia, 'The relationships of social participation and community ratings to health and health behaviors in areas with high and low population density', Social Science & Medicine 59 (2004), 2303–2312.

M. Gupte, 'Participation in a gendered environment: The case of community forestry in India', Human Ecology 32 (2004), 365–382.

K. S. Hanna, 'The paradox of participation and the hidden role of information - A case study', Journal Of The American Planning Association 66 (2000), 398–410.

B. Jacobs and N. Price, 'Community participation in externally funded health projects: lessons from Cambodia', Health Policy And Planning 18 (2003), 399–410.

R. James, 'Participation disadvantage in Australian higher education: An analysis of some effects of geographical location and socioeconomic status', Higher Education 42 (2001), 455–472.

M. Jennings, ' "We must run while others walk": popular participation and development crisis in Tanzania, 1961-9', Journal Of Modern African Studies 41 (2003), 163–187.

C. Y. Jim and S. S. W. Xu, 'Stifled stakeholders and subdued participation: Interpreting local responses toward Shimentai Nature Reserve in South China', Environmental Management 30 (2002), 327–341.

L. Kapiriri, O. F. Norheim and K. Heggenhougen, 'Public participation in health planning and priority setting at the district level in Uganda', Health Policy and Planning 18 (2003), 205–213.

D. Kayo, 'Power sector reforms in Zimbabwe: will reforms increase electrification and strengthen local participation?' Energy Policy 30 (2002), 959.

G. Khasankhanova, 'Public participation to improve water resource management in Uzbekistan', Water Science and Technology 51 (2005), 365–372.

A. Kohli, 'Peoples Participation, Local-Government and Rural-Development - the Case of West-Bengal, India - Westergaard, K', Pacific Affairs 61 (1988), 540–541.

S. Kumar, 'Does "participation" in common pool resource management help the poor? A social cost-benefit analysis of joint forest management in Jharkhand, India', World Development 30 (2002), 763–782.

C. Lado, 'Female Labor Participation In Agricultural Production And The Implications For Nutrition And Health In Rural Africa', Social Science & Medicine 34 (1992), 789–807.

C. Leeuwis, 'Reconceptualizing participation for sustainable rural development: Towards a negotiation approach', Development And Change 31 (2000), 931–959.

W. Lise, 'Factors influencing people's participation in forest management in India', Ecological Economics 34 (2000), 379–392.

T. Marsden and J. Murdoch, 'The shifting nature of rural governance and community participation', Journal of Rural Studies 14 (1998), 1–4.

K. Mathiyazhagan, 'Willingness to pay for rural health insurance through community participation in India', International Journal Of Health Planning And Management 13 (1998), 47–67.

J. Midgley, 'Projects with People - the Practice of Participation in Rural-Development - Oakley, P', Third World Planning Review 13 (1991), 212-214.

M. M. O. Mirghani and H. H. G. Savenije, 'Incorporation of people's participation in planning and implementation of water resources projects', Physics and Chemistry of The Earth 20 (1995), 229.

R. V. R. Mohan, 'Rural water supply in India: Trends in institutionalizing people's participation', Water International 28 (2003), 442–453.

S. G. Moon, S. C. Mayo and G. C. Mccann, 'Local Participation in Rural Development Planning in Korea - Its Functions and Dysfunctions', Sociologia Ruralis 8 (1968), 484–484.

P. Nanda, 'Women's participation in rural credit programmes in Bangladesh and their demand for formal health care: Is there a positive impact?' Health Economics 8 (1999), 415–428.

L. Pare, C. Robles and C. Cortez, 'Participation of indigenous and rural people in the construction of developmental and environmental public policies in Mexico', Ids Bulletin-Institute Of Development Studies 33 (2002), 83-+.

D. Parisi, D. K. McLaughlin, S. M. Grice, M. Taquino and D. A. Gill, 'TANF participation rates: Do community conditions matter?' Rural Sociology 68 (2003), 491–512.

E. Phimister, E. Vera-Toscano and A. Weersink, 'Female participation and labor market attachment in rural Canada', American Journal Of Agricultural Economics 84 (2002), 210–221.

J. Ploger, 'Public participation and the art of governance', Environment and Planning B-Planning & Design 28 (2001), 219–241.

C. Potter and R. Gasson, 'Farmer Participation in Voluntary Land Diversion Schemes - Some Predictions from a Survey', Journal of Rural Studies 4 (1988), 365–375.

J. N. Pretty and P. Shah, 'Making soil and water conservation sustainable: From coercion and control to partnerships and participation', Land Degradation & Development 8 (1997), 39–58.

S. Rajakutty, 'Peoples Participation in Monitoring and Evaluation of Rural-Development Programs - Concepts and Approaches', Journal of Rural Development 10 (1991), 35–53.
M. L. Santhanam, 'Citizen Participation in Rural-Development - Chaturvedi, Hr, Mitra, Sk', Journal of Rural Development 2 (1983), 490–491.
J. A. Seeley, J. F. Kengeyakayondo and D. W. Mulder, 'Community-Based HIV/AIDS Research - Whither Community Participation - Unsolved Problems In A Research-Program In Rural Uganda', Social Science & Medicine 34 (1992), 1089–1095.
E. D. Setty, 'Peoples Participation in Rural-Development - a Critical Analysis', Indian Journal of Social Work 46 (1985), 73–83.
P. N. Sharma, 'Community Participation For Forest Watershed Management In Laos', Journal Of Soil And Water Conservation 47 (1992), 499–504.
M. S. Sherraden, 'Policy Impacts Of Community Participation - Health-Services In Rural Mexico', Human Organization 50 (1991), 256–263.
D. M. Shucksmith, J. Rowanrobinson, C. T. Reid and M. G. Lloyd, 'Community Councils as a Medium for Public-Participation - a Case-Study in Grampian Region', Journal of Rural Studies 1 (1985), 307–319.
J. D. Sidaway, 'A voice for the excluded. Popular participation in development: Utopia or necessity? - Stiefel, M, Wolfe, M', Journal of Rural Studies 12 (1996), 211–211.
A. Smalley, 'Peoples Participation, Local-Government and Rural-Development - the Case of West Bengal, India - Westergaard, K', Journal of Contemporary Asia 17 (1987), 366–369.
D. V. Stefanovic, 'Participation in Urban and Rural-Development in Yugoslavia', Ekistics-the Problems and Science of Human Settlements 42 (1976), 235–236.
D. Storey, 'Issues of integration, participation and empowerment in rural development: The case of LEADER in the Republic of Ireland', Journal of Rural Studies 15 (1999), 307–315.
L. E. Swanson, 'Rural policy and direct local participation: Democracy, inclusiveness, collective agency, and locality-based policy', Rural Sociology 66 (2001), 1–21.
L. Tacconi and C. Tisdell, 'Rural-Development Projects in LDCS - Appraisal, Participation and Sustainability', Public Administration and Development 12 (1992), 267–278.
M. Tewdwr-Jones, 'Rural government and community participation: The planning role of community councils', Journal of Rural Studies 14 (1998), 51.
R. Thomas, 'Rural-Development Projects and Popular Participation in Zambia - Saasa, O', Canadian Journal of African Studies-Revue Canadienne Des Etudes Africaines 28 (1994), 176–177.
M. D. Turner, 'Political ecology and the moral dimensions of "resource conflicts": the case of farmer-herder conflicts in the Sahel', Political Geography 23 (2004), 863–889.
H. Valokivi, 'Participation and citizenship of elderly persons: User experiences from Finland', Social Work in Health Care 39 (2004), 181–207.
R. Vengroff, 'Popular Participation and Administration of Rural Development - Case of Botswana', Human Organization 33 (1974), 303–309.
M. Warner, ' "Consensus" participation: an example for protected areas planning', Public Administration And Development 17 (1997), 413–432.
T. Webler, S. Tuler and R. Krueger, 'What is a good public participation process? Five perspectives from the public', Environmental Management 27 (2001), 435–450.
T. Webler, S. Tuler, I. Shockey, P. Stern and R. Beattie, 'Participation by local governmental officials in watershed management planning', Society & Natural Resources 16 (2003), 105–121.
S. H. Williams, 'Citizen Participation in City and Regional-Planning - Effective American Methodology', Town Planning Review 47 (1976), 349–358.
K. White, 'Cooperatives and local development: Theory and applications for the 21st century', Professional Geographer 57 (2005), 144–146.
S. M. Zanen and W. T. Degroot, 'Enhancing Participation Of Local People - Some Basic Principles And An Example From Burkina-Faso', Landscape and Urban Planning 20 (1991), 151–158.

Chapter 16
The Decision Guidelines Making

The making of a set of guidelines usually arises after a previous rural analysis work, aiming to provide some guidance for development or to solve the problems identified in the analysis. Guidelines are simple directives that should be considered when detailed norms, laws or proposals are made. They represent the basic principles driving any further action. In formal terms, the guidelines should be flexible but not too vague.

Developing a set of rural development guidelines involves a combination of skills, abilities, experience and professional competence. The task of writing the guidelines can be initiated by a single person whom will supply a first, well organized draft or a three-tier scheme, based upon the previous rural analysis. This skeleton will then be used as the starting point for discussion. This frame can be modified, or a new scheme proposed from scratch, but it is advisable to have a kick-off document to center the discussion.

16.1 The Gap Between Scientific Analysis and Decision-making

There is a difference between a scientific based approach to management and actual decision-making. Scientists can provide the best available information and generate objective models, but it is up to the decision makers to follow such directions (or not). There are good reasons to consciously depart from the best scientific advice, such as different priorities, moral grounds, or simply the unwillingness of the people involved to carry out the proposed actions. In a democracy, the ultimate power belongs to the people. The analyst may be certain that a decision is wrong in the best interest of the majority, but all he can do is to note that and continue with his work if the majority prefers to do otherwise.

16.2 Decision-making Procedures and Timing

One key element that is all too often forgotten in planning is timing. Plans have a dynamic component that has to always be considered. First, if the analysis work takes too long, the final analyst report may be telling a history about the past, and not about the present. The danger of outdating the analysis is always present.

Problems with time also appear, in the sense of not taking into account that timing is as important as the action. A timely action can increase the benefits of the action itself, whereas an untimely action can not only be useless but harmful and contrary to the intended purpose. The examples are numerous: an action to provide maternity information in a remote area should be preceded or go along with gender and equality policies to be fully effective. Proposing a ski resort in a rural area with recent advanced agricultural practices may bring unwanted economic risks and adverse social effects, whereas after a few years of consolidation of the new economic activity, the same project could create full-scale economies with little social opposition.

16.3 Rural Planning, Regional Planning and Impact Assessments

The matching between the various levels of planning requires separate attention. Decisions and plans at a higher level may supersede rural plans, but consistent rural plans may have a down-top effect on legislation and planning.

Rural planning also relates with environmental impact assessment (EIA) in several ways. EIA is an activity with parallels to rural analysis, and some reports in one may be useful to another. Social impact assessment (SIA) also provides powerful tools to this part of the rural analysis. Contributions such as Becker and Vanclay' synthetic list of social impacts may be of use with the appropriate caveats.

References and Bibliography Used

J. Alwang, P. B. Siegel and S. L. Jorgensen, 'Seeking guidelines for poverty reduction in rural Zambia', World Development 24 (1996), 1711–1723.

C. J. Barrow, 'Social Impact Assessment: an Introduction'. (2000) London: Arnold.

H. Becker, and F, Vanclay. 'The international handbook of SIA'. (2003) Cheltenham: E Elgar.

B. Field and D. Hamilton, 'Pragmatic guidelines for determining road investment priorities in developing countries', International Journal of Sustainable Development and World Ecology 9 (2002), 223–231.

T. Pinto-Correia, R. Gustavsson and J. Pirnat, 'Bridging the gap between centrally defined policies and local decisions - Towards more sensitive and creative rural landscape management', Landscape Ecology 21 (2006), 333–346.

D. K. Sharma, D. Ghosh and J. A. Alade, 'A fuzzy goal programming approach for regional rural development planning', Applied Mathematics And Computation 176 (2006), 141–149.

I. G. Simpson, 'Objectives, Guidelines and Management-Training for Local Rural-Development Programs', Agricultural Administration 8 (1980), 43–57.

A. G. van der Valk and R. W. Jolly, 'Recommendations for research to develop guidelines for the use of wetlands to control rural nonpoint source pollution', Ecological Engineering 1 (1992), 115.

L. J. Webb and D. M. Smyth, 'Ecological Guidelines and Traditional Empiricism in Rural-Development Acknowledgments', Environmentalist (1984), 99–105.

Chapter 17
Public Participation and Information Technologies

Once the guidelines, reports, or documents have been produced, it is of the interest of the rural area to disseminate the results to the public. There are some documents that should remain private, such as the information given nominally, private information on for instance taxes or earnings, or social or personal preferences and large etcetera dictated by common sense. Whereas the database of a rural analysis/management project should not be considered public in any case, the reports and guidelines should be. It is therefore required to have this component in mind when writing the reports.

17.1 Effective Communication

While some parts of the reports might be very technical, an additional effort should be put to make the language accessible to any literate person. Pompous or verbose comments should always be avoided. This applies not only to English but to most other languages too: Latin-derived languages such as Spanish or French allow a far more complex syntax than English, being prone to fool the reader with a nice empty shell. Natives of such languages should be aware of this fact and avoid rhetorical statements. In general, each sentence should convey an idea as simply as possible and sentences should be put one after another in a meaningful, related way. Vain phrases such as 'a host of very complex dynamics might explain the process involved in this problem' can only generate discomfort in the reader, who would like to know more about those dynamics to understand the problem.

The language should be kept straightforward, and ideas should be conveyed as simply as possible. This applies to any piece of scientific literature, but especially to public communication. Redundant or rhetoric sentences or comments must be avoided altogether. Also, it is advisable to hide the literary style of the writer as much as possible: the language of a a rural analysis report should be closer to a microwave oven manual than to a novel, and should also maintain the same standards that of a manual in terms of being revised over an over to avoid any misunderstandings or wrong doings, and to be widely understood.

Materials and ideas should be organized in clear, short phrases. Itemized structure helps to keep the text clean and to differentiate between different ideas. If two items are close enough to be potentially confused by the reader one of them should be either removed or indented.

Diagrams require a certain degree of education to be fully understood and therefore should be kept to a minimum. Pictures and images should be easy to grasp, and not complicated layered images. Regarding to maps, they have to be as simple as possible if they are intended to illustrate rather than to provide planning information. A map such as the one in Fig. 17.1 is simple enough to be read and understood by any person.

If the figure is a simulation or a model (Fig. 17.2), this fact should be made explicit. Statements such as 'this image is not a picture but a computer-generated model' are key, no matter how obvious it might seem to an educated person. The same applies for false-color combinations of satellite imagery.

Ideally, a good communication campaign should cover several levels of literacy and education. In the first level, the central ideas should be emphasized without using technical terms. The second level might involve some more specific information, with the third and fourth being more complex, specialized levels. In all the three phases information technologies can help to disseminate the work. Typographic and web design strategies exist to fulfill the basic requirements for effective communication.

The clarity requirement is not only a tenet of communication. In rural analysis and management, it can make the difference between receiving useful feed back or a large number of questions; or between engaging people in the process or drawing a line between the planners and the locals. Finally, telecommunications networks are

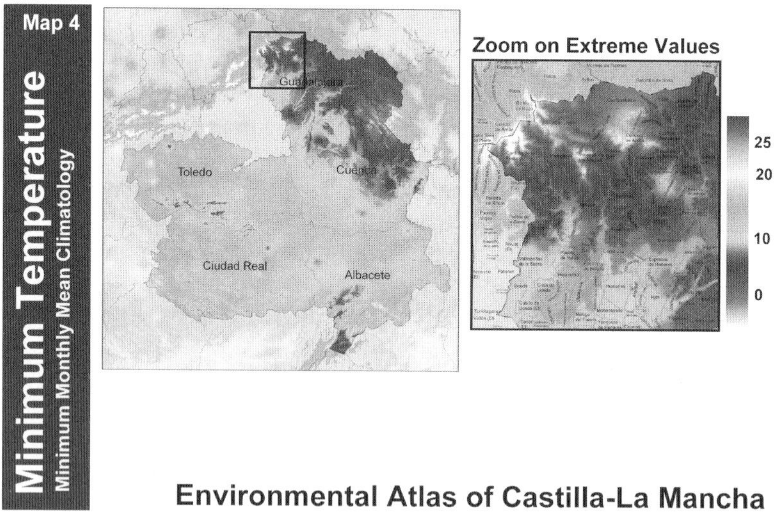

Fig. 17.1 An example of a simple, effective map. Minimum temperatures (climatology) of Castilla-La Mancha region, with a zoom in a special interest area (extremes in minimum temperature) (see color plate on following page 333)

Fig. 17.2 A shaded digital elevation model of a flood plain. Zamora, Spain. It should always make clear the difference between models, simulations, computer-generated graphics and reality

not only useful for rural development itself, but also to convey the rural project to the local population. An intelligent use of existing technologies can help the success of the project.

References and Bibliography Used

H. Albrecht, 'Extension Education and Rural-Development Vol, 1 - International Experience in Communication and Innovation - Crouch, Br, Chamala, S', Sociologia Ruralis 22 (1982), 73–74.

J. C. Allen, 'Telecommunications and Rural-Development - a Study of Private and Public-Sector Innovation - Schmandt, J, Williams, F, Wilson, Rh, Strover, S', Rural Sociology 58 (1993), 304–306.

S. Bandias and S. Ram Vemuri, 'Telecommunications infrastructure facilitating sustainable development of rural and remote communities in Northern Australia', Telecommunications Policy 29 (2005), 237–249.

H. Barrett, 'Communication and Rural-Development - Bordenave, Je', Convergence 10 (1977), 85–86.

N. Berkeley, D. Clark and B. Ilbery, 'Regional variations in business use of information and communication technologies and their implications for policy: Case study evidence from rural England', Geoforum 27 (1996), 75–86.

C. Courtright, 'Which lessons are learned? Best practices and World Bank rural telecommunications policy', Information Society 20 (2004), 345–356.

E. W. Coward, 'Impact of Communication on Rural Development - Investigation in Costa Rica and India - Roy, P, Waisanen, Fb and Rogers, Em', Rural Sociology 35 (1970), 441–441.

S. M. A. Eid and O. M. Hegazy, 'Investigation into Communication Needs for Rural-Development through Modeling Techniques', Resource Management and Optimization 2 (1983), 269–292.

K. Elmenoufi, 'Communication and Rural-Development in an Egyptian Village', Journal of Arab Affairs 4 (1985), 57–65.

J. H. Farrington, 'Reducing Isolation - Telecommunications and Rural-Development - Black, Jb', Scottish Geographical Magazine 103 (1987), 180–180.

S. Ghorpade, 'When Telephones Reach the Village - Telecommunications in Rural-Development - Hudson, He', Journalism Quarterly 62 (1985), 660–661.

D. Goldschmidt, 'Financing Telecommunications for Rural-Development', Telecommunications Policy 8 (1984), 181–203.

S. Grimes, 'Exploiting Information and Communication Technologies for Rural-Development', Journal of Rural Studies 8 (1992), 269–278.

C. A. Hollifield, J. F. Donnermeyer, G. H. Wolford and R. Agunga, 'The effects of rural telecommunications self-development projects on local adoption of new technologies', Telecommunications Policy 24 (2000), 761–779.

H. E. Hudson and E. B. Parker, 'Information gaps in rural America : Telecommunications policies for rural development', Telecommunications Policy 14 (1990), 193.

H. Lee, Y. Shi and S. M. Nazem, 'Supporting rural telecommunications: A compromise solutions approach', Annals Of Operations Research 68 (1996), 33–45.

H. Lee, Y. Shi, S. M. Nazem, S. Y. Kang, T. H. Park and M. H. Sohn, 'Multicriteria hub decision making for rural area telecommunication networks', European Journal Of Operational Research 133 (2001), 483–495.

G. Madden, S. J. Savage, G. Coble-Neal and P. Bloxham, 'Advanced communications policy and adoption in rural Western Australia', Telecommunications Policy 24 (2000), 291.

S. Mitchell and D. Clark, 'Business adoption of information and communications technologies in the two-tier rural economy: some evidence from the South Midlands', Journal of Rural Studies 15 (1999), 447–455.

M. Mueller, 'Electronic Byways - State Policies for Rural-Development through Telecommunications - Parker, Eb, Hudson, He', Journal of Communication 43 (1993), 147–159.

M. Mueller, 'Telecommunications and Rural-Development - a Study of Private and Public-Sector Innovation - Schmanot, J, Williams, F, Wilson, Rh, Strover, S', Journal of Communication 43 (1993), 147–159.

M. Oden, 'Information and communication technologies and rural development', Journal of the American Planning Association 70 (2004), 107–108.

S. Pal, 'Telecommunications for Rural-Development', Electronics Information & Planning 19 (1991), 30–33.

E. B. Parker, 'Communication Satellites for Rural-Development', Telecommunications Policy 2 (1978), 309–315.

C. Philo, 'Telecommunications in Rural England - Report to the Rural-Development Commission and Oftel - Ec-Transport-Planning-Group', Journal of Rural Studies 6 (1990), 347–348.

T. Poffenberger, 'Impact of Communication on Rural Development - Investigation in Costa-Rica and India - Roy, P, Waisanen, Fb and Rogers, Em', Journal of Asian Studies 29 (1970), 961–962.

R. Ramirez, 'A model for rural and remote information and communication technologies: a Canadian exploration', Telecommunications Policy 25 (2001), 315.

R. Ramirez and D. Richardson, 'Measuring the impact of telecommunication services on rural and remote communities', Telecommunications Policy 29 (2005), 297–319.

R. Richardson and A. Gillespie, 'Advanced communications and employment creation in rural and peripheral regions: A case study of the Highlands and Islands of Scotland', Annals Of Regional Science 30 (1996), 91–110.

G. M. Robinson, 'Telecommunications in Rural England', Journal of Rural Studies 7 (1991), 135–136.

P. C. Sharma, 'Integrated Communication - Bringing People and Rural-Development Together - Webster, Rl', Rural Sociology 42 (1977), 301–302.

L. F. Stephens, 'Integrated Rural-Development and Communication - Some Views - Coombs, Ph', Journalism Quarterly 54 (1977), 814–815.

M. A. Tarrant, C. Overdevest, A. D. Bright, H. K. Cordell and D. B. K. English, 'The effect of persuasive communication strategies on rural resident attitudes toward ecosystem management', Society & Natural Resources 10 (1997), 537–550.

T. Tsiligirides, 'Teleworking - An Information Technology Tool For Integrated Broad-Band Communication Development In Rural-Areas Of Europe', Journal Of Information Technology 8 (1993), 241–255.

G. Valentine and S. L. Holloway, 'A window on the wider world? Rural children's use of information and communication technologies', Journal of Rural Studies 17 (2001), 383–394.

M. E. Voland, 'Extension Education and Rural-Development - International Experience in Communication and Innovation, Vol 1, - Crouch, Br, Chamala, S', Rural Sociology 47 (1982), 576–577.

B. Wellenius, 'When Telephones Reach the Village - the Role of Telecommunications in Rural-Development - Hudson, He', Journal of Communication 36 (1986), 159.

M. D. Woods, 'Electronic Byways - State Policies for Rural-Development through Telecommunications - Parker, Eb, Hudson, He, Dillman, Da, Strover, S, Williams, F', Growth and Change 24 (1993), 298–299.

F. W. Young, 'Impact of Communication on Rural Development - Investigation of Costa-Rica and India - Roy, P, Waisanen, Fb and Rogers, Em', American Journal of Sociology 76 (1971), 1166–1168.

Chapter 18
Global Change Issues in Rural Planning

A growing concern on rural science is the impact of global change on rural communities. Earth is a dynamic planet and global changes have occurred in the past due to natural causes. The difference now is the pace of these changes. The scientific bases of climate change are well explained in the reports by the International Panel of Climate Change (IPCC). The latter IPCC report (2007) presents some novelties over the precedent reports. Now, there is overwhelming evidence of the human effects on climate. The best available scientific evidence indicates that this new acceleration is linked with human activities, namely greenhouse gas emissions from the industrial revolution, land use changes, and the exponential change of the planet's population.

It has to be made clear here that there is no scientific controversy on this topic and that dissonant voices are not based on the best scientific evidence, but in particular interpretations of a data selection. Even if all the climatologists in the world were wrong on this, the current consensus is well founded in view of all the available data and our methods of analysis.

Climate changes may affect some rural areas such as coastal or semiarid areas. A steady increase in sea level may flood large regions of the globe. The effect of phenological changes in the ecosystem might reduce the chances of adapting to the change by reducing the number of individuals and thus the gene pool. A low entropic state in that pool reduces the chances of mutations and adaptation, resulting in loss of biodiversity.

18.1 The Precautionary Principle

A central concept in global change research is the precautionary principle. It states that even if we were not completely certain that climate change is a real threat, the risk is too high to not take any proactive action.

The principle seems a like common sense statement, but has been widely criticized from several sides, not only from sceptical environmentalists but also from some scientists from the establishment. As with the concept of sustainable development, the definition is ample enough to be easily attacked. A common, non-partisan criticism is that the vast amount of uncertainty in our current climate simulations would make anyone think twice before starting costly, and at best useless changes in global economies. On the other hand, one might wonder if it is worth thc risk doing nothing considering the importance of the bet. Probably, it is not: cleaner sources of energy can only benefit human health, improve biodiversity, and can even boost technology and science by posing new urgent challenges to human ingenuity.

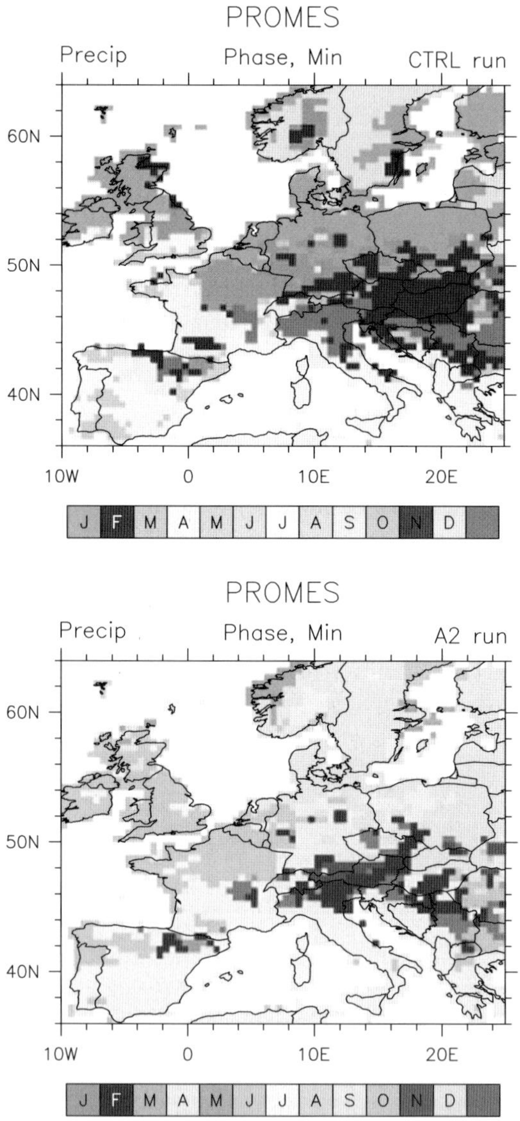

Fig. 18.1 Changes in the phase minimum of the annual cycle of precipitation in Europe for the period 2070–2100 under an increased greenhouse emissions scenario compared with 1960–1990, PROMES model simulation

18.2 From Assessment to Mitigation

Assessment of the effects of climate changes is a highly specialized task. It involves an extensive use of advanced, not-on-the-shelf technology. Even the interpretation of the models' outcomes requires a deep understanding of what is behind the equations. Pitfalls are common, and extreme care should be put into extracting the right conclusions. The uncertainty in the limits of the results, the probabilistic nature of simulations, and our still limited understanding of natural processes should be balanced when faced with future climate simulations.

Some derived plots from models' outputs can help to link simulations with actual rural issues, such as farming practices and water availability. Figures 18.1 and 18.2 show the predicted changes of the precipitation in Europe for an increased greenhouse effect scenario. This kind of plot may be more useful than seasonal averages for some agricultural applications. The onset of spring, and the time of arrival of the first rainfall of the year are critical for crops, and mid-term planning should take into account foreseeable changes.

Knowing the effects of likely global climate change is an assessment exercise that usually yields a mitigation strategy. This implies to combining a good knowledge of the current dynamics of the rural area and a global change simulation. Analyzing the

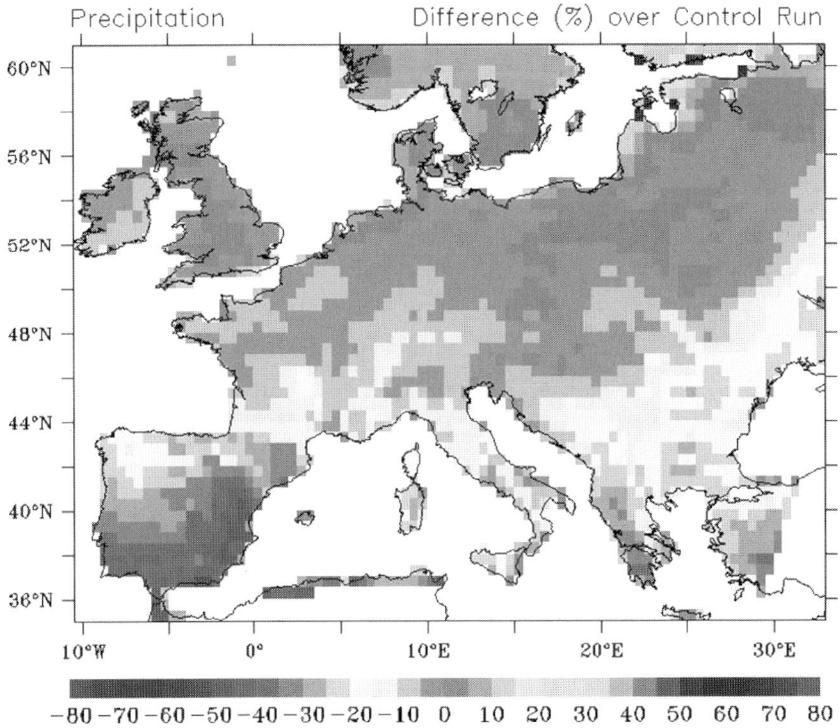

Fig. 18.2 Differences between future (2070–2100) and present climate precipitation

relationship between current climate conditions and current economy and society permits to infer the effects of new climate conditions (assessment) on the rural area. Mitigation consists of trying to avoid the unwanted effects of those changes by using the targeted policies.

References and Bibliography Used

M. F. Abu-Taleb, 'Impacts of global climate change scenarios on water supply and demand in Jordan', Water International 25 (2000), 457–463.

L. D. Brekke, N. L. Miller, K. E. Bashford, N. W. T. Quinn and J. A. Dracup, 'Climate change impacts uncertainty for water resources in the San Joaquin River Basin, California', Journal of the American Water Resources Association 40 (2004), 149–164.

Q. P. Chiotti and T. Johnston, 'Extending The Boundaries Of Climate-Change Research - A Discussion On Agriculture', Journal Of Rural Studies 11 (1995), 335–350.

Q. P. Chiotti and T. Johnston, 'Extending the boundaries of climate change research: A discussion on agriculture', Journal of Rural Studies 11 (1995), 335.

K. Clayton, 'Climate Change and World Agriculture - Parry,M', Journal of Rural Studies 7 (1991), 134–135.

R. Dautray, 'Les perspectives de l'energie nucleaire dans le cadre des changements climatiques: The future of nuclear energy in relation with climate change', Comptes Rendus de l'Academie des Sciences - Series IIA - Earth and Planetary Science 333 (2001), 811.

D. J. Davidson, T. Williamson and J. R. Parkins, 'Understanding climate change risk and vulnerability in northern forest-based communities', Canadian Journal Of Forest Research-Revue Canadienne De Recherche Forestiere 33 (2003), 2252–2261.

J. C. de Araujo, D. Doll, A. Guntner, M. Krol, C. B. R. Abreu, M. Hauschild and E. M. Mendiondo, 'Water scarcity under scenarios for global climate change and regional development in semiarid northeastern Brazil', Water International 29 (2004), 209–220.

P. Deshingkar, 'Climate change adaptation in India: A case study of forest systems in Himachal Pradesh', International Journal Of Environment And Pollution 9 (1998), 186–197.

T. Dockerty, A. Lovett, G. Sunnenberg, K. Appleton and M. Parry, 'Visualising the potential impacts of climate change on rural landscapes', Computers, Environment and Urban Systems 29 (2005), 297–320.

H. Eakin, 'Institutional change, climate risk, and rural vulnerability: Cases from central Mexico', World Development 33 (2005), 1923–1938.

J. M. Gibbons and S. J. Ramsden, 'Robustness of recommended farm plans in England under climate change: A Monte Carlo simulation', Climatic Change 68 (2005), 113–133.

C. Giupponi, M. Ramanzin, E. Sturaro and S. Fuser, 'Climate and land use changes, biodiversity and agri-environmental measures in the Belluno province, Italy', Environmental Science & Policy 9 (2006), 163.

T. Grothmann and A. Patt, 'Adaptive capacity and human cognition: The process of individual adaptation to climate change', Global Environmental Change-Human And Policy Dimensions 15 (2005), 199–213.

S. Hartley, R. C. Harriss and T. Blanchard, 'Urban Water Demand and Climate-Change', Natural Resources Forum 18 (1994), 55–62.

J. A. A. Jones, 'Climate-change mitigation and European land-use policies', Journal of Rural Studies 16 (2000), 259–260.

A. Kanudia and R. Loulou, 'Advanced bottom-up modelling for national and regional energy planning in response to climate change', International Journal of Environment and Pollution 12 (1999), 191–216.

V. V. N. Kishore, P. M. Bhandari and P. Gupta, 'Biomass energy technologies for rural infrastructure and village power–opportunities and challenges in the context of global climate change concerns', Energy Policy 32 (2004), 801–810.

H. Meinke and R. Stone, 'Seasonal and inter-annual climate forecasting: The new tool for increasing preparedness to climate variability and change in agricultural planning and operations', Climatic Change 70 (2005), 221–253.

R. P. Motha and W. Baier, 'Impacts of present and future climate change and climate variability on agriculture in the temperate regions: North America', Climatic Change 70 (2005), 137–164.

J. E. Olesen and M. Bindi, 'Consequences of climate change for European agricultural productivity, land use and policy', European Journal Of Agronomy 16 (2002), 239–262.

V. Perarnaud, B. Seguin, E. Malezieux, M. Deque and D. Loustau, 'Agrometeorological research and applications needed to prepare agriculture and forestry to 21st century climate change', Climatic Change 70 (2005), 319–340.

P. S. Ramakrishnan, 'Sustainable development, climate change and tropical rain forest landscape', Climatic Change 39 (1998), 583–600.

X. Ren, L. Zeng and D. D. Zhou, 'Sustainable energy development and climate change in China', Climate Policy 5 (2005), 185–198.

M. J. Salinger, M. V. K. Sivakumar and R. Motha, 'Reducing vulnerability of agriculture and forestry to climate variability and change: Workshop summary and recommendations', Climatic Change 70 (2005), 341–362.

W. D. Solecki and C. Oliveri, 'Downscaling climate change scenarios in an urban land use change model', Journal of Environmental Management 72 (2004), 105–115.

P. Vellinga and R. J. T. Klein, 'Climate-Change, Sea-Level Rise and Integrated Coastal Zone Management - an Ipcc Approach', Ocean & Coastal Management 21 (1993), 245–268.

H. D. Venema, E. J. Schiller, K. Adamowski and J. M. Thizy, 'A water resources planning response to climate change in the Senegal River Basin', Journal of Environmental Management 49 (1997), 125–155.

Chapter 19
Public Dissemination Tools: An Example for Inventories

This chapter shows an example of tools that can be used for the public dissemination element of a rural management project. It comprises of a set of maps done for an EU EQUAL project in Lleida, Spain (Figs. 19.1 to 19.23).

The objective of the map series was to help locals identify relevant aspects of their rural area in the inventory. While it may seem surprising at first glance, that for instance a map of the Geo-sites could be useful for this purpose, it turns out that people were able to point to other sites with similar geological merits by comparing with the existing places, thus identifying new potential sites. Maps of official minimum temperature and precipitation were fine-tuned by people in the mountain areas by indicating places with frequent departures from those averages. Wind speed maps were of use to the municipalities with an interest in attracting wind turbine initiatives, while climate classifications were found useful to orientate reforestation and to asses the impacts of climate change in the area. Maps of the femininity ratio (female population divided by male population) also identified some social issues regarding equal opportunities in rural populations.

Tools from public participation can emanate directly from a previous analysis phase. Thus, many of the maps presented here were also used in the analysis phase. The role of local knowledge is critical: regardless of the knowledge and expertise exhibited by the analyst when using geographical information, locals always know better. They can easily identify recently closed paths, inaccessible sectors, out-dated information and recent changes.

Public participation tools are not limited to this kind of information. They also include means to discuss management proposals (meetings, local interest groups), polls, lectures, Internet forums, or exchanges with the analysts and policy-makers.

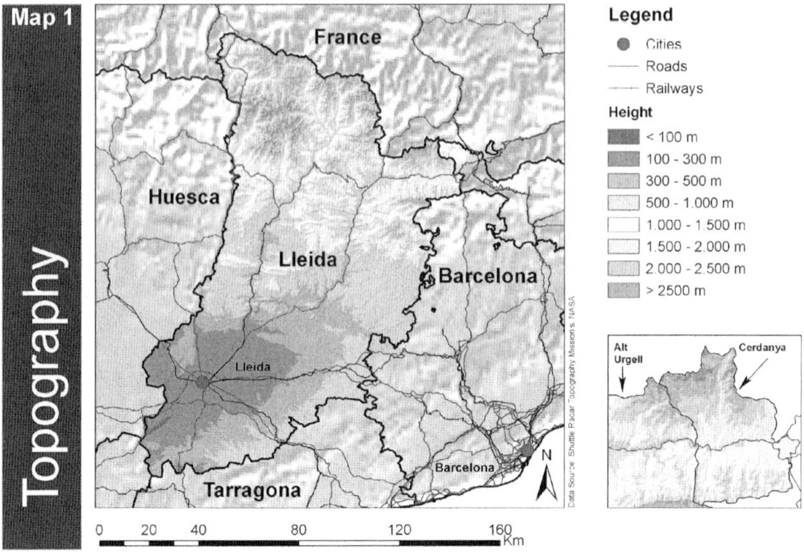

Fig. 19.1 Topography

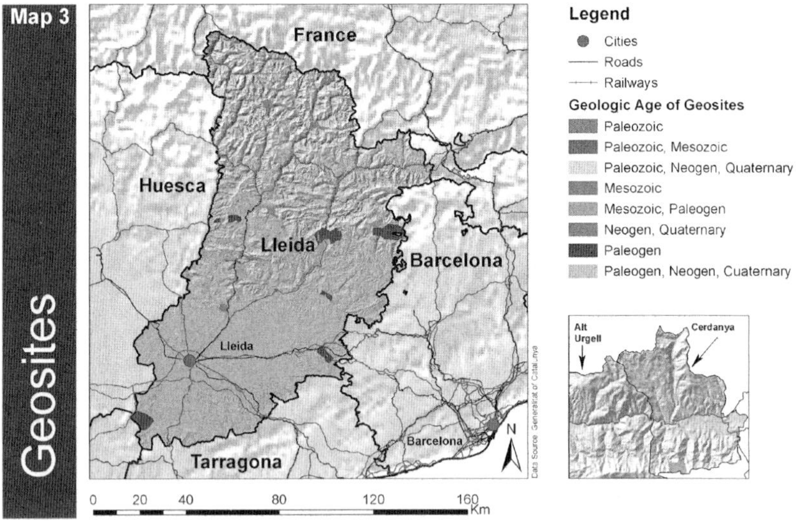

Fig. 19.2 Geosites

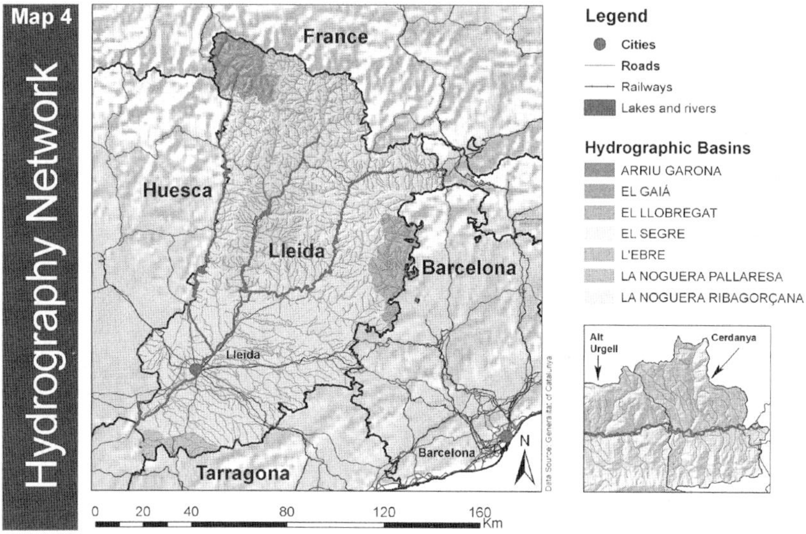

Fig. 19.3 Hydrography network

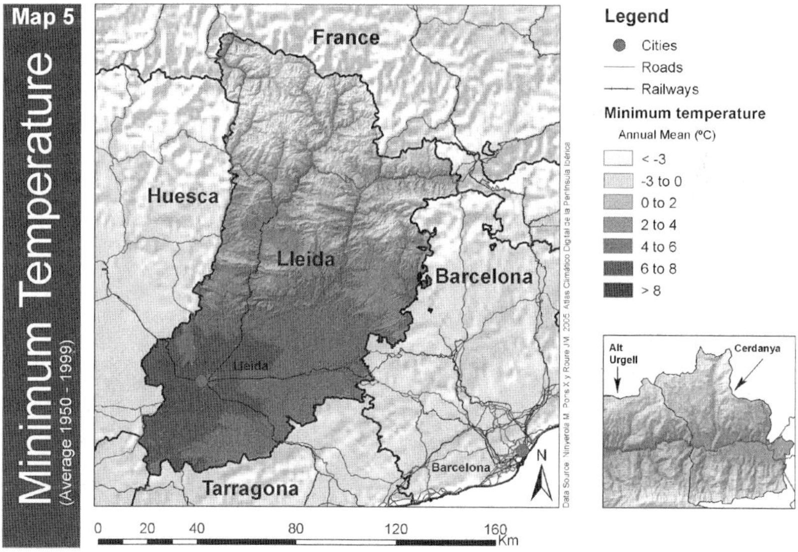

Fig. 19.4 Minimum temperatures (climatology)

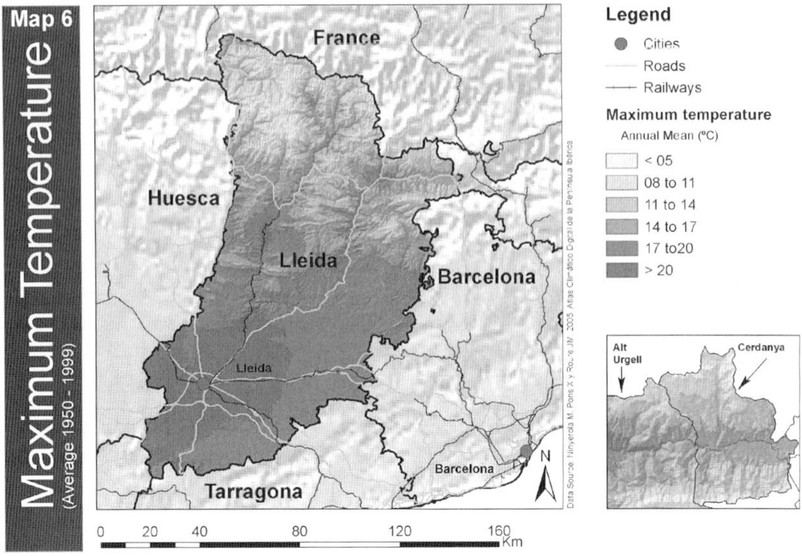

Fig. 19.5 Maximum temperatures (climatology)

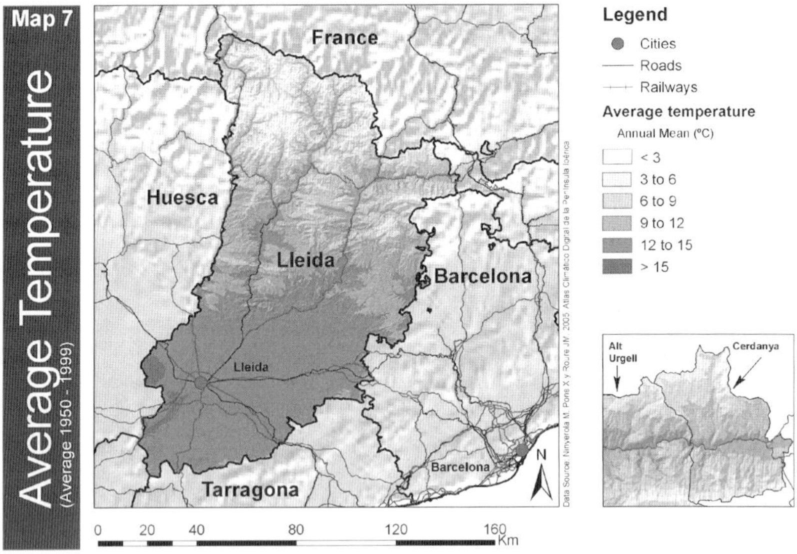

Fig. 19.6 Average temperature (climatology)

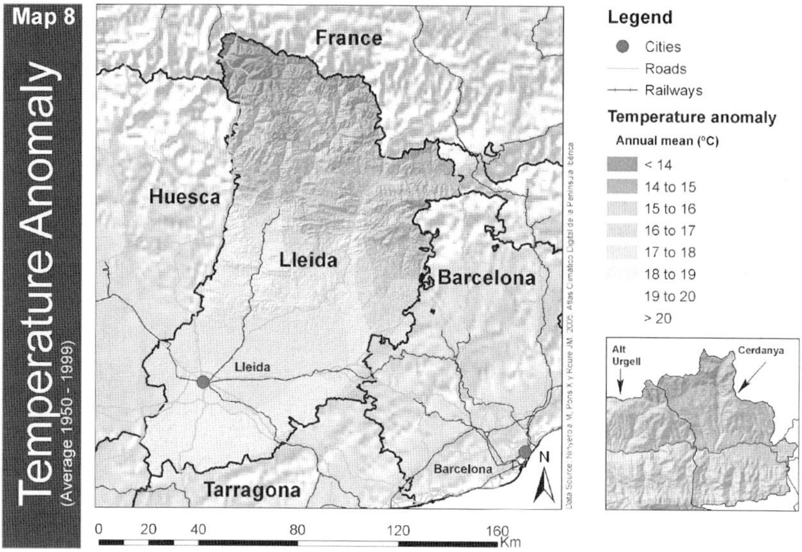

Fig. 19.7 Temperature anomalies

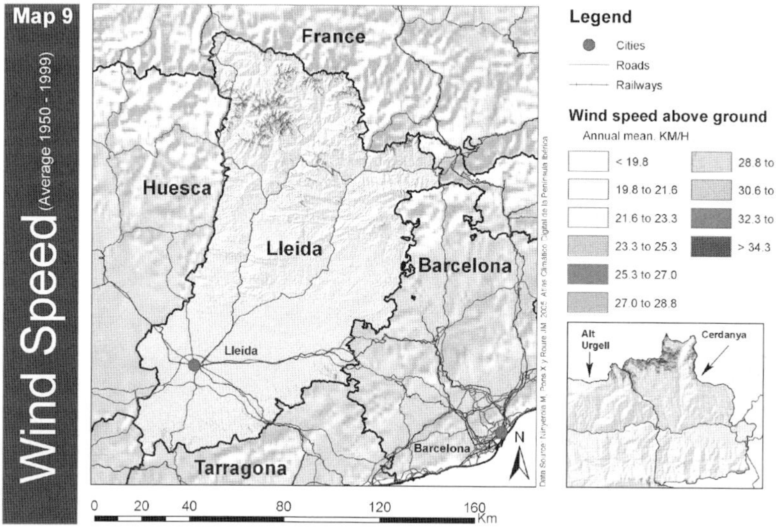

Fig. 19.8 Average wind speeds (climatology)

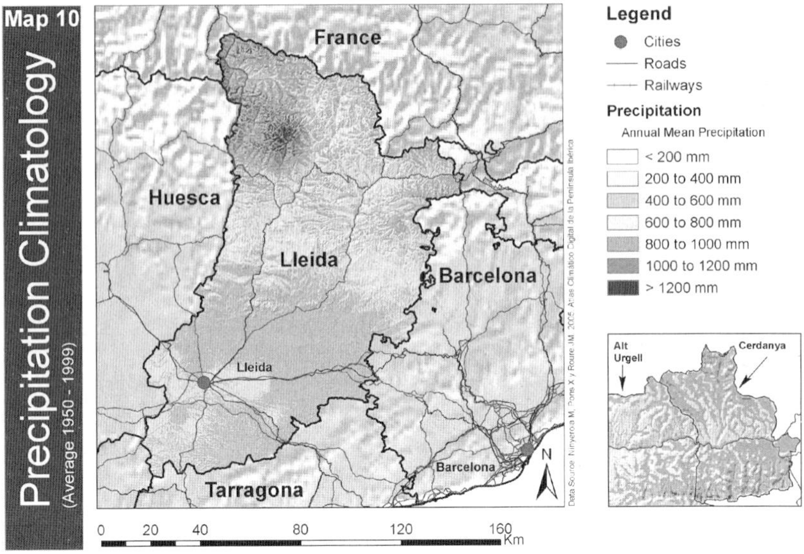

Fig. 19.9 Precipitation (climatology)

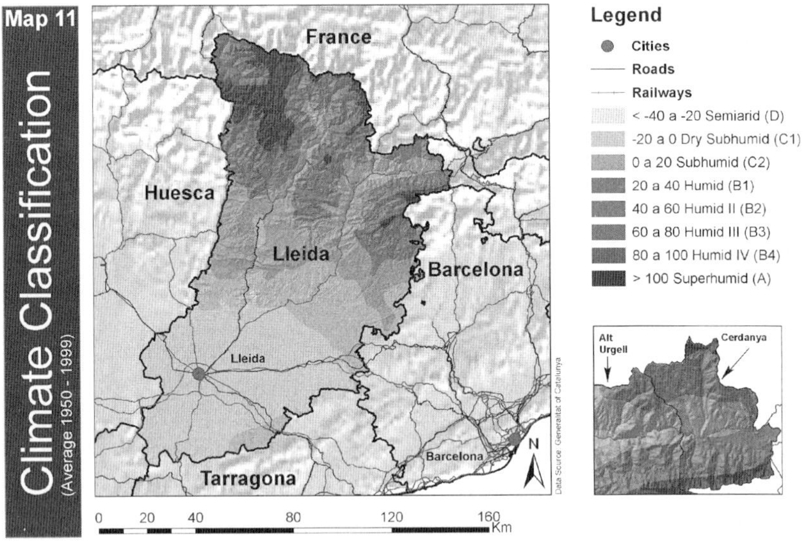

Fig. 19.10 Climate classification

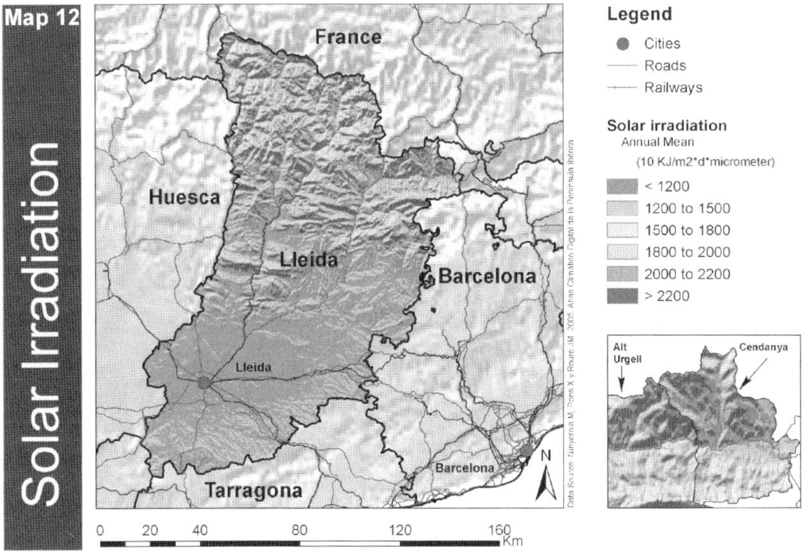

Fig. 19.11 Solar irradiation

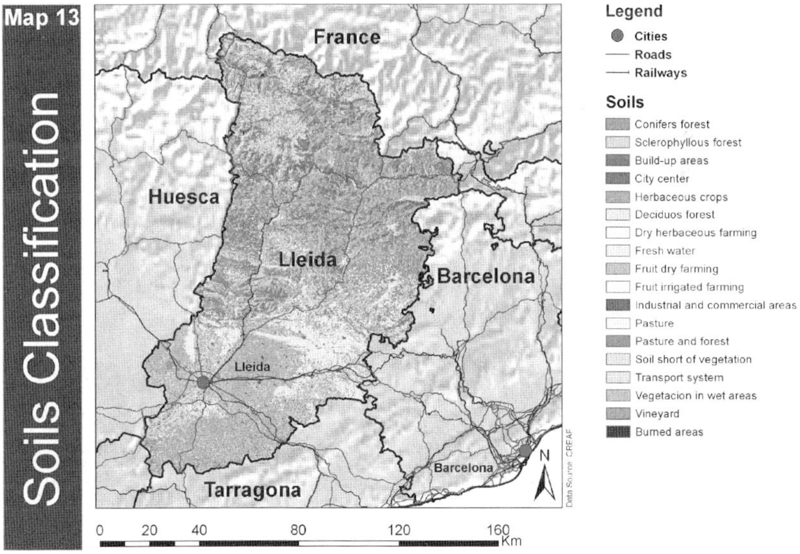

Fig. 19.12 Soils classification

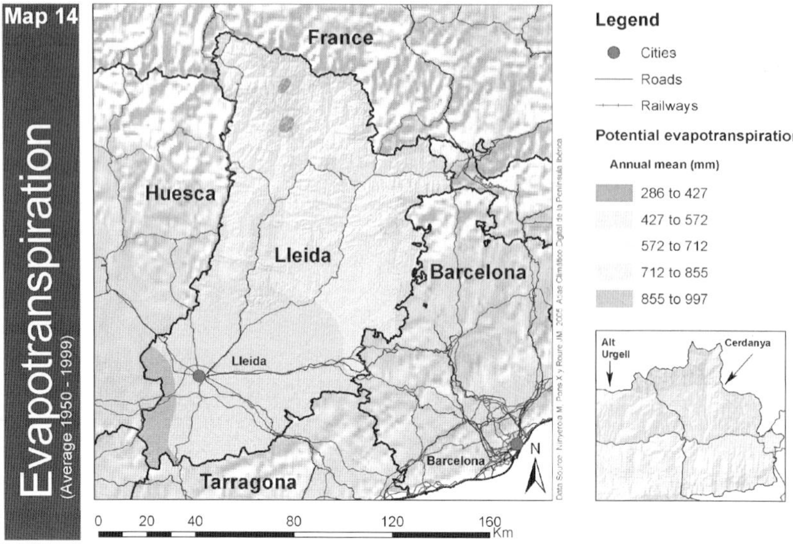

Fig. 19.13 Evapotranspiration

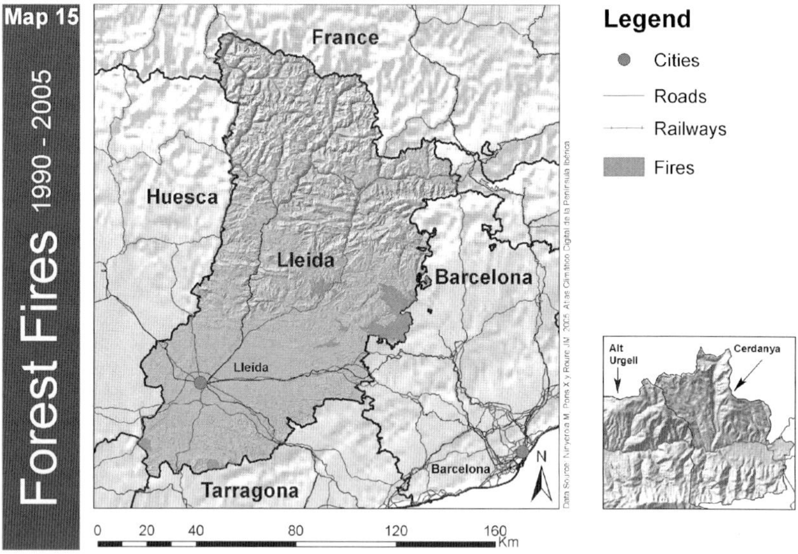

Fig. 19.14 Recent forest fires events

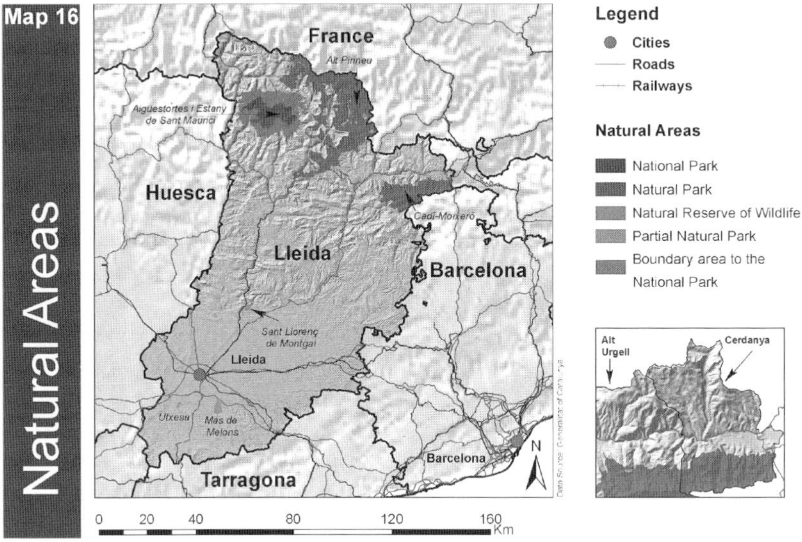

Fig. 19.15 Official natural areas

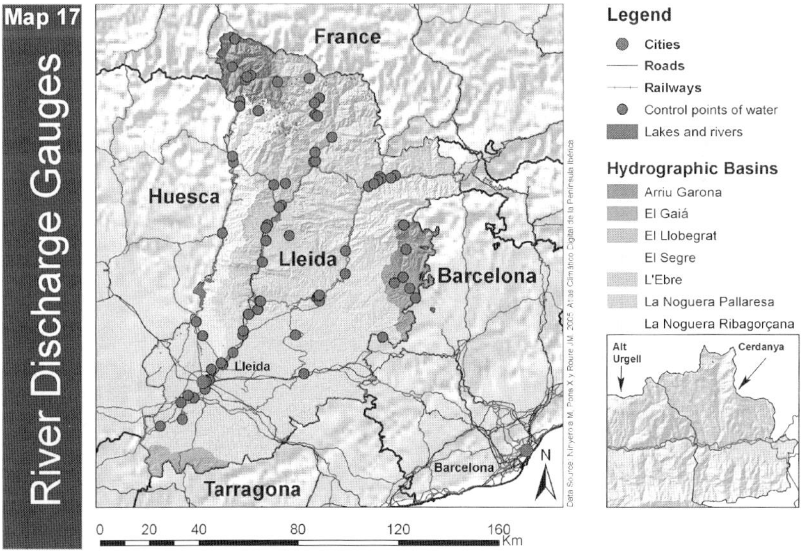

Fig. 19.16 River discharge gauge network

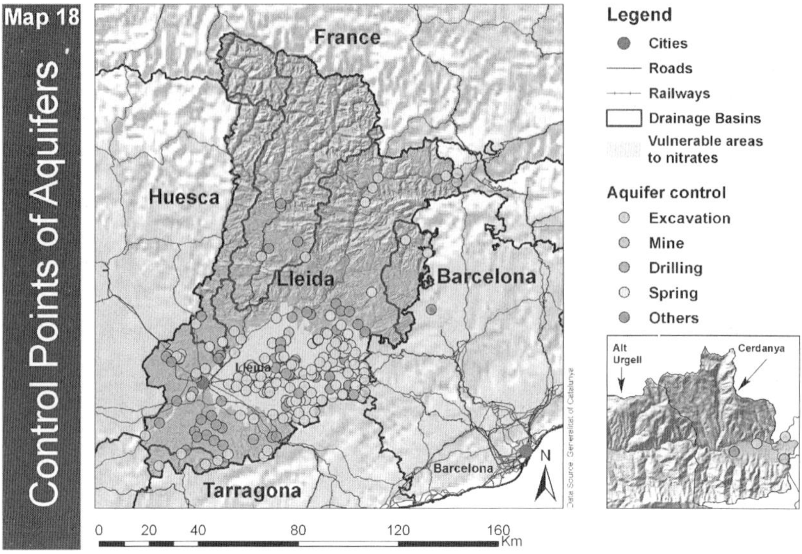

Fig. 19.17 Control points of aquifers

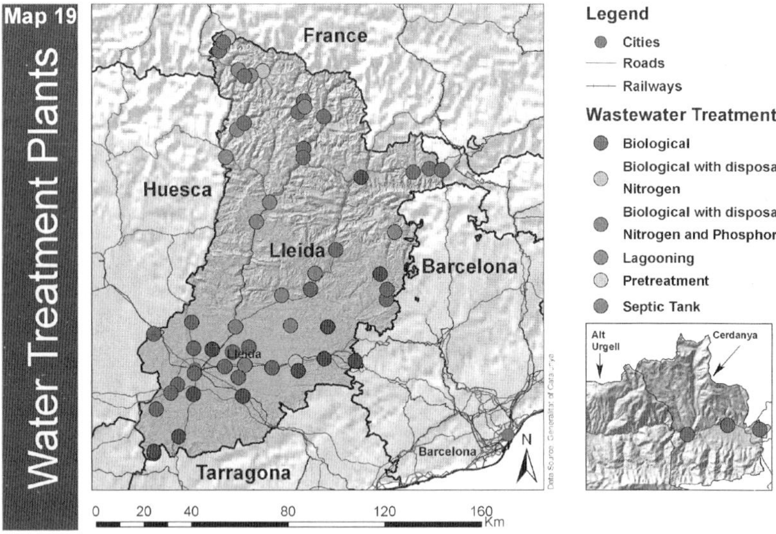

Fig. 19.18 Water treatment plants

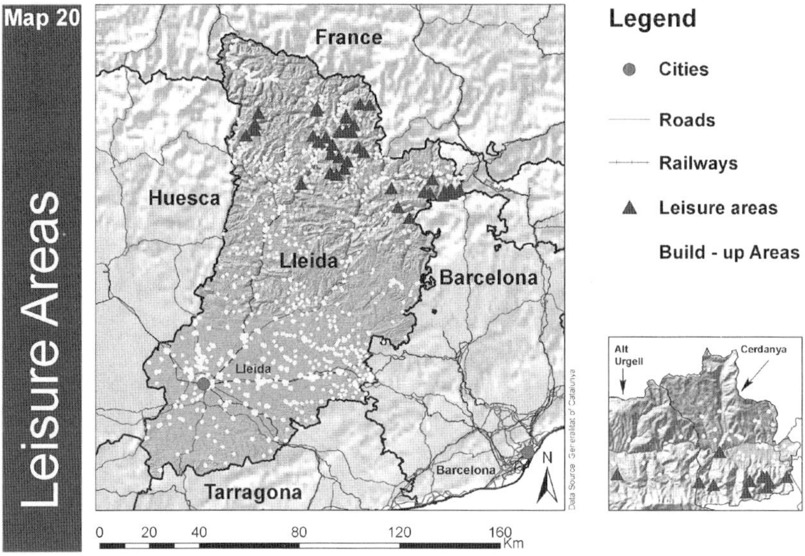

Fig. 19.19 Official leisure areas

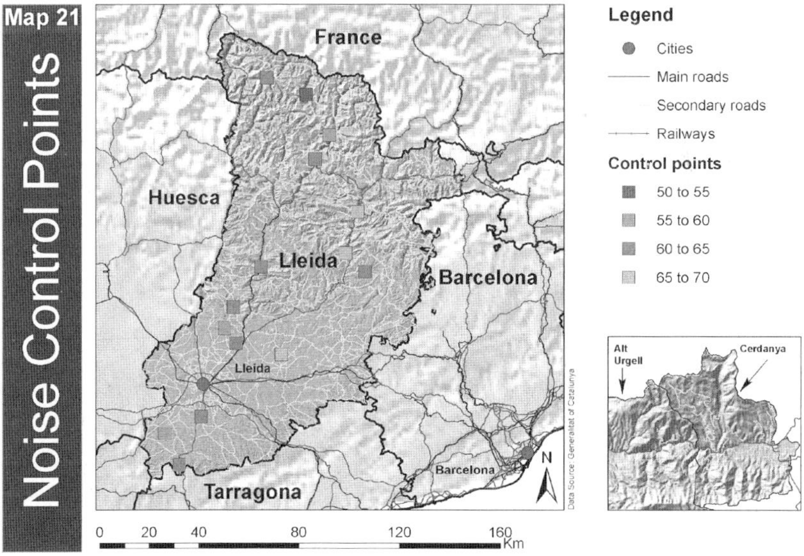

Fig. 19.20 Noise control points

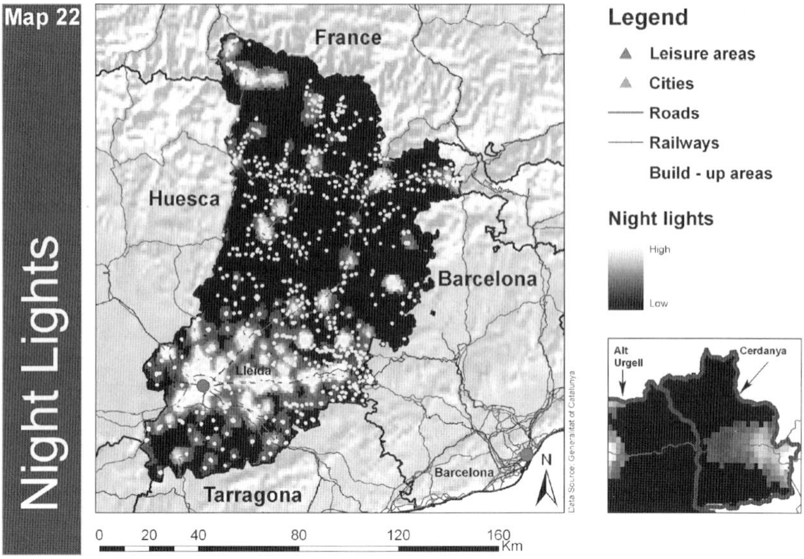

Fig. 19.21 Night lights

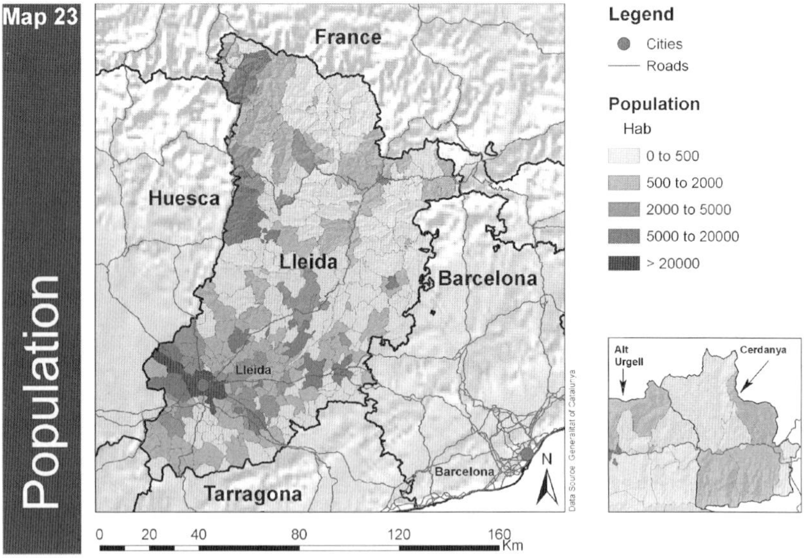

Fig. 19.22 Population

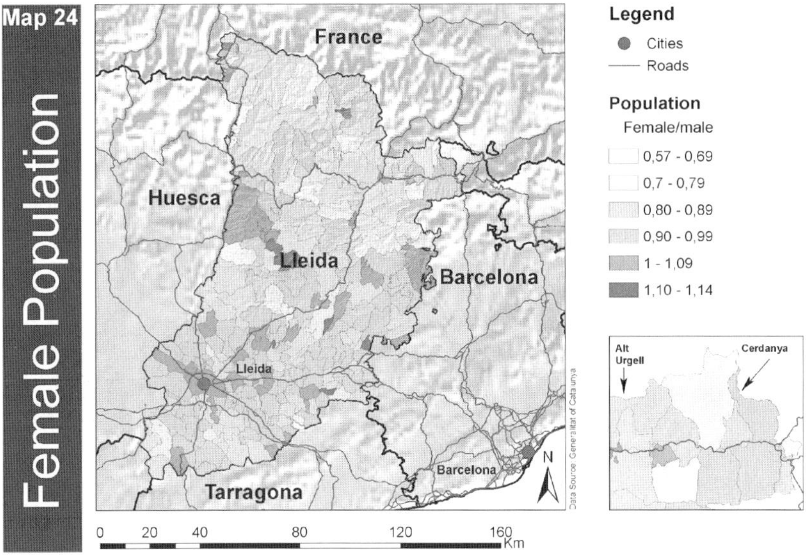

Fig. 19.23 Female population ratio

References and Bibliography Used

L. Kapiriri, O. F. Norheim and K. Heggenhougen, 'Public participation in health planning and priority setting at the district level in Uganda', Health Policy and Planning 18 (2003), 205–213.

G. Khasankhanova, 'Public participation to improve water resource management in Uzbekistan', Water Science and Technology 51 (2005), 365–372.

L. Pare, C. Robles and C. Cortez, 'Participation of indigenous and rural people in the construction of developmental and environmental public policies in Mexico', Ids Bulletin-Institute Of Development Studies 33 (2002).

J. Ploger, 'Public participation and the art of governance', Environment and Planning B-Planning & Design 28 (2001), 219–241.

D. M. Shucksmith, J. Rowanrobinson, C. T. Reid and M. G. Lloyd, 'Community Councils as a Medium for Public-Participation - a Case-Study in Grampian Region', Journal of Rural Studies 1 (1985), 307–319.

D. Storey, 'Issues of integration, participation and empowerment in rural development: The case of LEADER in the Republic of Ireland', Journal of Rural Studies 15 (1999), 307–315.

T. Webler, S. Tuler and R. Krueger, 'What is a good public participation process? Five perspectives from the public', Environmental Management 27 (2001), 435–450.

Chapter 20
A Rural Plan

20.1 An Example in a Formerly Depressed Area

This chapter describes the key points of a rural plan in a formerly depressed area of Western Europe. The area chosen corresponds to a part of 'Las Hurdes' shire, in the province of Cáceres, West Spain. The actual rural management and development plan was done in 1999 as a part of an independent academic project. What is presented here is a recreation of that work (the original document was written in Spanish) with new currently available tools. In the time elapsed, some of the proposals were carried out whereas others were not considered, or disregarder for a number of reasons.

There are many templates suitable for presenting a rural plan. Here, we have used a document formatted with the popular 'Report Template' of Microsoft Word. The major advantage of this layout is that it allows a quick appraisal by the decision makers, highlighting the main topics and organizing the information in a pleasant way (Figs. 20.1, 20.2 and 20.3). There are other templates well suited to those purposes available from StarOffice with Linux.

The synthesis report is just a minute part of the actual rural plan documentation, but it is useful to illustrate most of the topics covered by the whole study. What follows is a snapshot of the report is shown. Grey boxes are used in the figures to either describe the parts of the document or to clarify why something is done in that way and not in another. Development axis and actual proposals are also given in the same say way reported style. It has to be noticed that a rural plan report is not an academic document, so it should be easily readable by non-experts and formatting should allow a quick look of the contents. While empty terminology should be avoided, special care should be put in to using the same language of the decision-makers.

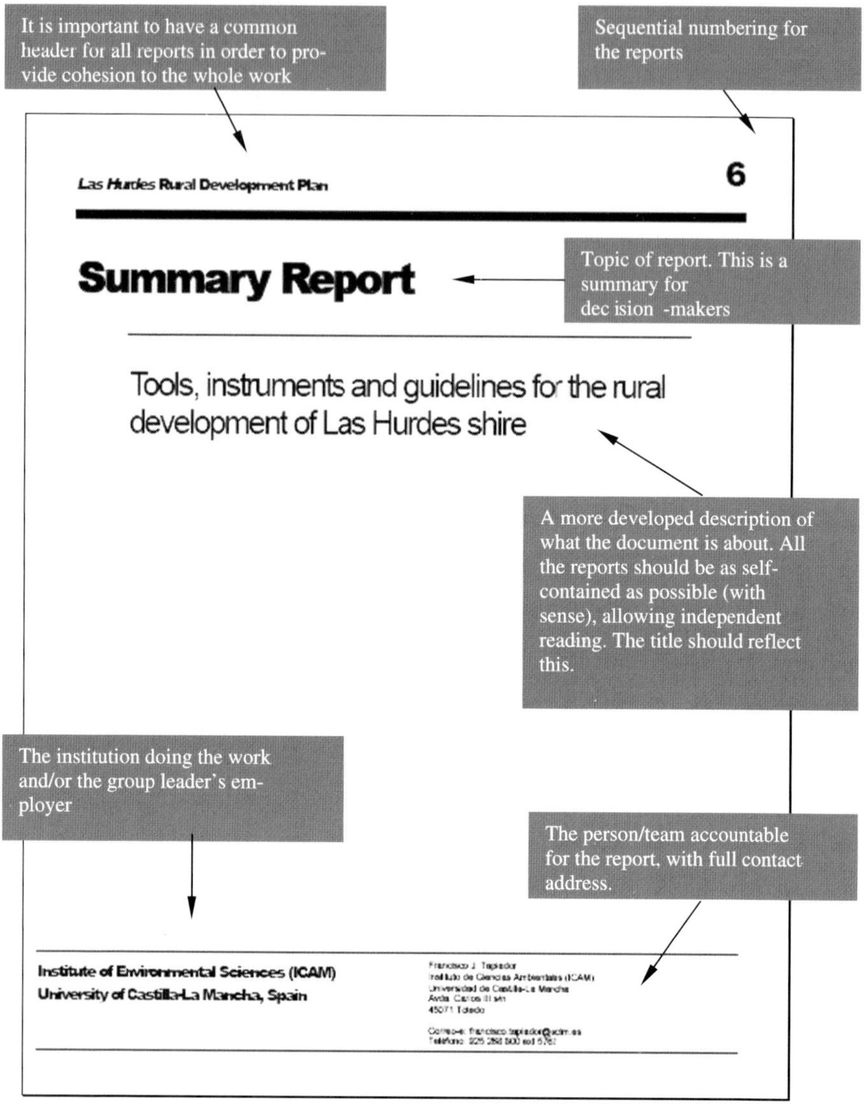

Fig. 20.1 Cover of a rural plan synthesis report, as delivered to the client

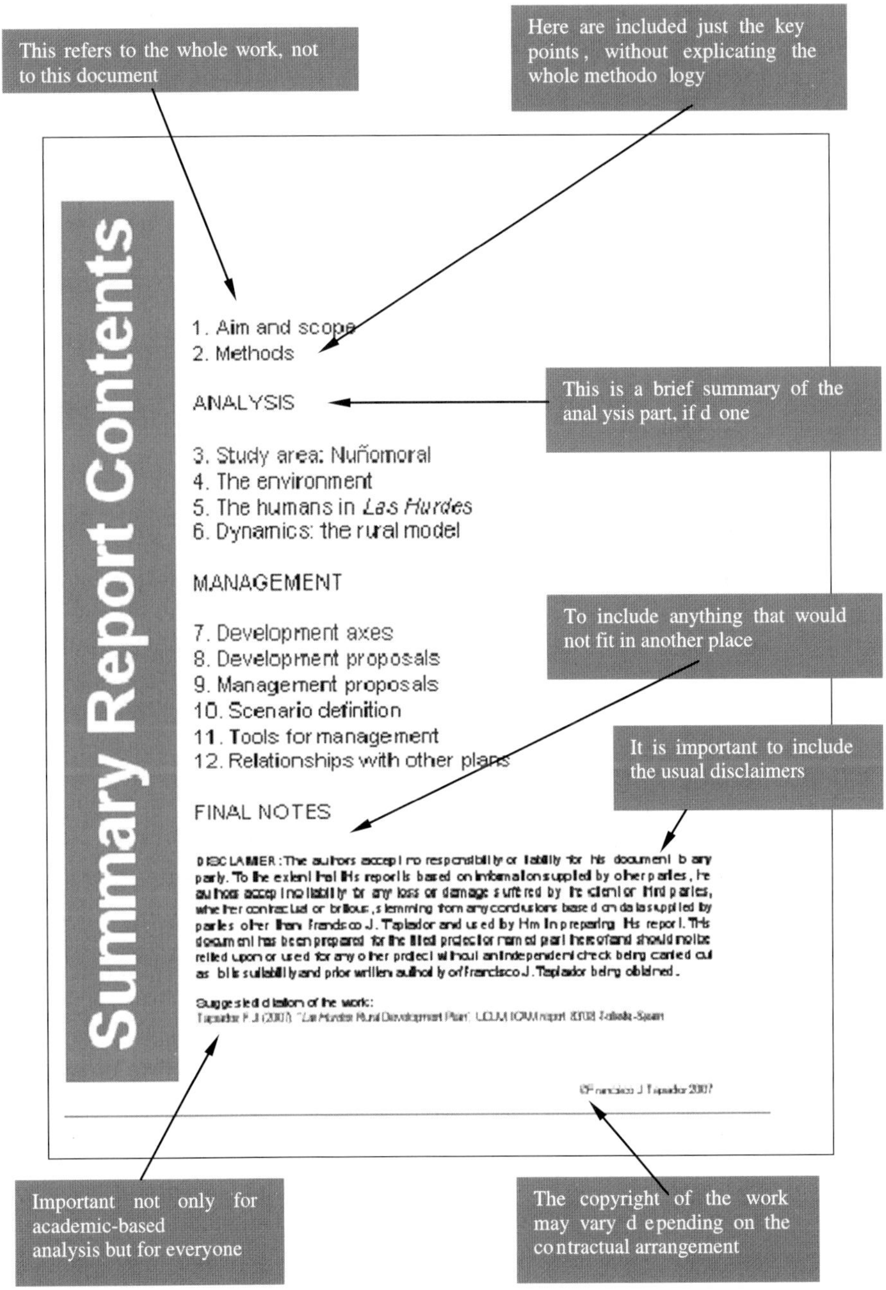

Fig. 20.2 The basic contents of a rural plan summary report. Items in blue are those being shown later in this chapter

Section 3 Study Area

Nuñomoral

A municipality with reduced accessibility

Within Las Hurdes, Nuñomoral presents several characteristics that make this municipality a singular area. First, it is located in a dead-end road. Accessibility through a narrow road is difficult in winter, and tourism activities should take into account that there not many passer-by people can be expected: Nuñomoral is the last village on the road.

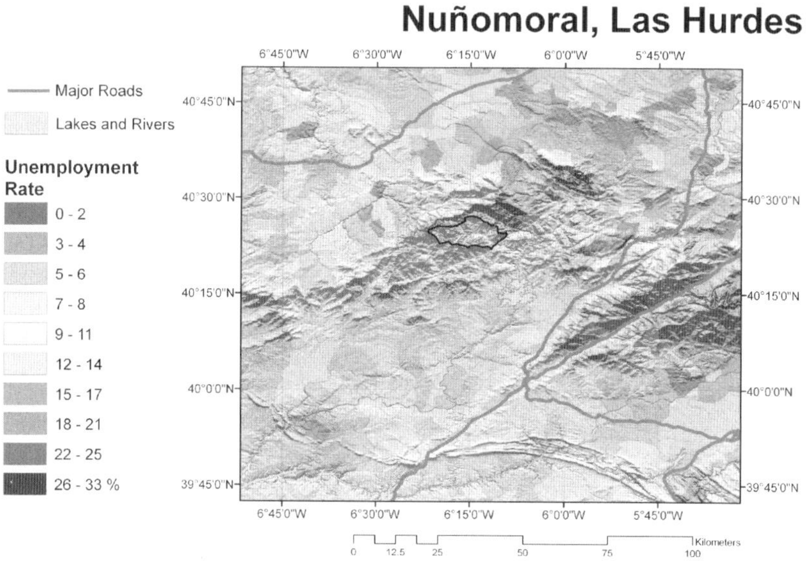

Figure 3.1 The municipality of Nuñomoral. Location, major roads and unemployment rate in 2005.

Fig. 20.3 The study area, Nuñomoral municipality

The rest of the chapter illustrates some of the key parts of the rural plan, including different formatting styles used to provide an idea of the kind of document that can be expected. In general, a rural plan should be concise, but explanations may be needed to justify the opportunity of some actions. It is important to realize that this document may be used by both planners, locals and regional authorities, decision-makers and the public, and as such should be easy to grasp without sacrificing the contents.

Section

7 Development Axes

Remoteness, Quality Products, Information

Key points for a sustainable development

The analysis work has identified three major axes to increase the development of the area deserving special attention. Those development directions are endogenous, so they may complement external help in terms of funding or inversion.

Remoteness

Increased accessibility is deemed as critical for the future development of the area. Developing the road network in terms of surface improvement, provision of private services such as local taxis, and the definition of a network of emergency and contingency routes are crucial to improving the quality of life of the inhabitants, attracting visitors, developing the local economies and ensuring a sustainable development. An inexpensive signposting policy can help to increase the appeal of the area to casual visitors, expanding the potential tourist market. Providing closed walking circuits from a few parking areas will also increase the interest over competing natural/rural areas. Increased accessibility will benefit local population by allowing improved health care, education, personal interactions and trade opportunities. Remoteness, however, can also be an asset for the area. Low accessibility is not an impediment by itself: Zermatt, the last village to access the Matterhorn mountain in the Alps also has a remote location, but even this factor and the limitations to road traffic have not impeded a thriving economic activity. In a far lower scale, El Gasco can transmute this weakness into an opportunity if it is capable of marketing the natural values of the area and the local productions, making it a desirable location for a certain type of tourist.

(continued)

(continued)

Produce quality products

The analysis has identified a potential for producing high-quality, high added-value agrarian productions in the area. Both the climatology and the historical exploitation have produced varieties of organic food that may be of commercial interest. Thus, goat cheese, Iberian ham, cherries, olive oil, mushrooms, truffles, venison meat, honey, bee glue or royal jelly could be better marketed. The secluded character of many plots in areas with little or no pollution make the production valuable for some markets, while the intrinsic quality of the production well beyond agribusiness standardization, increases the value for organic consumers and gourmet markets. Creating a trade-mark for these productions is a must. Identifying and differentiating the production from other areas is important, protecting quality with a legal brand. Producers should organize themselves with the help of commercial and technical assistance to improve both the packaging and the post-production stages, inserting those products into the market chain at regional, national and even international level.

Improve information and information access

Increased networking between local entrepreneurs and potential clients is also an identified development axe. Access to information technologies would have a tremendous impact in terms of social structure and economic activity.

Section
8 Development Proposals

Specific actions
Proposals for developing Nuñomoral

Action #1 Improve accessibility and promote a remoteness idea

★ Roads are instrumental in enabling visitor accessing to the area and to improve the quality of life of the local population. Nonetheless, a certain idea of remoteness can help to market the values of the municipality. A delicate balance between not deterring potential visitors and enhancing the remote rural character of the area should be sought.

★ Foster a local taxi service to service the local population, with an emphasis in elderly population.

★ Improve signposting to direct visitors to special interest areas in a quick way.

★ Create ample parking areas for visitors.

★ Prepare circular walks in natural areas, enabling people to spend from an half hour to several hours in natural environments.

Action #2 Use financing instruments

★ Apply for EU financing instruments for rural improvement: PRODER, LEADER, FEOGA, etc.

★ Use transversal financing tools (e.g. EQUAL programs) to improve rural development as a whole.

(continued)

(continued)

Action #3 Promote energetic independence

★ Promote the use of solar energy and small wind turbines where applicable.

Action #4 Develop a green guide of the area

★ Using an in-depth scientific study of the analysis phase as a basis, a green guide should be produced to disseminate the natural aspects of the valley.

Action #5 Identify places of scientific interest

★ The analysis part has identified El Gasco *Volcano* (actually a possible meteorite impact) as a potential scientific place of interest. Access to this resource should be improved, but at the same time ensuring the preservation of this place. Other areas can also be of scientific interest (botanical, geomorphologic, cultural, etc.)

Action #6 Develop high-quality local productions

★ Goat cheese, Iberian ham, cherries, olive oil, mushrooms, truffles, venison meat, honey, bee glue or royal jelly could be better marketed, emphasizing the genetic singularities.

★ The secluded character of many plots in areas with little or no pollution makes those productions valuable. The intrinsic quality of the productions, well beyond agribusiness standardization, increases the value for organic consumers and gourmet markets.

★Create a trade-mark or protected denomination for local productionsm, and promote a 'rural inn' scheme.

★Improve the organization of local producers with the help of commercial and technical assistance to improve both the packaging and the postproduction stages.

★Insert natural products into the market chain at regional, national and international level using specialized fairs and the instruments and means provided by the ministries.

Action #7 School pooling

★ An effective school transportation system may require to pool resources between municipalities. Developing a location strategy for facilities is also important. This plan should take into account not only the current situation but also the objective potentialities of each area.

Action #8 Promote adventure sports initiatives

★A potential resource for the area is the rough landscape. Adventure sport promoters should be made aware of what the area has to offer to them. To that end, areas with little use or too remote can be dedicated to extreme sports without affecting rural life.

Action #9 Improve health care, with emphasis in elderly population

★Health care should be improved by dedicating more means of access, but also by improving information.

Action #10 Increase access to information technologies

★ Increase networking activities between local entrepreneurs and potential clients using information technology.

★Full access to information technology has a tremendous impact in terms of social structure and economic activity.

(continued)

(continued)

Action #11 Localized preservation of habitats

★Some habitats deserve special protection. This is applicable to both human habitats (traditional houses) and to ecosystems in the area.

Action #12 Attract returning emigrants

★Returning migrants can help to improve the living conditions of the area. New economic resources and new generations would benefit the area. Promoting the sense of place and the regeneration of second residential homes for natives are sensible measures to attract additional resources.

Action #13 Reopen natural spaces

★Natural spaces lacking current economic activity should be used. This applies to remote woods or places no longer in agrarian or forestry use.

Action #14 Analyze the viability of new services and activities

★New services or activities such as hunting, fishing, local private security for large properties, adventure expeditions, environmental lectures, university field trips, services for naturalist or cultural scholars should be promoted by the public sector. While it goes beyond the scope of this document to analyze the economical viability of such proposals, the area has the potential to sustain some of those activities.

Action #15 Recover the rivers and streams

★Rivers and streams such as the Malvellido River present valuable scenery. Recovering the areas nearby, improving accessibility and signalling viewpoints would contribute to the appeal of the area. Natural and environmental issues are also of interest.

(continued)

Action #16 Promote local crafts as an economic complement

★Specialized fairs could be planned to fill the periods when agrarian labour is less required and an extra income could be available, or when more visitors are expected (spring and autumn). Wicker and tin crafts, ceramics, pottery, traditional hives, musical instruments, embroidery, are all complementary activities that, in some cases, can even be a primary occupation in the area.

★Promote craft associations and joint activities.

Action #17 Use local specialization to allocate new services

★Local specialization should be used for sensible resource management. A swimming pool in one village can be complemented with a small sport centre or private gym in another, a special care unit in a different, nearby area with a dispensary.

Additional Reading

A. W. Gilg and M. Kelly, 'Rural planning in practice: The case of agricultural dwellings', Progress in Planning 47 (1997), 75-&.

A. W. Gilg and M. P. Kelly, 'The delivery of planning policy in Great Britain: Explaining the implementation gap. New evidence from a case study in rural England', Environment and Planning C-Government and Policy 15 (1997), 19–36.

Z. W. Guo, X. M. Xiao, Y. L. Gan and Y. J. Zheng, 'Landscape planning for a rural ecosystem: case study of a resettlement area for residents from land submerged by the Three Gorges Reservoir, China', Landscape Ecology 18 (2003), 503–512.

M. R. Perez, M. G. Zhong, B. Belcher, X. Chen, M. Y. Fu and J. Z. Xie, 'The role of bamboo plantations in rural development: The case of Anji County, Zhejiang, China', World Development 27 (1999), 101–114.

R. Raja, C. E. Sooriamoorthi, P. Kanniapan and T. Ramachandran, 'Energy planning and optimization model for rural development - A case of sustainable agriculture', International Journal of Energy Research 21 (1997), 527–547.

G. Signorello and G. Pappalardo, 'Domestic animal biodiversity conservation: a case study of rural development plans in the European Union', Ecological Economics 45 (2003), 487–499.

E. L. Webb and Thiha, 'Integrating social preference in GIS-aided planning for forestry and conservation activities: A case study from rural SE Asia', Environmental Management 30 (2002), 183–198.

Appendix
Templates for a Rural Project

This appendix comprises of a set of templates that can be used in a rural project, including a rural development project checklist. These forms are the basis of any rural analysis, and all nine templates should always be included in the final report.

The nine templates are:

1. A *checklist of the phases* of the project. The person responsible for every phase should be recorded with the date the task finished.
2. The *environmental indicators* form
3. The *social indicators* form
4. The *economic and institutional* indicators form
5. The *environmental elements* analysis form.
6. The *anthropic elements* analysis form
7. The *cross-relationships* form; this is an extended version of the two previous forms, so only half of the table has to be completed. Sometimes, filling this form in will involve revisiting the previous ones. This is not unusual and does not indicate mistakes. It is however a beneficial consequence of considering the full spectrum of relationships under new light
8. The *rural landscapes typology* form. The team should choose three or more cells to characterize the landscape/s of the rural area.
9. A *checklist of the report* chapters.

The idea behind these forms is not to substitute a proper analysis, but to not forget essential aspects of the work: having the nine forms filled does not guarantee a brilliant final study/project, but shows that time has been devoted to the most relevant aspects of the rural area.

PROJECT FORM 1

Task	Responsible Person	Done
1- Client specifications, aims and objectives		
2- Problem definition		
3- Budgeting and agreement on a preliminary study		
4- Field work, phase 1		
5- Need and requirements of specific analyses		
6- Preliminary feasibility study. Basic indicators.		
7- Project definition meeting and budgeting. Client approval		
8- Acceptance. Commissioning of external specific analyses		
9- Database, archives and information sources identification and exploration		
10- GIS data model definition, version 1		
11- Field work, phase 2		
12- Meeting with the social agents, clients, expert panel and team		
13- Additional information gathering		
14- Integration of data from specific analyses fieldwork		
15- GIS redefinition and GIS data model, version 2		
16- Data processing		
17- Cross-relationships identification		
18- Models development (optional)		
19- Preliminary analyses		
20- Users and agents feedback		
21- Proposals and scenario development		
22- Third field work visit		
23- Draft report and tools for public participation		
24- Final meeting with the client		
25- Public presentation and participation		
26- Integration of results from public participation		
27- Delivery of the final report		

PROJECT FORM 2

Theme	Sub-theme	Environmental indicator	Value
Land	Agriculture	Arable and Permanent Crop Land Area	
		Use of Fertilizers	
		Use of Agricultural Pesticides	
	Forests	Forest Area as a Percent of Land Area	
		Wood Harvesting Intensity	
	Desertification	Land Affected by Desertification	
	Urbanization	Area of Formal and Informal Settlements	
Oceans, Seas and Coasts	Coastal Zone	Algae Concentration in Coastal Waters	
		Percent of Total Population Living in Coastal Areas	
	Fisheries	Annual Catch by Major Species	
Fresh Water	Water Quantity	Annual Withdrawal of Ground and Surface Water as a Percent of Total Available Water	
	Water Quality	BOD in Water Bodies	
		Concentration of Faecal Coliform in Freshwater	
Biodiversity	Ecosystem	Area of Selected Key Ecosystems	
		Protected Area as a % of Total Area	
	Species	Abundance of Selected Key Species	

PROJECT FORM 3

Theme	Sub-theme	Social Indicator	Value
Equity	Poverty	Percent of Population Living below Poverty Line	
		Gini Index of Income Inequality	
		Unemployment Rate	
	Gender Equality	Ratio of Average Female Wage to Male Wage	
Health	Nutritional Status	Nutritional Status of Children	
	Mortality	Mortality Rate Under 5 Years Old	
		Life Expectancy at Birth	
	Sanitation	Percent of Population with Adequate Sewage Disposal Facilities	
	Drinking Water	Population with Access to Safe Drinking Water	
	Health care Delivery	Percent of Population with Access to Primary Health Care Facilities	
		Immunization Against Infectious Childhood Diseases	
		Contraceptive Prevalence Rate	
Education	Education Level	Children Reaching Grade 5 of Primary Education	
		Adult Secondary Education Achievement Level	
	Literacy	Adult Literacy Rate	
Housing	Living Conditions	Floor Area per Person	
Security	Crime	Number of Recorded Crimes per 1,000 Population	
Population	Population Change	Population Growth Rate	
		Population of Formal and Informal Settlements	

PROJECT FORM 4

Theme	Sub-theme	Economic and Institutional Indicator	Value
Consumption and Production Patterns	Material Consumption	Intensity of Material Use	
	Energy Use	Annual Energy Consumption per Capita	
		Share of Consumption of Renewable Energy Resources	
		Intensity of Energy Use	
	Waste Generation and Management	Generation of Industrial and Municipal Solid Waste	
		Generation of Hazardous Waste	
		Management of Radioactive Waste	
		Waste Recycling and Reuse	
	Transportation	Distance Traveled per Capita by Mode of Transport	
Institutional Capacity	Information Access	Number of Internet Subscribers per 1000 Inhabitants	
	Disaster Preparedness and Response	Economic and Human Loss Due to Natural Disasters	

PROJECT FORM 5

ENVIRONMENTAL ELEMENTS	GEOLOGY	MINERAL RESOURCES	SOILS	GEOMORPHOLOGY	METEOROLOGY	CLIMATE	HYDROLOGY	LAND COVER	PLANTS AND ANIMALS DISTRIBUTION	NATURAL RESOURCES	SCENIC VALUES	NATURAL RISKS
GEOLOGY												
MINERAL RESOURCES												
SOILS												
GEOMORPHOLOGY												
METEOROLOGY												
CLIMATE												
HYDROLOGY												
LAND COVER												
PLANTS AND ANIMALS DISTR.												
NATURAL RESOURCES												
SCENIC VALUES												
NATURAL RISKS												

A	DIRECT EFFECT
B	VARIABLE EFFECT
C	INDIRECT OR WEAK EFFECT
-	UNCOMMON OR LONG TERM EFFECT

PROJECT FORM 6

ANTHROPIC ELEMENTS	HABITAT	AGRARIAN LAND	SOCIAL STRUCTURE	LEISURE	UTILITIES AND SERVICES	TRANSPORT AND ACCESSIBILITY	ENERGY	INDUSTRIAL ACTIVITIES	GENDER RELATIONSHIPS	PUBLIC HEALTH ISSUES	POLITICS	RURAL TRADITIONS AND FOLK	RURAL TOURISM
HABITAT													
AGRARIAN LAND													
SOCIAL STRUCTURE													
LEISURE													
UTILITIES AND SERVICES													
TRANSPORT AND ACCESSIBILITY													
ENERGY													
INDUSTRIAL ACTIVITIES													
GENDER RELATIONSHIPS													
PUBLIC HEALTH ISSUES													
POLITICS													
RURAL TRADITIONS AND FOLK													
RURAL TOURISM													

A	DIRECT EFFECT
B	VARIABLE EFFECT
C	INDIRECT OR WEAK EFFECT
-	UNCOMMON OR LONG TERM EFFECT

PROJECT FORM 7

CROSS RELATIONSHIPS

ON THIS ELEMENT

THE EFFECT OF THIS ELEMENT

		1 Geology	2 Mineral Resources	3 Soils	4 Geomorphology	5 Meteorology	6 Climate	7 Hydrology	8 Land cover	9 Ecology	10 Natural resources	11 Scenic values	12 Natural and environmental risks	13 Habitat	14 Agrarian land	15 Social structure	16 Leisure	17 Utilities and services	18 Transport and accessibility	19 Energy	20 Industrial activities	21 Gender relationships	22 Public health issues	23 Political issues	24 Rural culture, traditions and folklore	25 Rural tourism
1	Geology																									
2	Mineral resources																									
3	Soils																									
4	Geomorphology																									
5	Meteorology																									
6	Climate																									
7	Hydrology																									
8	Land cover variety																									
9	Ecology																									
10	Natural resources																									
11	Scenic values																									
12	Natural and environmental risks																									
13	Habitat																									
14	Agrarian land																									
15	Social structure																									
16	Leisure																									
17	Utilities and services																									
18	Transport and accessibility																									
19	Energy																									
20	Industrial activities																									
21	Gender relationships																									
22	Public health issues																									
23	Political issues																									
24	Rural culture, traditions and folklore																									
25	Rural tourism																									

A DIRECT EFFECT B VARIABLE EFFECT C INDIRECT OR WEAK EFFECT - UNCOMMON OR LONG TERM EFFECT

PROJECT FORM 8

RURAL LANDSCAPE MATRIX

PRIMARY FACTOR

		1 Geology	2 Mineral Resources	3 Soils	4 Geomorphology	5 Meteorology	6 Climate	7 Hydrology	8 Land cover	9 Ecology	10 Natural resources	11 Scenic values	12 Natural and environmental risks	13 Agrarian land	14 Habitat	15 Social structure	16 Leisure	17 Utilities and services	18 Transport and accessibility	19 Energy	20 Industry	21 Gender relationships	22 Public health issues	23 Political issues	24 Rural culture, traditions and folklore	25 Rural tourism	Base Geosystem
1	Geology																										Abiotic
2	Mineral resources																										Abiotic
3	Soils																										Abiotic
4	Geomorphology																										Abiotic
5	Meteorology																										Abiotic
6	Climate																										Abiotic
7	Hydrology																										Abiotic
8	Land cover variety																										Abiotic
9	Ecology																										Biotic
10	Natural resources																										Biotic
11	Scenic values																										Mixed
12	Natural and environmental risks																										Anthropic
13	Habitat																										Anthropic
14	Agrarian land																										Anthropic
15	Social stucture																										Anthropic
16	Leisure																										Anthropic
17	Utilities and services																										Anthropic
18	Transport and accessibility																										Anthropic
19	Energy																										Anthropic
20	Industry																										Anthropic
21	Gender relationships																										Anthropic
22	Public health issues																										Anthropic
23	Political issues																										Anthropic
24	Rural culture, traditions and folklore																										Anthropic
25	Rural tourism																										Anthropic

PROJECT FORM 9

Part		☑☒	Responsible Person
1	*Aim and description of the project*		
2	How commissioned the analysis		
3	Aims of the analysis		
4	Projected budget and final costs		
5	Timetable and schedule, both projected and final		
6	Team composition, work breakdown and responsibilities		
7	Notes from the team coordinator		
8	Notes from the team members		
9	*Location map of the area in the region and country*		
10	Detailed local maps used in the analyses		
11	Situation map		
12	Basic indicators		
13	*Measurement methods used*		
14	Measurement protocols		
15	Methods		
16	*Database description and access directions*		
17	GIS description		
18	*Environmental elements description and maps*		
19	(one epigraph for element)		
20	*Anthropic elements description and maps*		
21	(one epigraph for element)		
22	*Geographical modelling (optional)*		
23	Model diagram		
24	Geographical model description. Rationale of the elements used		
25	Parameterizations used		
26	Scenario definition		
27	*Synthesis*		
28	Generalities		
29	Landscapes analysis		
30	Landscape report		
31	Results of the modelling		
32	Notes for development and management		
33	*Final notes*		
34	*ANNEXES*		
35	Public information documentation		
36	Additional resources		

List of Figures

List of Tables

Index

Color Plates

Plate 1 Integrated DEM and satellite images fusion. Royal Site of 'La Granja' in the Central System, Spain. Snow appears in the top of the mountain, whilst the dam and the village are clearly visible in the plain. Forest and roads are also apparent. The landforms of these products can be calculated using automated procedures (see Fig. 3.5 on p. 43)

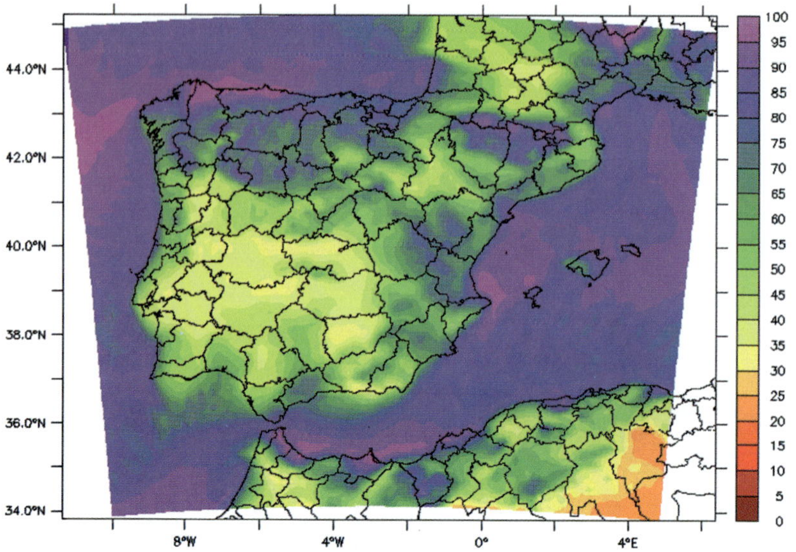

Plate 2 An example of high-resolution weather forecast. Surface Relative Humidity (%) in Spain for the 15/08/2005 (http://momac.uclm.es) (see Fig. 3.6 on p. 44)

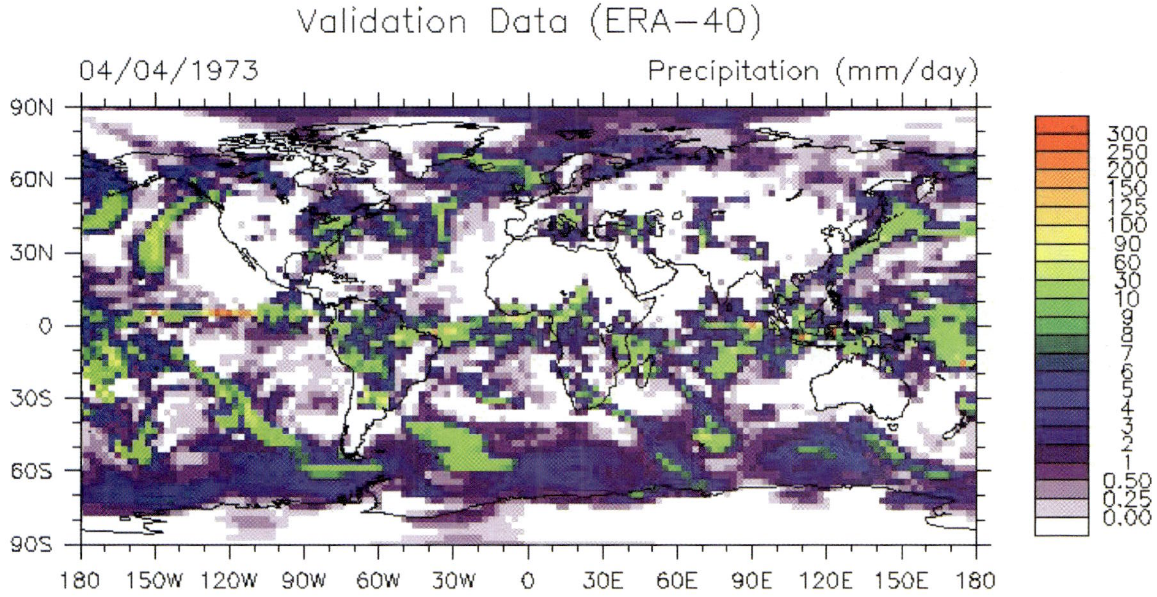

Plate 3 Merged satellite and model global precipitation estimation (see Fig. 3.9 on p. 50)

Plate 4 Comparison between IKONOS and SPOT satellite images for the same area. Color combination in the SPOT image has been chosen to enhance irrigated land (see Fig. 3.10 on p. 53)

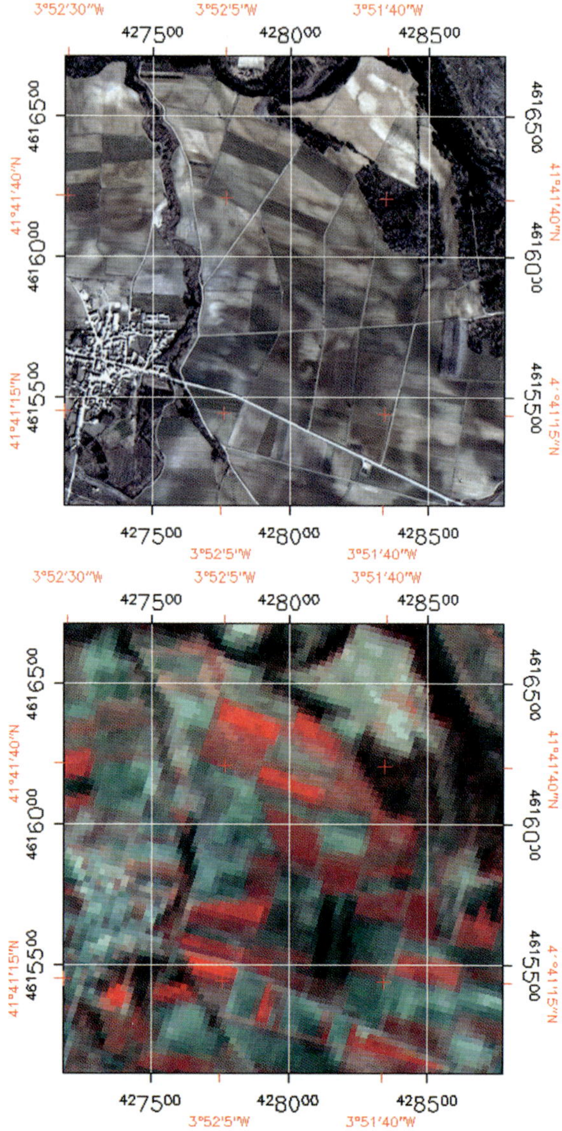

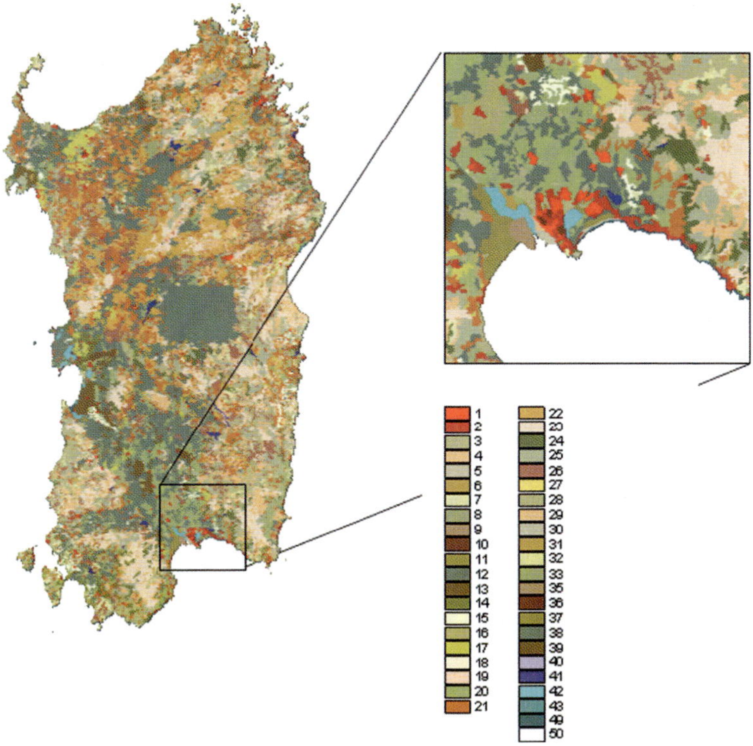

Plate 5 CORINE-2000 land use map example (Sardinia). Key corresponds with the three levels version of CORINE-2000 land cover classification (see Fig. 3.11 on p. 54)

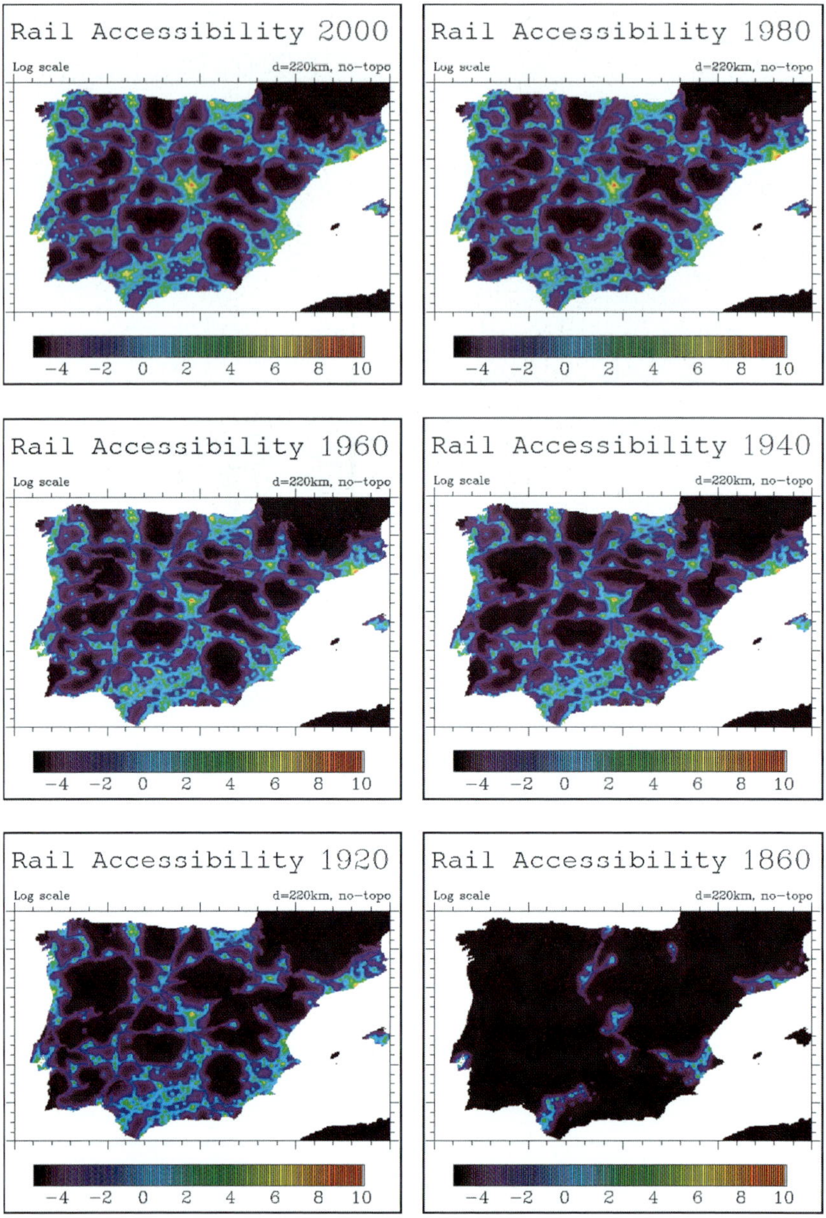

Plate 6 Changes in the Iberian rail accessibility. Large rural areas still present a low accessibility by train (see Fig. 5.5 on p. 91)

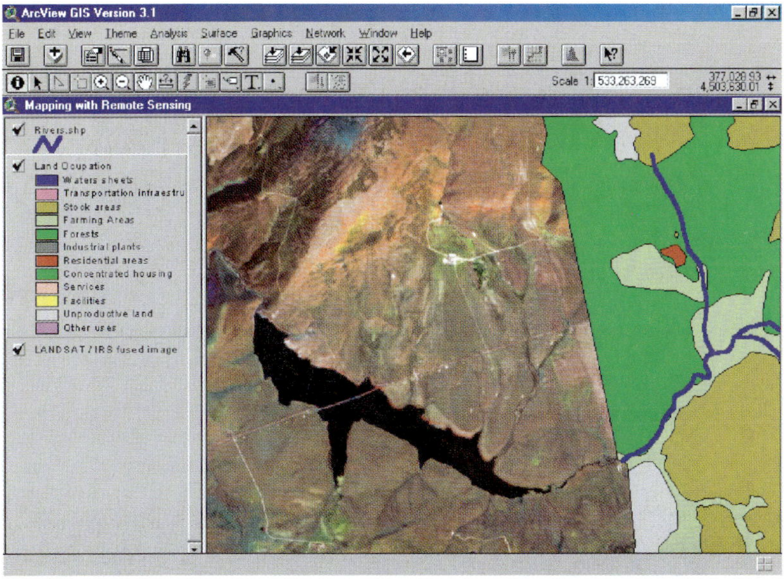

Plate 7 Integration of GIS and remote sensing for water management (see Fig. 5.7 on p. 113)

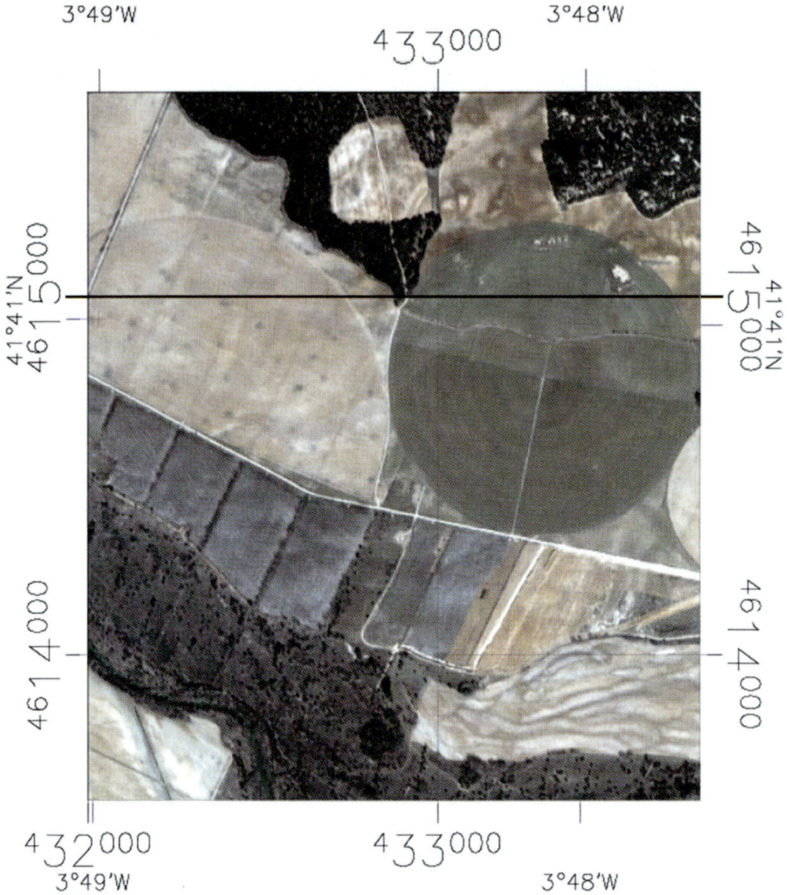

Plate 8 IKONOS composite image, showing the capability of this satellite to show small features in the crops. Trees are also clearly visible in the image (see Fig. 5.8 on p. 115)

Plate 9 A false-color LANDSAT image. Irrigated land appears in reds, woods in dark green and rivers in dark (see Fig. 5.9 on p. 116)

Plate 10 A potential 14/15 landscape (Mainly a social landscape with a leisure component). Peri-urban settlements and part-time agriculture plots near a large city (see Fig. 7.1 on p. 134)

Plate 11 An example of the graphical documentation required for the landscape report. Type 3.1 (or 10.3) landscape. Map by Endoxon AG (see Fig. 7.2 on p. 136)

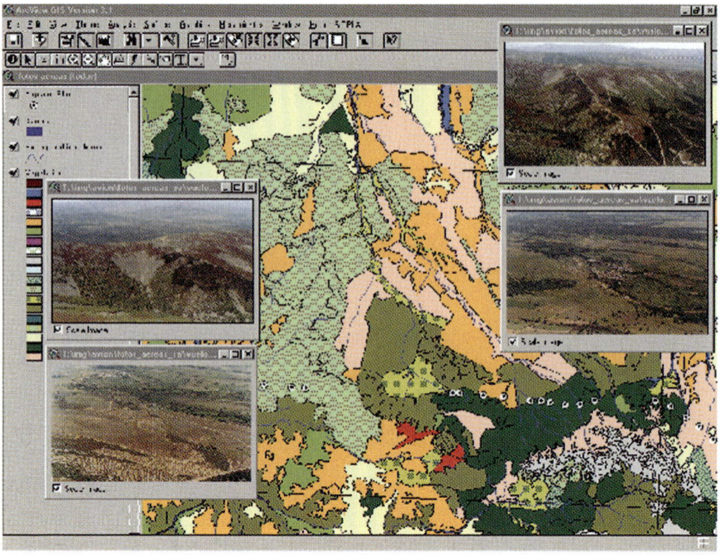

Plate 12 GIS support for oblique aerial photographs of the landscapes. Photographs are linked to specific coordinates in a land-use map to help interpretation (see Fig. 7.3 on p. 137)

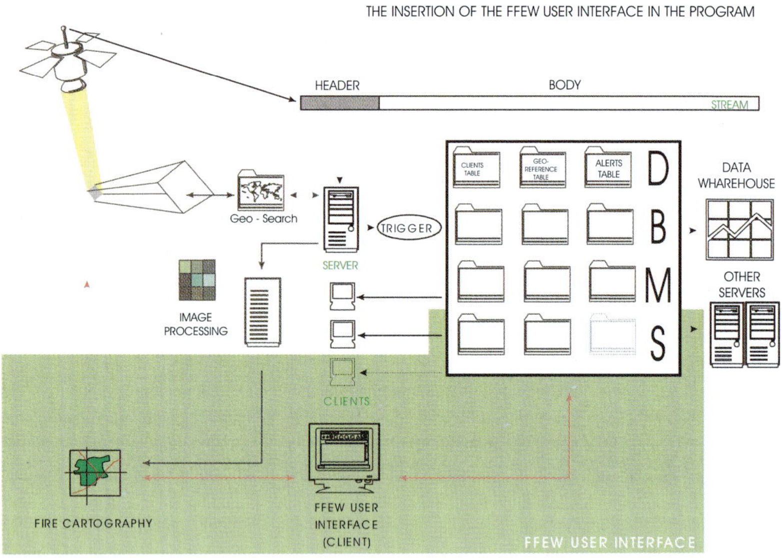

Plate 13 An expert system example: expert system for forest fires early warning. A system of this kind can automatically calculate contingency plans and emergency routes for the population, and are an excellent support for decision-making (see Fig. 11.1 on p. 248)

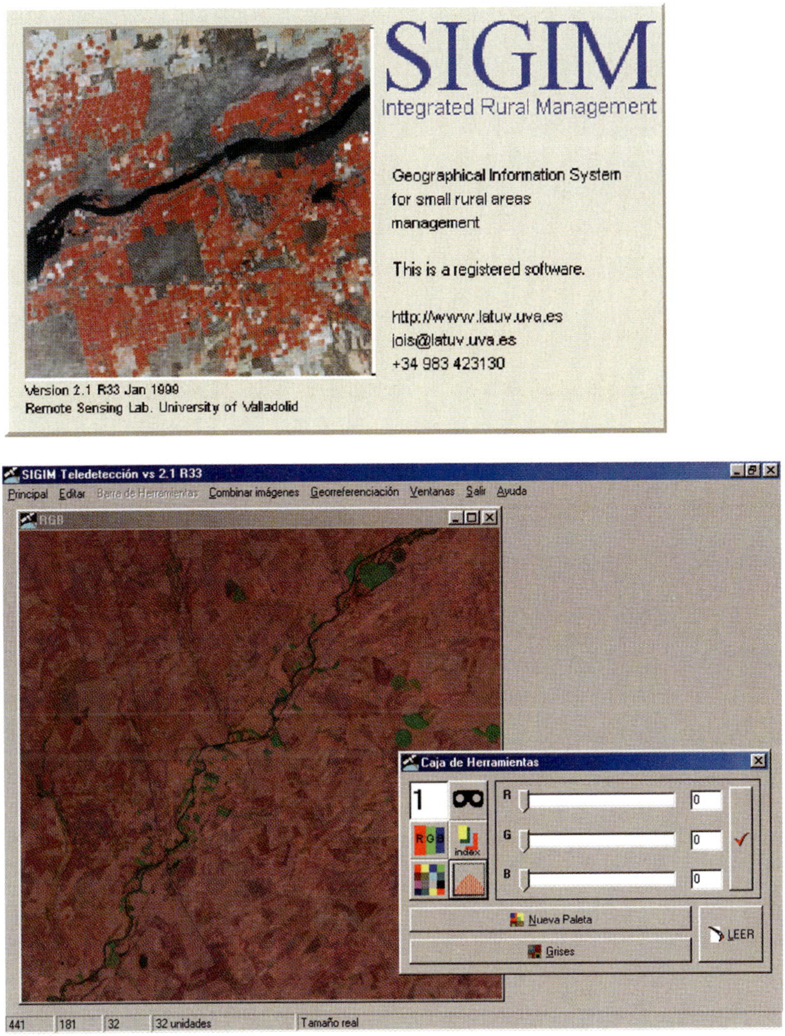

Plate 14 SIGIM welcome pop-up screen (translated to English; original version in Spanish) and the remote sensing module, featuring masking, band combination, georeferencing, vegetation indexes calculations and thresholds tools over a LANDSAT image (see Fig. 11.2 on p. 249)

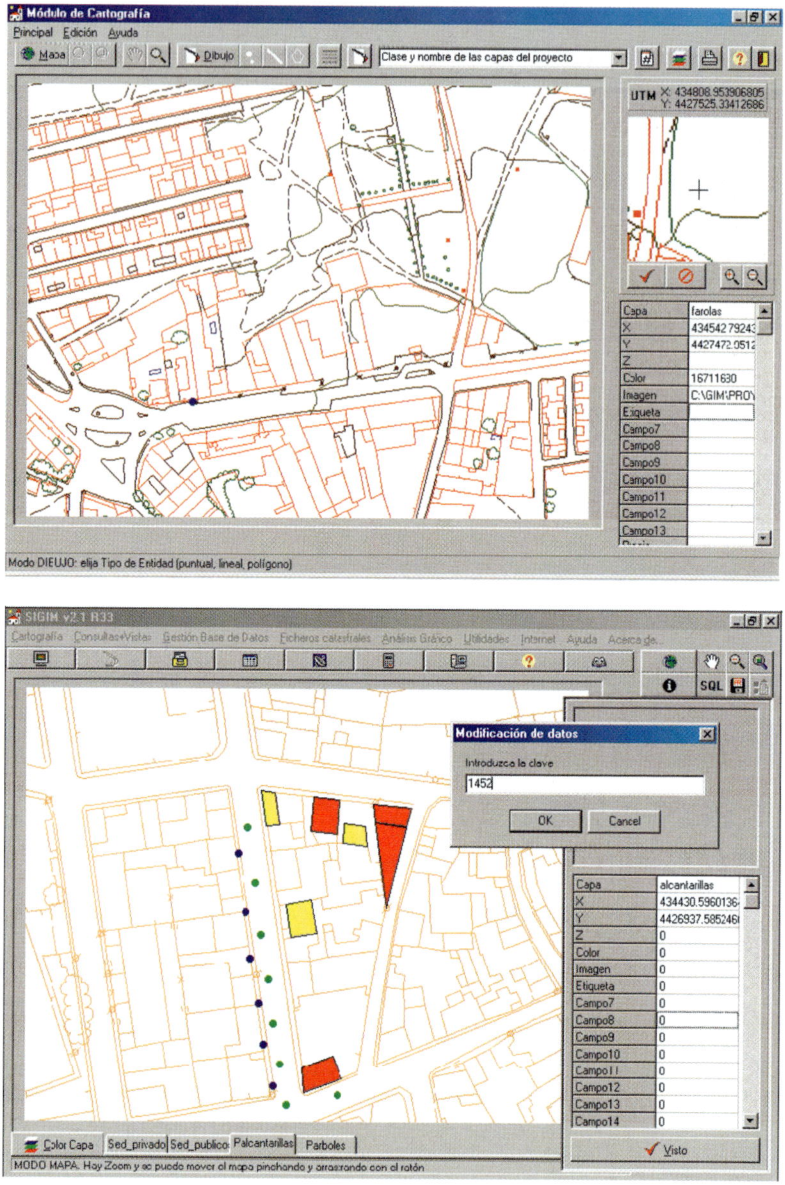

Plate 15 SIGIM cartographical module, with the basic cartography layout and the georeferencing tools (see Fig. 11.3 on p. 250)

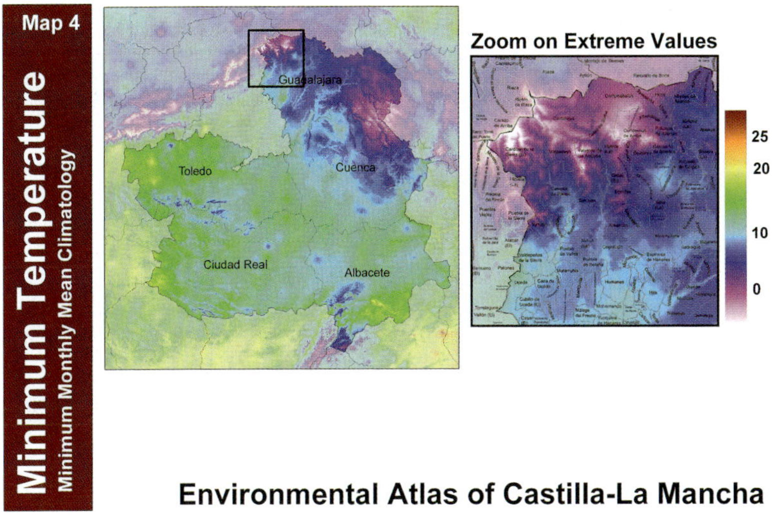

Plate 16 An example of a simple, effective map. Minimum temperatures (climatology) of Castilla-La Mancha region, with a zoom in a special interest area (extremes in minimum temperature) (see Fig. 17.1 on p. 280)

Printing: Krips bv, Meppel
Binding: Stürtz, Würzburg